To Michael
Love
Mother & Mike
"2008"

City of Heroes

OTHER BOOKS BY RICHARD N. CÔTÉ

Mary's World:

Love, War, and Family Ties in Nineteenth-Century Charleston

Accustomed to affluence and opportunity in the South's Golden Age, Mary Motte Alston Pringle (1803-1884) represented the epitome of Southern white womanhood. Her husband was a wealthy rice planter who owned four plantations and 337 slaves. *Mary's World* illuminates in lavish detail the world and psyche of this wealthy, well-educated, well-intentioned woman, her family, and their slaves before, during, and after the Civil War. **"An unparalleled window into the mind and heart of one of South Carolina's most interesting women." — George C. Rogers, Jr., Ph.D., Distinguished Professor Emeritus, University of South Carolina.**

Theodosia Burr Alston:

Portrait of a Prodigy

For Vice President Aaron Burr, providing his daughter, Theodosia, with an extraordinary education was much more than just a lifelong obsession. By the time she could walk, Burr had envisioned an incredible goal for her and crafted a master plan to achieve it. From her youth through her marriage in 1802, Theodosia was groomed and educated to become a female Aaron Burr and take her intended station in life: nothing less than president, queen, or empress. **"Côté has produced a fascinating portrait of a talented, enigmatic young woman." —Walter Edgar, Ph.D., author of *South Carolina: A History*.**

Strength and Honor:

The Life of Dolley Madison

Born a Quaker farm girl in the North Carolina wilderness, the heroism Dolley displayed during the War of 1812, and the integrity that characterized her entire life, made her an extraordinary role model, and she became one of the most-acclaimed women in America. Based on more than 2,000 of her letters, this intimate portrait explores the life of a vivacious, dedicated woman, who triumphed over adversity and tragedy to help build the American republic and define the role of First Lady. **"Cote's reinterpretation of her life provides a very human profile of a legendary historical figure." — American Library Association *Booklist*.**

City of Heroes

The Great Charleston Earthquake of 1886

Richard N. Côté

Mt. Pleasant, S.C.

Publisher's Cataloging-in-Publication Data
(*Provided by Quality Books, Inc.*)

Côté, Richard N.
City of Heroes : the Great Charleston Earthquake of 1886
by Richard N. Côté--1st ed.
p. cm.
Includes bibliographical references and index.
LCCN 2005906172
ISBN-10: 1-929175-45-0 (trade hardcover)
ISBN-13: 978-1-929175-45-1 (trade hardcover)
ISBN-10: 1-929175-46-9 (trade paperback)
ISBN-13: 978-1-929175-46-8 (trade paperback)

1. Charleston Earthquake, S.C., 1886. 2. Earthquakes —South Carolina—Charleston—History—19th century. 3. Charleston (S.C.)—History. I. Title.

F279.C457C68 2006 975.7'91503
QBI06-600114

BISAC categories:

HIS036040	History/United States/19th century
NAT023000	Nature/Natural Disasters
NAT009000	Earthquakes

First edition, first printing, August 2006.

This book was printed in the United States of America on archival-quality paper that meets the guidelines for performance and durability of the Committee on Production Guidelines for Book Longevity for the Council on Library Resources.

Corinthian Books
483 Old Carolina Court
Mt. Pleasant, SC 29464
www.corinthianbooks.com
(843) 881-6080

Dedication

This account of the victims and heroes of the Great Charleston Earthquake of 1886 is dedicated to Elizabeth ("Liz") Jenkins Young, a pioneer and driving force for historic preservation, a fluent speaker of Gullah, and since 1944, Charleston's first and foremost female tour guide.

The 1886 Earthquake Prayer

Oh mah Gawd an mah Father,
Ain yuh feel how dis earth do tremble like
Jedgement Day?
Come down heayh, Lawd
An help yo poor people in dere trial and trib'lation.
But oh do, Massa God, be sho and come Yoself,
And doan sent yo Son,
Caus dis ain' no time fuh Chillun.

—Prayer of a woman on the
Santee River, South Carolina

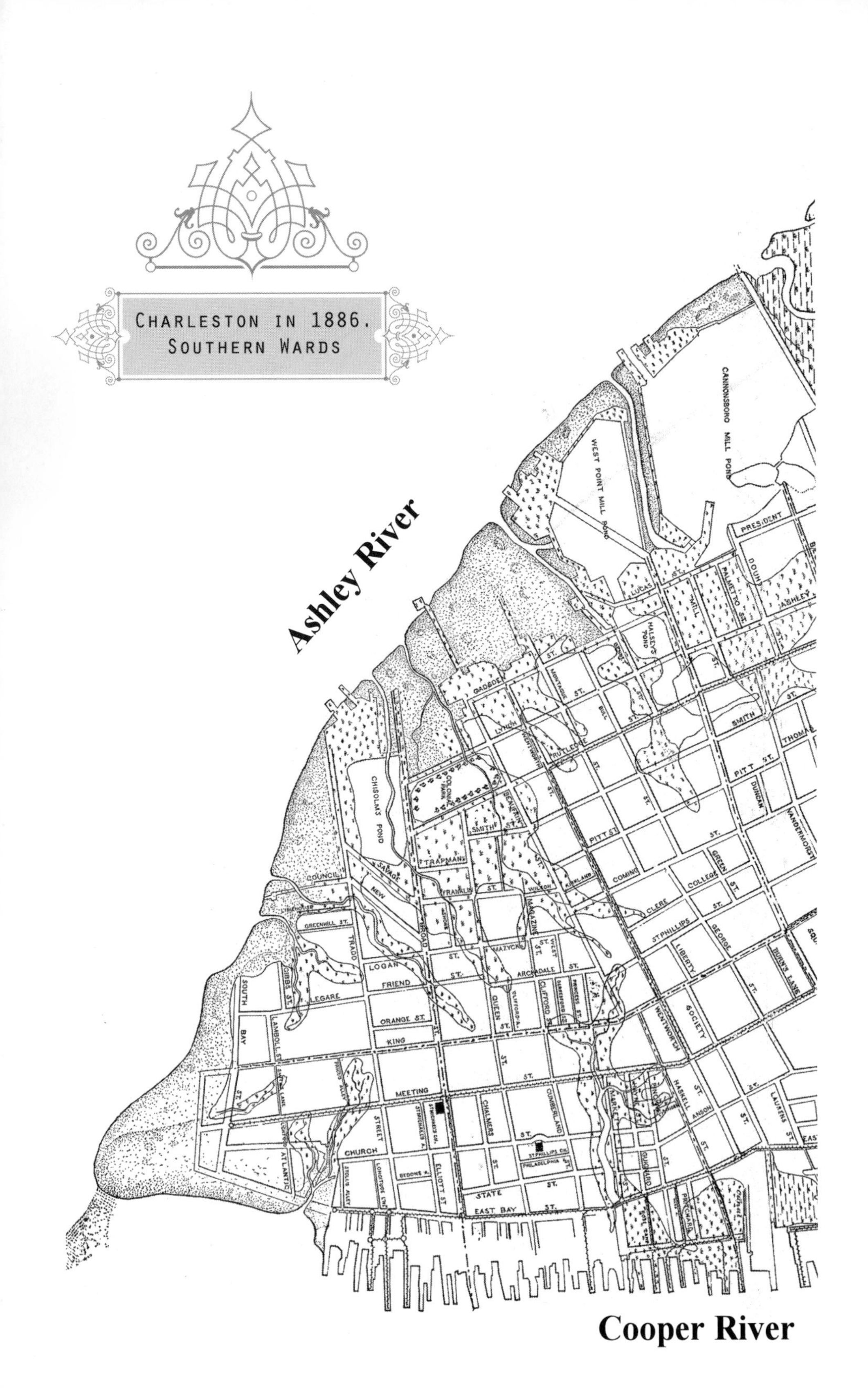

Charleston in 1886.
Southern Wards
Ashley River
Cooper River
CANNONSBORO MILL POND
WEST POINT MILL POND
PRESIDENT
LUCAS
MILL
PALMETTO ST.
ASHLEY
HALSEY'S POND
GADSDEN
MONTAGUE ST.
BULL
SMITH
LYNCH
RUTLEDGE
PITT ST.
CHISOLM'S POND
COLONIAL PARK
SMITH ST.
DUNCAN
COMING
GREEN
COLLEGE
TRAPMANN
SAVAGE
COUNCIL
NEW
BROAD
FRANKLIN ST.
WILSON
MAGAZINE
CLERE
ST PHILLIPS
GEORGE
LIBERTY
BURN'S LANE
GREENHILL ST.
TRADD
LOGAN
FRIEND
MAZYCK
ARCHDALE
SOUTH BAY
GIBBS ST.
LEGARE
ORANGE ST.
QUEEN
CLIFFORD ST.
PRINCESS ST.
WENTWORTH
SOCIETY
LAMBOLL ST.
KING
MEETING
HASSELL
ANSON
LAURENS ST.
MARKET
CHALMERS
CUMBERLAND
CHURCH
STREET
ATLANTIC
ELLIOTT ST
STATE
EAST BAY
ST. PHILLIPS CH.
PHILADELPHIA ST.

Charleston in 1886.
Northern Wards
Cooper River
Town Creek
Cannonsboro Mill Pond
Grove
Moultrie
Huger
Congress St.
Mount St.
Fishburn St.
Line
Ave.
Race
St.
Poinsett St.
Butler St.
President
Spring St.
Bee St.
Doughty
Palmetto St.
Ashley
Bogard Lane
Sheppard
Park
Cannon
Rock
Laurel
Percy
Smith
Marion Place
Morris
Felix St.
Jasper
Thomas St.
Radcliffe
Coming
St. Phillips St.
Pitt St.
Duncan
Vanderhorst
Warren
King
Green
George
Colu mbus
Line
Cooper
Jackson
Harris
Johnstone
Hill
Hanover
Blake
The Mall
America
Drake
Amherst
Meeting
John
Hudson St.
Marion Square
Burns Lane
Charlotte
Henrietta St.
Elizabeth St.
Alexander
Washington St.
Laurens St.
East Bay St.
Washington St.
Vernon
Marsh
Inspection
Concord
Wharf
Gas Works
Bay St.
Mary

A Note to Readers

Unless otherwise indicated, all structures, streets, parks, and places mentioned are in the city of Charleston, South Carolina, and all cities are in the State of South Carolina.

When the text cites "the newspaper," or a quotation is not referenced by a source citation, the source is the 1886 and 1887 issues of the Charleston *News and Courier*, the city's largest-circulation newspaper at the time. If the article was published in the *News and Courier* in other years, a complete source citation is provided.

Where direct quotations from any source have been used, the original spelling, capitalization, and punctuation have been preserved.

All dollar amounts are expressed in 1886 gold dollars. If a sum appears in parentheses following a dollar amount, that sum is the approximate value in 2006 dollars: 1886 dollars multiplied by an inflation factor of 20.1.

Contents

Chronology

1886	
August 27-29:	Several small earthquake shocks are felt in Summerville and Charleston.
August 31:	At 9:51:12 p.m., the first shock of the magnitude 7.3 earthquake hits Charleston and Summerville. Three more follow before midnight.
September 1:	Eight aftershocks. All hospitals are destroyed. All communications connections to the outside world are severed.
September 2:	The U.S. Geological Survey dispatches a scientific team to investigate the earthquake.
September 3:	Acting Mayor William E. Huger calls a special meeting of city council.
September 4:	The city council establishes the Executive Relief Committee (E.R.C.) to provide food and shelter for 40,000 homeless Charlestonians.
September 6:	One hundred tents are received from the federal government and are erected in public parks.
September 7:	Mayor Courtenay returns to Charleston from his European vacation.
September 8:	The E.R.C. starts to erect wooden booths to shelter the homeless. Mass food distribution begins.
September 10:	The E.R.C. declares that it has sufficient tents and ceases building more booths.
September 15:	The E.R.C. starts accepting applications for rebuilding aid.
September 17:	Courtenay seeks donations nationwide to assist the earthquake victims.
October 5:	Courtenay ends solicitations for further relief funds.
October 9:	The E.R.C. ends mass food distribution, as demand has dropped to just a few hundred people, who can be cared for by existing agencies.
November 27:	The Stockdell Report provides detailed damage information for 8,000 structures.
1887	Aftershocks continue for the entire year.
March 21:	The E.R.C. makes its final charity payments and ceases operation.
July 25:	The E.R.C. submits its final financial report to the city council.
October 31-November 5:	Mayor Courtenay declares the earthquake recovery essentially complete. The city celebrates Galaweek to commemorate the end of the ordeal.

PREFACE

There are no living survivors of the Great Charleston Earthquake of August 31, 1886. No one remembers its effects on the city, the State of South Carolina, and the rest of the Eastern seaboard. Yet the seismic history of the eastern United States is dominated by the 1886 earthquake. The earthquake was devastating, lethal, and terrifying; the recovery process was inspirational. No one who lived through it ever forgot it.

Some native-born Charlestonians have heard stories about it from elderly family members. Most of those who have adopted the Lowcountry as their home only know what the guides tell them on the carriage tours, when earthquake-bolt gib plates (end caps) are pointed out as interesting local architectural curiosities. In all, probably not more than one local resident in a hundred is aware of the magnitude of the disaster that took place 120 years ago, or the likelihood of a recurrence.

Paul Pinckney, a descendant of a South Carolina family that produced two signers of the United States Constitution, was a young man when the 1886 earthquake shook the South Carolina Lowcountry like an alligator rolling its prey. Twenty years later, he wrote a letter to the editor of the *San Francisco Chronicle.* It was published on May 6, 1906, eighteen days after San Francisco was scourged by its own massive earthquake and three days of fire, which killed hundreds of people, leveled thousands of buildings, and left tens of thousands homeless. Pinckney hoped to inspire the San Franciscans with the story of his own astounding seismic experience—and the triumphant rebirth of his native Charleston.

August 31, 1886, he recalled, had been exceedingly hot, and the evening was unusually sultry, with such a profound stillness in the air that people commented on it. "The temblor," he wrote, "came lightly, with a gentle vibration of the houses as when a cat trots across the floor." Then Pinckney went on to calmly describe the most terrifying night anyone then alive had experienced.[1] The most powerful earthquake ever to strike east of the Mississippi had unleashed an immense burst of energy that had been accumulating deep beneath the earth for hundreds of years. Ten seconds after it started, at 9:51 p.m., Charleston was in chaos. Within forty-five seconds, most of the damage was done. Within minutes, the entire population of both places, save for the dead, injured, or those too infirm to move, had fled into the streets. Within a few days after the first shock, at least 40,000 of the city's 60,145 residents were sleeping in the streets and public parks every night.[2]

The 1886 earthquake killed and injured more people than all previous American earthquakes combined. The city's official death record named eighty-three victims. Based on new research presented here, the casualty count is now known to be at least one hundred twenty-four dead and one hundred thirty-nine severely injured. Countless others died or were hurt and went uncounted. Total casualties—those who died or were injured by falling debris, premature childbirth, or exposure to the elements, and those who committed suicide or were driven insane by the terror—were likely as many as five hundred souls. The earthquake shook an immense area of about 2,500,000 square miles, and was personally experienced by two-thirds of all Americans. It reached as far north as Toronto, Canada and Somerset County, Maine; south to Key West, Florida and Havana, Cuba; west to Omaha, Nebraska; and east to Long Island, New York and Bermuda.[3]

Sixty-seven percent of Charleston's brick buildings were severely damaged or destroyed. Every building in the city and in Summerville, twenty-two miles to the northwest and within several miles of the earthquake's twin epicenters, was affected. Of fourteen thousand chimneys in Charleston, scarcely one hundred escaped damage.[4] The damage to Charleston's structures alone was estimated by the city's assessor at $5 to $6 million ($100.5 to $121.2 million in 2006 dollars).[5]

The aftershocks eventually trailed off, but they never stopped. A systematic review of newspaper reports for 1886 through 1889 identified 522 aftershocks following the August 31, 1886 event.[6] In 1959, a chimney was knocked down, plaster fell, and a ceiling cracked in Summerville. In 1974, an outside brick wall was separated from a North Charleston house, and a five-hundred-ton machine tool jumped around on its bed at Ladson, near Summerville.[7] On November 19, 2005, yet another small earthquake was felt in Summerville.[8]

The earthquake that struck seconds after 9:51 p.m. on August 31, 1886, was devastating beyond imagination. For South Carolinians born before that time, "*de quake*" was the most frightening natural event they ever experienced. Yet in many ways, the disaster relief and rebuilding process was even more amazing than the violent geological event itself. Despite its imperfections, Charleston's recovery from a natural catastrophe that touched every one of its more than 60,000 citizens remains unsurpassed in its speed, efficiency, frugality, and humanity. Its successful aspects are presented here as The Charleston Disaster Recovery Model.

Most important was the example set by the civic leaders, the residents of Charleston and Summerville, and the tens of thousands of generous people—including countless benefactors in the North—who made the recovery possible. In the fourteen months that followed the earthquake, a legion of heroes, male and female—some prominent and many whose names will never be known—stepped forward to carry out the most rapid and inspiring rebirth any disaster-stricken city in America has ever experienced. This is their story.

A view of Charleston from St. Michael's steeple

1

As When a Cat Trots Across the Floor

From the fatal consequences of earthquakes, we are happily exempt.
—David Ramsay
South Carolina historian, 1808

Summer 1886

In 1886, earthquakes were not a hot topic of conversation in the elegant, two-hundred-year-old port city of Charleston, South Carolina. But something suspicious was definitely underway—and under foot. In the spring, a small tremor knocked several books off the shelves at the library of the College of Charleston one afternoon. Other than the librarian, few people were aware of the event.[1]

During the sitting of the federal court in June, there was a "decided rattling of the sashes in the court room which excited much observation, and which was thought by some to have been produced by some other cause than passing wagons or a boiler explosion."[2] Even though there was no logical explanation for the odd event, it was largely ignored. Some time later, a Charleston man recalled that while sitting in a rocking chair in his library that June, he was "slightly thumped by an earth movement."[3] It did not make much of an impression on him at the time.

Another June tremor passed virtually unnoticed by everyone except John Rugheimer of Summerville, a peaceful pineland village of 1,840 souls twenty-two miles northwest of Charleston. He had mentioned it to others, but, aside from his attractive young wife, no one paid any particular attention, and the old Bavarian tailor's concerns were forgotten.[4] During July, there were no reports of tremors, and those of June slipped quickly from mind.

In August, Thomas Turner, the fifty-three-year-old, British-born president of the Charleston Gas-Light Company, had vacated his residence at 152 Calhoun Street in the city and taken refuge from the steamy Lowcountry summer in a one-story bungalow in Summerville. The village, with its efficient South Carolina Railway connections, was rapidly becoming a bedroom suburb of the "City by the Sea," as Charleston was affectionately known. With Turner was his wife, Ellen; his sister-in-law, Sarah Quarmby; his niece, Ellen Quarmby; and Susan Gaillard, their black servant.[5]

On the morning of Tuesday, August 24, between eight and nine o'clock, Turner and his family were startled and frightened by a tremendous noise that resembled the explosion of a steam boiler or the discharge of a large cannon. His house was shaken to its foundations, and kitchen utensils were thrown from the shelves, but no real damage was done.[6] Taking the extraordinary event in stride, Turner heeded the familiar long, shrill whistle, boarded the 8:05 a.m. commuter train, and left the Summerville depot for his office on Meeting Street in Charleston's downtown business district.

On Friday, August 27, at about 1:30 a.m., Summerville resident Eugene Tighe had been up extremely late and was preparing for bed when a jarring shock below him caused a pitcher to rattle in its basin, and he felt as if his whole house was being lifted from the ground and then thumped heavily back into position.[7] Unlike Tighe, most of the villagers slept on blissfully, and the early morning shock evidently passed unnoticed in Charleston.[8]

The events later that morning caught the attention of everyone in Summerville. At 8:30 a.m., shortly after the commuter train had departed for Charleston with its usual load of businessmen, "without the slightest premonition of disturbance or danger, the shock of an earthquake was distinctly felt. It produced, of course, the utmost consternation, as it lasted for several seconds. People left their houses and ran out into the street to avoid the imminent crash of a falling house or a roof tumbling in on one of the inmates." The shock was particularly severe near the railroad depot, where articles were shaken off the shelves of several stores.[9]

Every house in the village had been mildly shaken. A reporter from the *News and Courier* interviewed a local resident and wrote, "A rumbling sound was first heard in a northeasterly direction from the town, and . . . the sound was followed by an explosion resembling that of a cannon at a distance." Some imaginative residents thought a meteor smashing into the earth had caused the tremor, but since there had been no sudden blaze of light in the sky or on the ground, the theory was discarded.[10]

When John B. Gadsden, son of John Gadsden, principal of the high school in Summerville, returned home that afternoon, his excited family greeted him with the news of the earthquake. Because the motion of the commuter train and its distance from Summerville at the time had shielded Gadsden from feeling the vibrations, he was naturally skeptical. He shared the strong "doubting Thomas" opinion of many who had not felt the shock. These incredulous villagers tried to find logical answers. "Some boiler at one of the numerous phosphate works that surround us has burst," and "someone was blowing up trees with dynamite" were a few of the numerous guesses.[11] Other than store goods and plates falling, there was no significant damage in the village.

Few outsiders took Summerville's earthquake reports seriously. Colonel John H. Averill, master of transportation of the South Carolina Railway, and intendent (mayor) of Summerville, noted that the local telegraph operator had reported the event to Charleston as an earthquake shock and was laughed at.[12]

Although the Charleston *News and Courier* published accounts of the event the next day, few in the city paid them any great heed, and fewer yet had personally felt anything. When a Charleston resident telegraphed the news of the Summerville tremors to the *Atlanta Constitution*, the editors made note—and also scoffed. They wrote on August 28, referring to the previous day, "A decided sensation was caused in Summerville by a shock of earthquake about 8:30 this morning. The shock was preceded by a dull rumbling followed by a sound as of a cannon shot fired at a distance."[13] Later in the same issue, they quipped, "A telegram from Charleston announces an earthquake near that place yesterday. This is merely imagination. It was simply the shock accompanying the announcement that Atlanta had won yesterday's [baseball] game from Memphis, ensuring it the pennant."[14]

The *Constitution* soon had something bigger to excite its readers. On August 27, catastrophic earthquakes struck on the far side of the Atlantic, and the newspaper's perspective changed from mocking to serious when the news reached Atlanta. On August 29, a commentator from the *Constitution* wrote, "There is something out of joint somewhere. Within the past two days seismatic [*sic*] disturbances have been reported from various quarters of the globe. In Greece and Egypt the shocks have been of alarming violence. But the remarkable feature of the business is that almost simultaneously with the earthquake in the old world we have had a touch of it here. South Carolina has been shaken up and a slight tremor has been felt in Georgia. Earthquakes are no new thing in this country, but the Atlantic slope has enjoyed comparative immunity for a long time. It is to be hoped that the recent manifestations do not mean anything serious."

Then, with uncanny insight, the commentator noted, "We have not built our cities or towns with earthquakes in view.

They have no place in our domestic economy, as it were, and if they are to become common, everyday affairs, it is to be feared that the bottom will drop out of our artificial civilization almost before we know it. Further developments will be awaited with intense interest."[15] The wait was eerily brief.

Between 4:45 and 5:30 a.m. on Saturday, August 28, Summerville felt a shock that started every dog barking and aroused the entire village. In Charleston, some noticed a slight rocking of houses and rattling of windows, but most were unaware of it.[16]

John B. Gadsden, the Summerville skeptic who had dismissed the previous vibrations, quickly changed his mind. He was lying in bed awake, planning his day, when he heard what sounded like a terrible explosion at the southeast corner of his house. This was quickly followed by the sensation that someone under his bed was trying to tip the bed on its back. The timbers in his house creaked and shivered as if they were haunted. The noises gradually passed away, and everything was quiet.[17]

About 1:45 p.m. that same day, a stronger shock in Summerville flung a bed against a wall. In other houses in the village, windows and crockery were broken.[18] Gadsden heard "detonations in the distance which seemed to come from deep below us."[19] Several aftershocks were also felt that day as far away as Augusta, Georgia, and Wilmington, North Carolina.[20] In Augusta, houses swayed, bells tolled, pictures were thrown from the walls, a valuable French mirror was smashed to atoms, a woman fainted from fright, and a baby was thrown from a couch to the floor, suffering scratches and bruises.[21] This shock and the one on August 27 were "also felt by several persons in Columbia [South Carolina], who recognized their character, and commented upon them at the time and afterwards without knowing that similar tremors had been observed elsewhere in the state."[22]

The next morning there was not a skeptic in Summerville, but the Charleston papers treated the subject lightly.[23] It was said that on August 28, an Italian workman in Summerville warned, "Two little shake; big one come soon." If he did, it is doubtful that many would have heeded his prophecy. As far as most South Carolinians were concerned, earthquakes—of which few had any memory—were little more than a temporary curiosity.

In Summerville, the tremors continued. Thomas Turner, by now an earthquake veteran, was reading on the piazza of his home on Sunday, August 29, between 3:00 and 4:00 p.m. when he felt several vibrations moving his chair. Then he saw the posts of the piazza move from their vertical position, as if the foundation of his house had been pushed with considerable force.[24] In Camden, a small town about 115 miles northwest of Charleston, some residents also had the novelty of hearing the rumbling of an earthquake on the same day. The *News and Courier* said of Camden, "One gentleman reports it as having shaken his windows pretty badly, but I can hear of nothing more serious than a severe shaking up of everything."[25]

On Monday, August 30, the earth was mercifully quiet. After five days of mysterious and unsettling seismic events, the villagers were greatly relieved, and began to speak of their

Houses on South Battery before the earthquake

"late disturbance" as something that had finally faded away.[26] They had no way of knowing that death and destruction were silently stalking them several miles below the surface of the earth.

On Tuesday night, August 31, 1886, everything in Charleston and the surrounding area seemed normal. The morning had been hot, sultry, and uncomfortable. As evening arrived, the breeze from the rising tide, which usually cooled the city, was absent. Not a whiff of air stirred the sails in the harbor, and the sky was clear. The setting sun sank quickly, with a glow only briefly noticeable. The first inkling of something unusual came when the nearly full moon arose. It was surrounded by mist, and some noticed that the moonlight seemed odd. Then just after dark, a peculiar sulphurous odor permeated the edges of the city. The people heading home or sitting on their piazzas to escape the stifling heat in their houses attributed the rank smell to the fertile, pungent pluff mud of the nearby marshes.[27]

Just before ten o'clock, at the *News and Courier*'s office at 19 Broad Street in the heart of the city, Assistant Editor Carlyle "Carl" McKinley was editing the September 1 edition. It was late, and McKinley's tired eyes tried to fend off sleep as he proofread columns of tedious statistics. The paper's editor-in-chief, Capt. Francis W. Dawson, had finished his day and was reading in his elegant home on Bull Street.

A few doors down the street from the newspaper office, Julian M. Bacot, an up-and-coming young attorney, had been working late. After briskly walking the eight blocks to his home on Coming Street, he sat down at 9:40 p.m. to make the daily entry into his diary. A man of precise and methodical habits, he dutifully dipped his steel-tipped pen into an antique pressed-glass inkwell and started writing, even though the only interesting thing that had happened was a slight shower around dinnertime.

A few blocks away, on Meeting Street, Sarah Middleton, a twenty-five-year-old black woman, was walking wearily along the sidewalk on the east side of the Guard House, the main police station. She had just finished her work as a washerwoman for one of Charleston's white families and was on her way home. She tried to ignore the pain in her wrists just below her thumbs, which grew more intense each day. Wringing out the clothes seemed to aggravate it. Her hands itched from the soap, her back hurt, and she was exhausted. Susan Middleton, a relative, had joined her on the walk home, and the two young women found temporary relief from the drudgery of their lives by giggling about the foibles of their employers and the charms of the handsome young coachman they had seen on the street the day before.

Several blocks to the southwest, near Savage Street, "Jumbo," a large, friendly dog belonging to the Chisolm family, would not stop his frantic barking. Despite the irritating noise, Charlestonians prided themselves on their good manners, and all the neighbors ignored him.[28]

Forty-seven-year-old Edward Laight Wells, a prosperous cotton broker and prolific author, had been reading at his home on the southeast side of the peninsula. A few minutes before ten, he put down his leather-bound volume of English verse and walked to the window that offered him a beautiful view of the harbor. He described the water as looking like a mirror, "reflecting & multiplying the many lights that twinkled on their surface, which hung from the rigging of the vessels at anchor opposite the windows of my house. The stars were shining brightly & a soft tender light enveloped everything. It was a charming night," he wrote, "such a one as might perhaps had suggested to the sentimental thoughts of lovers' guitars serenading fair ladies' chambers."[29]

On the southern part of the peninsula, near White Point Gardens, Laura A. Bennett, a black woman, had just gone to bed. The baby that was soon due was kicking hard, and the heavily pregnant woman was very uncomfortable. The usual worries ran through her head. Would the child be healthy? How long would she have to endure labor? Where would the money come from to support this new life? Feeling bloated and miserable, she fell asleep.

A few blocks away, off Church Street, in a particularly notorious alley known for its brothels, two prostitutes were plying their trade. The elder one was counting the days until she could start a business of her own. She had carefully put aside a large portion of her income toward her ultimate goal—to set herself up as a madam. Dutifully making sounds of pleasure, she envisioned the huge red velvet sofa that would grace the parlor where her clients would sit.

An inbound passenger train from the upstate had recently pulled into the South Carolina Railway terminal just north of Marion Square, its brakes screeching and axles smelling of hot grease, carrying its usual load of vacationers returning from the mountain resorts of North Carolina. After helping his passengers disembark, Harvey G. Senseney, the train's affable baggage master, hopped onto a horse-drawn trolley car belonging to the Enterprise Railroad and headed home to his apartment on northern King Street.[30] He had been delayed when a well-dressed old man with an ear trumpet insisted that his prized leather trunk had been lost. Despite Senseney's best efforts to find the luggage while shouting to make himself understood, the elderly gentleman had been mollified only when his distracted son appeared and explained that the trunk had already been loaded onto a waiting carriage.

The South Carolina Railway's northbound night train, with Mr. Burns at the throttle and his black fireman, Mr. Arnold, shoveling coal into the boiler's furnace, had left Charleston, belching smoke. The inbound passenger express train from Columbia, which normally arrived at 9:10 p.m., had left the capital an hour late and was rushing towards home, trying to make up time. The people who had gathered to meet friends and family members at the train terminal in Charleston sat quietly or talked with their neighbors. Among them was Louis A. Beaty of suburban Mt. Pleasant, the editor and publisher of the *Berkeley Gazette*, who passed the time by chatting with fellow businessmen.[31] Nine-year-old Bertha Cade and her sister played nearby, waiting with their father for their mother, who was coming in on the train.

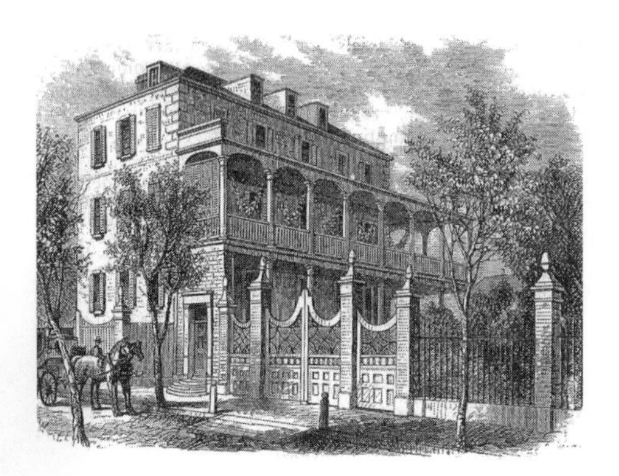

The Simmons-Edwards House, a typical Charleston mansion before the earthquake

In his lodgings on lower Meeting Street, in the heart of the ancient peninsular city, Robert Alexander, a genial twenty-four-year-old London native, was changing into his nightclothes, preparing for bed. An analytical chemist attracted to the city by its burgeoning phosphate trade, he had been in Charleston for only five months. The well-liked young man had recently bought a small steam pleasure yacht and had taken his first trip in it the previous evening.

At The Moultrie House, a popular resort on nearby Sullivan's Island, mothers tucked their children into bed, opening the windows wide to catch the cool, onshore sea breezes. In another part of the building, young adults and their parents were dancing at the hop being held there that evening.[32]

Miles out to sea, breezes ruffled the silver-grey hair of Charleston's highly regarded mayor, William Ashmead Courtenay. Moderate in height and solid of build, the fifty-five-year-old vigorous, handsome, successful businessman was enjoying a luxurious voyage on board the Cunard Line's state-of-the-art ocean liner, the R.M.S. *Etruria*. The sleek, imposing, black-hulled ship with twin red-and-black stacks was returning to New York from Liverpool, England, after stopping to pick up passengers at Queenstown, the chief port of Courtenay's ancestral home, Ireland.[33]

Courtenay's vacation abroad had been prompted by the need for a long-delayed rest. The previous year, a massive hurricane packing 125-mile-per-hour winds at its peak had mercilessly pummeled Charleston without warning. Although, miraculously, there were no fatalities, the city's waterfront, a mainstay of its economy, had been destroyed, and the lower quarter of the city was flooded. Losses were estimated at $2 million (approximately

William Ashmead Courtenay

$40.2 million in current dollars), and that was only the cost of the buildings, not their contents.[34] Nevertheless, as a matter of civic pride, Mayor Courtenay and the city council declined all offers of outside help, vowing to rebuild Charleston through sheer willpower using their own resources.

The mayor had thrown himself into the recovery and repair operations with such zeal that by the end of December 1885, he was totally exhausted. "I am admonished to seek rest, through some measure of relief, from the overwork incident to my present arduous duties," he told the city council in his annual report.[35] With the hurricane reconstruction substantially completed, and with half of his second term behind him, Courtenay thanked the council members and the people of Charleston for their numerous acts of confidence, consideration, and kindness, and then asked to be released from serving the rest of his term, effective March 1, 1886.[36]

Appalled at the thought of losing their extraordinary leader, the city council conferred anxiously with him in January. They proposed that instead of resigning, he take a leave of absence for as long as necessary to regain his health. To their great relief, at the council meeting on April 27, the mayor agreed to serve out the rest of his term after returning from a therapeutic European vacation.[37] He departed in early May. By August 31, 1886, rested and refreshed, Courtenay was returning home to resume his civic duties.

Life aboard the *Etruria* was relaxing and elegant. On her upper decks, Courtney was surrounded by the cheerful banter of his 550 fellow first-class passengers. Most of the remaining 800 passengers were poor Irish immigrants, who were kept

The R.M.S. Etruria *taking on passengers for New York*

out of sight, three decks below their wealthy shipmates. Courtenay would have been further soothed by the genteel clinking of sterling silverware on handsome English china plates, the unmistakable ring of crystal wineglasses, the pop of champagne corks—the *Etruria*'s wine cellar held 1,100 bottles of bubbly and 850 bottles of other wines—and the graceful waltzes played by the ship's tuxedoed dance band. The only motion he felt aboard the 501-foot-long luxury liner was the faint vibration of the powerful 14,500-horsepower steam engine four decks below him.

The *Etruria* held the Blue Riband speed record for a transatlantic crossing. She could maintain a sustained speed of more than nineteen knots, making her the fastest liner on the New York-Liverpool run. Indeed, she was faster than any warship in the Royal Navy. On August 31, with a bone in her teeth, she was two days out of Queenstown and four days from New York City, where Courtenay, who in private life was the Charleston agent for the Clyde & Company shipping line, would spend Sunday night, September 5. On Monday, if his schedule held, he would transfer to a Clyde Line coastal steamer and set sail for Charleston.

Near Asheville, North Carolina, 250 miles northwest of Charleston, on the evening of August 31, the Reverend Anthony Toomer Porter, the influential and outspoken fifty-seven-year-old rector of Charleston's Holy Communion Episcopal Church, was seated in the parlor of his vacation home on a mountainside in the Great Smoky Mountains, a favorite summer destination for affluent Charleston families. From his invalid wife's bedchamber a floor above, he heard strange noises. They continued so long that he went to the foot of the stairs and called to her nurse, "Why are you moving the furniture in my wife's room?" The nurse replied, "We thought you were moving the drawing-room furniture, for we heard the same noise; I thought it strange."[38]

Next, Porter heard what sounded like the wheels of several vehicles driving up the mountainside very fast—something completely unlikely at that time of night. Then came the sound of many railroad cars—but there was no railroad in the mountains. Immediately after the sounds ceased, one corner of his house was thrown up, landing with a thud moments later.

At that point, Porter realized they were in the middle of an earthquake. Rushing to his wife's side, he instructed the rest of the family to flee the house, but the shocks subsided. The neighbors were wild with excitement, and he soon learned that the earthquake had been felt more in the valley below than up where he was. At Asheville's city hall, vibrations in the tower had caused the bell to toll, but most residents, though alarmed, believed that the seismic shock was a local event. After a few hours of hurried conversation, Porter and his neighbors disbanded and went back to bed.

The next day, Wednesday, aroud noon, one of Reverend Porter's neighbors received a frightening telegram from Columbia, South Carolina, about halfway between Asheville and Charleston. It read, "We are all safe, but poor Charleston," and nothing more.[39]

Concerned for the safety of his aged mother, an aunt, and a niece, who had stayed behind in his Charleston house, and worried about the condition of his church and its large private school, Porter wrote, "We all began at once to telegraph Charleston, but received no response."[40]

It was not until eleven o'clock that night that he and his neighbors began to receive a reply. A half-sentence would come through the telegraph—then silence. Then a few more words—and more silence. After several hours of incessant telegraphing, the anxious group patched together enough information to draw a frightening conclusion: Charleston had been destroyed.

2

TERROR ON BROAD STREET

Great God the Earth Quake is Upon us!
—Julius M. Bacot

Charleston, South Carolina
August 31, 1886

For the residents of Charleston, Tuesday, August 31, had been a beautiful day. For the night shift at the *News and Courier*, the city's largest-circulation daily newspaper, it was utterly boring. Nothing of any significance had happened. The most urgent concern debated by the city council at its latest meeting had been whether to eliminate the "pernicious odors" from marshland known as the Cannonsborough ponds by using the city's garbage to fill in the marsh. The practice, which would eventually create new land to build upon, was approved, with the stipulation that no offal or dead animals from the slaughterhouses of the city market would be buried there and that each load of garbage would be covered with six to eight inches of dirt to minimize new odors.[1] For lack of anything more interesting, Wednesday's lead story was to be "The Trade of the Year. Charleston's Firm Stand in 1885-86."

The September 1 issue would feature the newspaper's annual trade review, designed to celebrate the city's progress in the previous twelve months (not that there had been much) and promote economic development in the next. The front page was devoted to the volume of trade in cotton, cotton goods, rice, turpentine, rosin, phosphate rock, fertilizer, lumber, and railroad ties shipped from Charleston in the preceding year. By 9:30 p.m., half of the Wednesday morning edition had been written, edited, set in type, proofread, and locked in to its page frames, ready for printing.

Even the weather was unremarkable. August had been hot and fairly dry, and the last day of the month was no exception. The high temperature reached eighty-nine degrees, the low had been seventy-five, and there had been virtually no precipitation, save for a brief shower around 6:00 p.m.[2] The evening sea breezes, which normally refreshed the peninsular city, were notably absent. In the harbor, the water was like glass. When the bells of St. Michael's Episcopal Church chimed the third quarter after nine o'clock, their familiar tones spread a blessing and a feeling of reassurance over the city, as they had since their installation in 1764. Charlestonians went to bed early and arose equally early. Many slept on cots or in hammocks strung across their second- and third-floor piazzas. Mosquito nets let the breezes through and kept people from being eaten alive by insects.

The streets were silent and virtually empty that night, save for the night watchmen, who set out on foot and horse patrols from the massive police station known as the Main Guard House. Nearly half a century old, the monumental building stood at the heart of the venerable, proud old city: the intersection of Broad and Meeting streets. Across the street to the north stood the Charleston County Courthouse, built in 1753 as the provincial capitol for the South Carolina colony. After the Revolution it continued to be used as the State House until it burned in 1788 and was rebuilt as the Charleston District Court House. To the northeast stood City Hall, built between 1800 and 1804 as a branch of the Bank of the United States. And to the east stood St. Michael's, constructed between 1752 and 1761. It was in the midst of this scene of calm, silence, and stability that the greatest earthquake in the recorded history of the eastern United States struck.

Carlyle "Carl" McKinley

Most Charlestonians were bathing or were already in bed by 9:51 that night. The newspaper editors, compositors (typesetters), and pressmen of the *News and Courier*, however, were working their usual evening shift so that the twelve-page newspaper could be printed and delivered to its subscribers shortly after dawn. Carlyle McKinley, known to all as Carl, the assistant editor, was editing the final copy to be typeset for the Wednesday morning edition.[3]

A talented and insightful writer, the friendly Georgia native had worked his way up through the newspaper ranks. As a young man, McKinley had been a Confederate soldier, volunteering to serve when most of his University of Georgia classmates signed up for duty. After the war ended, he returned to the university and graduated. Restless and seeking meaning in his life, he then enrolled in the Presbyterian Theological Seminary at Columbia, South Carolina, and graduated in 1874. Still searching for a calling, he abandoned the ministry, and after a brief stint as a teacher, he became the Columbia correspondent for the *News and Courier* during Reconstruction. He remained a journalist, a chronicler of South Carolina disasters, a poet, and a vitriolic commentator on racial issues for the rest of his life.[4]

The newspaper was printed at a plant on Elliott Street, but McKinley and his editorial colleagues worked in an impressive, three-story structure at 19 Broad Street, a block away. McKinley was at his desk on the second floor when his attention was diverted by a sound that appeared to come from the floor below. His first thought was that someone was rolling a heavy iron safe or a heavily laden hand truck. The sound lasted for two or three seconds, and the building trembled slightly. It was as if a horse-drawn wagon or trolley car had rumbled by on the street outside.

Suddenly recalling a small tremor from the previous Friday, his colleagues sprang to their feet, shouting, "What was that? An earthquake?" In seconds, they had their answer. In his account, published over the next three days, McKinley described the deadly and disorienting pandemonium that swiftly overwhelmed the city.

> The long roll deepened and spread into an awful roar, that seemed to pervade at once the troubled earth and the still air above and around. The tremor was now a rude, rapid quiver, that agitated the whole lofty, strong-walled building as though it were being shaken—shaken by the hand of an immeasurable power, with intent to tear its joints asunder and scatter its stones and bricks abroad, as a tree casts its over-ripened fruit before the breath of a gale.[5] From the first to the last it was a continuous jar[ring], adding force with every moment, and, as it approached and reached the climax of its manifestation, it seemed for a few terrible seconds that no work of human hands could possibly survive the shocks. The floors were heaving underfoot, the surrounding walls and partitions visibly swayed to and fro, the crash of falling masses of stone and brick and mortar was heard overhead and without, the terrible roar filled the ears and seemed to fill the mind and heart, dazing perception, arresting thought, and for a few panting breaths, or while you held your breath in dreadful anticipation of immediate and cruel death, you felt that life was already past and waited for the end, as the victim with his head on the block awaits the fall of the uplifted ax.[6]

For a few moments, the tremors diminished, giving the employees trapped inside the *News and Courier* building a ray of hope, which quickly vanished. Eight seconds later, the quake resumed in full force. No one expected to escape, but the survival instinct took over, and there was a sudden mass dash for the door. Before the newsmen reached the open air, they froze in fear, convinced they would be crushed or buried beneath the sinking roof and the falling walls.[7] In desperation, M. J. Flynn, a compositor, jumped from the window of the composing room into the side alley, sustaining serious injuries to his head, arm, and shoulder.

The News and Courier *building*

As the earth's waves rolled through town, they repeatedly lifted the ground and the buildings a foot high, and then dropped them back again a few seconds later. As one rolled underneath St. Michael's, the bells rang for the last time that night, and the handsome bell tower clock stopped at 9:51 p.m.[8] In less than a minute, the horrendous shaking stopped. Sensing that this might be their only chance to escape, the remaining newspaper staff finally scrambled down the stairway to the door.

As they ran outside, they clawed their way over a pile of masonry wreckage and rushed to seek refuge in the center of the street, as far away as possible from any building that might collapse upon them. Looking back at their offices, the men were shocked to see that the debris pile at their feet had been created when the handsome granite façade of their building separated from the front wall. The cornice from the roof had been torn off and had crashed down into the street.[9] The pile of masonry created by the roof coping, the multi-tiered cornice, and the wide decorative frieze consisted of large granite blocks, some of which weighed several tons. The quake had ripped the massive stones off the forty-five-foot-high building and flung them ten to fourteen feet into the street as if they were children's blocks. Some stones hit the street with enough force to break the underground water mains.[10]

On Broad Street, the first things that confronted Carl McKinley were anguished cries of pain and fear, the terrified prayers and wailings of women and children, and the shouts of bewildered, excited men. The nightscape before him was as bizarre as the event itself. Fallen bricks filled the street for a quarter of a mile from Meeting Street to the Old Exchange Building, which housed the U.S. Post Office.[11] A noxious cloud of whitish, dry, stifling dust, created by powdered lime and mortar from the shattered masonry, filled the air as high as the housetops. Through this dense fog of dust, the gaslights that usually illuminated the streets flickered feebly, causing the quake survivors to stumble over piles of bricks or become entangled in broken telegraph wires.

A few doors west of the shattered *News and Courier* office, W. W. Smith's stencil factory at 27 Broad Street and a cigar and tobacco shop at 25 Broad Street were nearly unrecognizable. The cornices of both brick buildings had been shaken free and slammed into the ground. Then the front wall of the second and third floors peeled off and crashed down, fully exposing all four front rooms above, which served as apartments for several families. Amazingly, the ground floor storefronts, with their extensive glass windows, were virtually undamaged.

Up and across the street, the conglomerated buildings at 60-64 Broad Street were heavily damaged. These structures made up the Confederate Home, organized in 1867 as the Home for Mothers, Widows, and Daughters of Confederate Soldiers of Charleston. The damage from the 1885 cyclone had scarcely been repaired by the time the earthquake tore off parts of the roof and threw down the gables at the southern and eastern sides, terrorizing more than one hundred frail, aged widows.[12]

Throughout the maze of shredded buildings and toppled telegraph poles, men and women rushed or staggered in every direction. Injuries were common. Samuel Hammond was thought to be fatally wounded when he fell from a third-story window on Broad Street. He broke both hips, both legs, and his left arm. When help arrived, Hammond said that he did not know whether he had jumped from the window or was thrown out. Despite his shattered limbs, he had somehow crawled out from the rubble-strewn sidewalk to the middle of the street. Dr. Peter Gourdin DeSaussure heard Hammond's heartrending shrieks of pain, rushed up, and took the gravely injured man in his arms. Fearing that his life was at an end, Hammond begged the doctor to seek his own safety. Both men survived.

W. W. Smith's stencil factory, 27 Broad St.

As McKinley walked through the streets, he absorbed everything he saw. Because it was evening when the quake occurred, some survivors were in nightclothes, some were partially clothed, and some were almost naked. Everyone was crazed with fear and excitement. A husband supported his shocked wife, whose arms hung listlessly by her sides. Her head fell back on her spouse's shoulder, and she moaned repeatedly as he tried to whisper something encouraging to hide his own panic. A few steps away, under the gas lamp, a woman lay prone, face up and motionless, on the pavement. The stunned crowd that had gathered in the street passed her by, and no one paused to see whether she was dead or alive.

The scene was one of utter chaos, but the veteran newsman reported it all in extraordinary detail. "A man in his shirt sleeves, with blood streaming over his clothing from a wound on his head, moves about among the throng without being questioned or greeted; no one knows which way to turn, or where to offer aid; many voices are speaking at once, but few heed what is said; you take note of all these things as one in a dream. The reality seems strangely unreal; and through it all is felt instinctively the presence of continuing, imminent danger, which will not allow you to collect your thoughts or do aught but turn from one new object to another."

A sudden flare of light illuminated the gloom of the chalk-filled air. A cry of "Fire!" rang out from the crowded street. McKinley and some of the crowd rushed towards the spot. On his way, McKinley spotted a man lying doubled up, silent and helpless, against a wall, but all thoughts of providing aid vanished when a low ominous noise signaling another tremor began. As the sound grew louder and nearer, like the growl of a wild beast swiftly approaching its prey, the crowd began a frenzied rush for open space.

A strange fragility hung in the air. "Shattered cornices and copings, the tops of the frowning walls, lie piled from both sides to the center of the street," McKinley wrote. "It seems that a touch now would send the broken masses left standing down upon the people below, who look up to them and shrink together as the tremor of the earthquake passes under them, and the mysterious reverberations swell and roll along like some infernal drum-beat summoning them to die. It passes away, and once more is experienced the blessed feeling of deliverance from impending calamity, which, it may well be believed, evokes a mute but earnest offering of mingled prayer and thanksgiving from every heart in the throng."

After the second severe shock passed at 9:59 p.m., the dust literally began to settle, revealing devastation. "The real force of the shock and severity of its destruction just began to dawn upon the people," McKinley wrote. Anxious to find his family, he cautiously made his way up Broad Street toward Meeting, following the rubble-filled streets toward his home on Coming Street. He could not abandon his journalistic instincts completely, however, and the record of his observations later brought the horror of that night vividly to life for his readers.

Along the way, McKinley saw that all the main thoroughfares reflected the destruction he had witnessed near his office. "Broad Street was instantly filled with men and women and children in all conditions of dress," he wrote. "Men in their shirt-sleeves, women and children in their night clothing, just as they rushed from their beds to escape impending destruction. There was a scene of the wildest confusion for a few minutes, but the exertions of a few cool-headed men soon brought affairs to a state of order, and had the people stationed in the safest possible condition." In the dark shadows of the night, he thought he saw a glimmer of hope. He wrote, "St. Michael's steeple towered high and white through the gloom, seemingly uninjured." Sadly, daylight would shatter this illusion. Pervasive cracks ran through the walls of the building. The steeple's tower had separated from the main sanctuary and had sunk eight inches into the earth. It looked as if the entire structure might collapse or have to be pulled down.

St. Michael's Church

The effect of heavy porticoes—common in Charleston—on the buildings to which they were attached soon became apparent. Turning onto Meeting Street, McKinley passed between city hall and the county courthouse, which did not seem to be heavily damaged. However, the morning light would show city hall cracked, with its rear wall bulging out. Across Meeting Street, large cracks appeared below the second-story windows of the courthouse, and its pediment had been thrown onto Broad Street.[13]

A block further up Meeting Street stood the Fireproof Building, designed in 1822 by Robert Mills to provide a secure repository for public records. Built of stone and iron with a copper roof and massive twin Doric porticos, the immensely solid building housed the records of the Register of Mesne Conveyances (register of deeds), the probate court, and various county offices. To make it less vulnerable to fires, the architect had persuaded the city to create a block-long park, known as Washington Park, as a firebreak between the building and the rest of the city. By the time McKinley reached it, the Fireproof Building had lost brownstones from its north and south pediments, but the rest stood solid, and the adjoining park was starting to fill up with dazed citizens seeking protection from falling masonry.[14]

Opposite the Fireproof Building stood the remains of the Hibernian Society Hall. Built in the Classical Revival style to resemble a Greek temple, the handsome forty-five-year-old hall had not withstood the onslaught of the earthquake. It was now one of the most severely damaged buildings in the city. It housed the Hibernian Society, which was founded in 1791 as a social club that raised funds for the relief of distressed immigrants. Since its founding, the society had been an ecumenical expression of Irish heritage, electing Catholic and Protestant presidents in alternate years. Loved by all white Charlestonians, the hall hosted numerous social gatherings and patriotic meetings.

On the night of the earthquake, the roof of its handsome portico on the east face crashed to the ground, completely demolished. In its fall, the portico roof shredded the outer four of its six massive, richly fluted and stuccoed brick columns and decapitated the two remaining center ones, throwing their Ionic capitals to the ground. In the process, the corners were ripped off the building, which also sustained cracks and fissures.[15]

Hibernian Society Hall

The Charleston Hotel during the earthquake

However the hall's badge of honor—a rendering of an Irish harp on the Meeting Street face—was left unscathed on its nakedly exposed front wall.

As he looked north up Meeting Street, piles of debris lined both sides of the street as far as McKinley could see. When he passed the imposing Charleston Hotel, the city's largest, he felt another tremor and watched as hotel guests who had stayed inside now fled the building. In the initial shock, the center portion of the parapet of the hotel's block-long Corinthian colonnade had hurtled to the sidewalk below, crushing the two ornate gas lamps that flanked the handsome entrance door.

Inside, Thomas H. Tolson, a hotel guest from Baltimore, had a near-death experience during the initial shock. That evening, he had been sitting outdoors, opposite the hotel, talking with an old gentleman. About 9:45, Tolson said good-night to his companion, returned to the hotel, and went up to his third-floor room. As he was lighting his gas lamp, he heard a strange noise, as if he had broken something. After inspecting the room and finding nothing damaged, he went to place his hat on a bureau.

The shock came without warning, and the lights went out. Desperate to keep from falling, Tolson flung out his hands in the darkness and clutched the window frame. The terrified guest thought the hotel was being lifted up and swung backward and forward a distance of fifteen or twenty inches with each vibration. Galvanized by panic, Tolson managed to get out into the hotel corridor and grope his way toward the street in utter darkness, amid falling plaster. When he reached the ground floor, the air was filled with plaster dust, and there was a dim light from gas lamps outside. All around him was a terrible roaring and moaning, and the din was heightened by the sound of falling timbers.

Frustrated at finding the front door of the hotel closed, Tolson groped about, trying to find the knob. As he was struggling with the door, tons of bricks from the upper part of the building fell in front of him. Had the door been open as he tried to escape, Tolson would have been crushed to death. Stunned by his narrow escape, he ran out through the heaps of masonry, falling twice while trying to get to the middle of the street.[16]

John and Mary Robinson

One young couple, John Webb Robinson, twenty-six, and his talented, petite wife, Mary Allston Phillips Robinson, twenty-two, had married less than a month before the earthquake. As did many Charleston newlyweds, John and Minnie, as she was called, chose to live in a boarding house after their marriage. They made their home at 6 Glebe Street, a handsome Georgian double house built c. 1770 as the parsonage for the rector of St. Philip's Church. By 1886, the building had become known as Mrs. Eason's boarding house. The young couple was jolted awake by the first shock, which threw a treasured wedding present—a heavy, ornate French mantel clock—to the floor. Its glass face cracked, but it kept running, and the Robinsons also survived.[17] Others were similarly fortunate. In one home during the initial shock, a man sprang from his bed, seized his child from its cradle, descended two flights of stairs, and slowly made his way down a long hallway in the darkness. As he went out the door to reach his garden, he narrowly escaped being crushed by a falling cornice.[18]

Those who immediately fled their homes were often killed or injured by falling debris. The walking wounded, dazed and moaning from being struck by bricks and masonry, were all over the streets. At the corner of Meeting and Market streets, a black woman lay unconscious. A policeman reported that he had seen two dead bodies in King Street, north of Broad. Above the city market, turkey vultures (dubbed "Charleston eagles" by the aristocratic Charlestonians), which lived by scavenging scraps from the butchers, perched ominously, sensing death in the air.[19]

"Charleston eagles" view the wreckage

On the south side of Market Street at Meeting, Henry Steitz, an importer and wholesale dealer in foreign and domestic fruit, had a full-blown disaster on his hands. His three-story building directly across from Market Hall had lost its cornices and virtually all of the front face of its top two stories, exposing the residential quarters above. All of the masonry had crashed to the street, covering the Market Street streetcar rails and leaving the top stories of the building as exposed as a child's life-sized dollhouse. In a bedroom in the now open-air top-floor apartment, a reporter noted, "a woman was seen to deliberately take down her mosquito net, put out the light, and leave the room. This was long after all the inhabitants had fled."

Steitz's fruit store

While Carl McKinley was making his way home, collecting his vivid impressions of the catastrophe, his boss, Capt. Francis W. Dawson, editor and publisher of the *News and Courier*, was on the second floor of his substantial masonry house when the first shock hit. Formerly the managing editor of the Charleston *Mercury*, Dawson and B. R. Riordan, former managing editor of the Charleston *News*, had purchased the *News* from its financially troubled owners shortly after the Civil War ended. Soon thereafter the *Mercury* closed its doors, and in 1872, Dawson and Riordan bought out their remaining competitor and renamed the merged enterprise the *News and Courier*.[20]

Dawson, a native of England, a devout Roman Catholic, and a former officer in the Confederate Navy, was a hard-working, seasoned businessman and a trained journalist. He soon made the *News and Courier* one of the most-respected and profitable newspapers in the southern states. By 1870, he had purchased an impressive house on Bull Street.[21] Just after the earthquake hit, he later told one of his reporters, "The house seemed literally to turn on its axis. The first shock was followed by a second and third, less severe than the first. The air was filled with the cries and shrieks of women and children. From every side in that quiet neighborhood came the cry, 'God Help us!' 'God save us!' 'Oh, my God!' It was worse than the worst battle of the war."

"When the first agony was over," he continued, "it was found that the ceiling of every room in the house was cracked, the big cistern was broken apart, the huge tank in the attic was pouring its flood of water into the bedrooms. In the parlors the statues all had been wrenched from their bases and thrown to the floor. In the hall the massive lamp had actually been turned around. In front of the house was a large porch, with heavy pillars and solid marble steps; all this was swept away as though it had been shaved off with a razor. The beautiful house was a wreck."

In another part of the city, young Harriet Kinloch Smith was being terrorized. Her father had been downstairs shutting up the house for the night, and Harriet and her brother, Willie, had gone to their rooms. As Harriet lit her lamp and went to her dressing table, there was a deafening noise, and the house rocked like a cradle. At the same moment, the staircase outside her room crashed to the ground. As she watched in horror, part of the northeast wall fell in, and she was trapped. Believing that she would be killed if she moved, Harriet stayed quiet, holding onto the bedstead until the shaking stopped.

Hearing Willie's frantic cries of "Harriet, Harriet!" she ran out of her room to find the stairs almost gone and the entire north wall fallen in. Losing her balance, she began to skid down a heap of bricks and mortar, but Willie caught her, saving her from serious injury. Reassured that she was all right, her father immediately took his son and daughter and ran out into the street to confront a scene of unimaginable horror. Crowds of half-dressed people roamed the area in shock under a sky lurid from the glare of immense fires. Bricks fell like hail in the frequent aftershocks, and terrified people moaned, sang, and prayed aloud.[22]

As with the Smith household, domestic tranquility at the Robson residence on Coming Street was shattered when the

Falling chimneys and cornices caused most of the early deaths

shock threw their home into the air and smashed it back to earth. The south wall fell, taking with it two piazzas, burying the family beneath the wreckage. Elizabeth Moffett, the Robson's neighbor, wrote to her daughter, Anna, "Our chimneys with most of those around sh[e]ared off to the roof—houses cracked & precious lives lost—the Robsons' piazzas torn off clean—Ainsley & his two sisters Mary & Sallie buried in the ruins—they, surrounded by the timbers, [were] but only bruised; he [was] struck upon the head—his sister held out her hand to him, & prayed with him. He was conscious when taken out but died in a few minutes." [23] He was trapped in the wreckage for three-quarters of an hour before he died. Twenty-eight-year-old Ainsley H. Robson had lived his entire life in Charleston. His funeral was held in the beautiful flower garden adjoining the house the next day, and he was buried in Magnolia Cemetery.

The Moffetts all survived, but their badly damaged house was nearly burned down by an oil lamp, which was upset by an aftershock and set fire to an upstairs room. After extinguishing the blaze with water-soaked blankets, the family spent the night sleeping on the ground floor piazza, ready for flight. Their neighbors, the Mazÿcks, slept in their stable. [24] "I never before, except during the Yankees' visit in Fairfield [South Carolina], experienced such terror," Elizabeth Moffett wrote. "I was thoroughly demoralized, though outwardly calm, but God grant I may never again know such an experience." [25]

Several blocks southwest of Broad Street, the venerable Miles Brewton House, a magnificent Palladian mansion at 27 King Street completed in 1769, had survived the earthquake, although the cataclysm gave its residents a severe scare. Susan Pringle, a third-generation step-niece of Miles Brewton, the wealthy merchant and slave trader who had commissioned the house to be built, was fifty-seven years old at the time. She was living in the house with her widowed brother, William Alston Pringle, the city's municipal court judge, then sixty-four. Their brother,

Susan Pringle

Motte Alston Pringle, lived nearby. William had been sick for several weeks and had gone to bed early.

Shortly after 9:00, Susan also began to prepare for bed. As she stood by her dressing table reading a book, the huge brick mansion rocked, swayed, and trembled like a house of cards about to fall. Susan could hardly stand, but somehow, she made it to William's room. The door was jammed shut, so the panic-stricken woman dashed through the drawing room, where the chandelier was literally flying in circles on the ceiling, to reach a second entrance to William's bedroom. When she reached her brother, he was sitting up in bed, nearly imprisoned by heavy, fallen bookcases. Having been reunited, Susan and William stood together in the bedroom waiting to see what would happen next.

After the second shock came, their brother Motte arrived to see if they were safe. When the quake finally subsided, Susan and William hurried downstairs, having gathered up money and important papers, and sat on chairs by the garden door, ready to escape into the street at a moment's notice. William, who was too sick to go any further, urged Susan to leave the house for a safer place, but she refused, insisting that if he were to be killed, then she would stay and be killed, too.

At daylight, the siblings took stock of the situation. Despite fallen chimneys and damage to the roof, their house was in good shape. There were no cracks in the walls, and the foundation and lower stories were intact. However, the outbuildings, with the exception of the old kitchen, were in ruins.[26] In the servants' courtyard, the water from the well "came up like a waterspout, overflowed the yard, and deposited six inches of sand for a distance of twenty steps around the well."[27] On Wednesday, Motte Pringle would find his aunt, Emma Alston, wandering near her house on the battery with her children. Her home, which was devastated by the earthquake, was not fit for habitation, so Motte brought his aunt and her family to stay with William and Susan until other accommodations could be found.

Death played no favorites during the earthquake. Chance and luck ruled the night. The wicked and the virtuous, the

The Main Guard House

black and the white, survived or died according to the roll of unseen dice. A block to the east of the Pringle residence, Robert Alexander, the amiable twenty-four-year-old analytical chemist from London, was in his room in at 20 Meeting Street.[28] When the first shock hit, he ran out the door and was struck by falling bricks. His horribly mangled body was taken to the City Hospital, where he died. His funeral was conducted at the Citadel Square Baptist Church on September 2, and he was interred in a donated plot at Magnolia Cemetery, far from his family and home in England.[29]

Washerwoman Sarah Middleton, who, moments before the quake, was laughing with her relative Susan at the end of a long day, was also one of the first to die. In the dark of night, illuminated only by the gas light at the southwest corner of Broad and Meeting streets, her lifeless body was pulled from the debris of the Main Guard House about one o'clock in the morning. She had been passing by the massive, two-story building as the upper edge of its east wall crashed to the ground, crushing her to death.[30] Sarah's grieving husband, Jacob, later retrieved her body, and she was taken to Beaufort, South Carolina, for burial.

Inside the Guard House, police Pvt. Ashley B. Haight was sleeping soundly in the policemen's dormitory on the second floor when the quake hit.[31] Despite the fact that his wife was alone at home with five small children, Haight loyally remained at his post and carried out the instructions of his chief. When he finally had the opportunity to go home, he had to extricate his family from the ruins of their house. Two of his children had been hurt by falling debris.

At 100 King Street, just below Broad, parts of a falling wall struck and killed three young black people, all neighbors, in a single deadly stroke. The victims included Grace Fleming, a nineteen-year-old nursemaid; Alexander "Aleck" Miller, a laborer; and twenty-three-year-old Joe Rodolf, also a laborer.[32] A few days later, aftershocks and a heavy rain brought the rest of the wall crashing down, this time causing no injuries.

Father Patrick L. Duffy, a much-loved, thirty-five-year-old Catholic parish priest who lived at Bishop Henry P. Northrop's Broad Street residence, was talking with his friend, Father Schachte, on the side porch when he heard a sound like thunder approaching. "In a few seconds the shock was upon us," Duffy wrote. "The other reverend gentleman [Father Bagley, who was inside] rushed for the door to the street but fortunately it was jammed by the debris, for the front gable and porch of our grand old building went crashing to the ground. If they had stepped out they would have been crushed beyond recognition.... After the first and severest shock I passed out over the piled up brick and lumber which had fallen, and received the next shock in the street amid the cries of the dying and the shrieks of the frightened people." Ignoring his bruises and the tears in his clothing, Duffy set off into the night to minister to his flock.

A few blocks away, at 119 Meeting Street, Irishman Maurice J. Lynch was standing in the front door of his son's saloon. He was enjoying the night air and totaling up the profits in his head when a stone column snapped and the large granite block over the saloon door fell and crushed him, breaking his legs. A friend rushed to the bishop's residence and left a frantic message that a parishioner was mortally injured and desperately needed a priest. The bishop was out of town, but minutes later, Father Duffy returned and learned of Lynch's fate. Duffy had already been busy that night, stumbling in the darkness across piles of brick and stone and fighting his way through snarled thickets of telegraph wires, hearing confessions on the streets and consoling the injured.[33] He rushed to the saloon, knelt over the dying man, removed, kissed, and replaced his priest's stole and read Lynch the Last Rites of the Roman Catholic Church. Father Duffy recalled, "As I stepped back and removed my stole, I met my friend, Capt. Dawson, of the *News and Courier*, who was hurrying to his office. Our escapes were similar and the rents [rips or tears] in our clothing identical."

Maurice Lynch died of his injuries just before midnight, and became the third of twenty-seven victims (seven white, twenty black) entered into the "Return of Deaths Within the City of Charleston" for the evening of August 31, 1886.[34] Eight more (three white, five black) were added the next day, and the official list of earthquake-related deaths within the city limits grew to eighty-three (twenty-two white, sixty-one black) by the end of October. Later inquiries would determine that the actual death and injury toll was far higher.

Many of the injured undoubtedly died later, but they were not recorded on the city's official rolls. About the same time that saloonkeeper Lynch was fatally injured, two Good Samaritans near the Pavilion Hotel, at the corner of Meeting and Hasell streets, heard piercing cries for help. They threaded their way through the debris to the Lazarus building at 66 Hasell Street, halfway between Meeting Street and Maiden Lane. There they extricated a white man and woman, half-buried in the ruins. The injured couple was placed in a horse-drawn wagon volunteered by a Mr. Pickett and taken to the City Hospital.[35]

The daughter of Mrs. M. J. Williams, made a drastic choice. To save herself from the shaking building, she jumped from the second story of her residence at the corner of Wentworth and Meeting streets, severely injuring her spine. Opposite the Pavilion Hotel on the corner of Meeting and Hasell streets, a brick struck Mrs. E. Galliot, a black woman, and her head was badly injured. Her daughter suffered the same fate. Mrs. Robert Martin, wife of a shoe merchant in Market Street, near King, was also badly hurt. Hundreds of such stories filled the newspaper in the following days, with only the names differing.

Carl McKinley's journey home would have taken him past the train station, but he had no way of knowing the dramas that were unfolding there. As the earthquake began, the people who had gathered to meet the incoming Northeastern Railroad train felt a rumbling and vibration such as a heavy train would cause and called, "There's the express!" The noise died out for an instant, but it quickly recommenced, this time getting louder and shaking the building more as it approached. Shouts of "The train is running away and will crash through the building!" were heard, but no train could cause what quickly took place. The long, low station began to tremble and then to vibrate. The ceiling appeared to lower itself several inches as the structure bowed and quivered. The building seemed to stand on both ends at once, and it began to shake. The agitation was so violent that people were thrown off their feet as they tried to get to the street.[36]

When the first shock wave hit, nine-year-old Bertha Cade and her older sister Bessie, daughters of Henry L. Cade, a well-known Charleston building contractor, were waiting at the station with their father for their mother, who was returning from the mountains. The large crowd stampeded wildly for the door, separating Bertha from her family. Whether Bertha was crushed by the crowd or struck by a falling fragment from the building is unknown, but she was severely injured. By the time her father found her, she was unconscious and remained that way for three days, except for a few moments when she complained of violent pains. She died on September 3. Had everyone stayed inside, injuries would probably have been few and minor, as the building itself lost only two joists and a little plaster.[37]

The fleeing crowd escaped the terrors inside the train station only to find the world outside in chaos. Housetops were tumbling in every direction. Carriage horses, whose drivers had run away in fright, dashed down streets in a tangle of hubs and wheels—a regular stampede, which threatened another form of death to those who sought safety in the street. Cobblestones in the streets seemed to shrink and surge and vibrate.[38]

Those who stayed in the station anxiously awaiting the train from Columbia, which was late, were elated when the train arrived. People scrambled aboard to find out if their loved ones and friends had been injured by the shocks. To the amazement of those who were waiting, the passengers on the train, which had been only three miles north of Charleston when the earthquake hit, had not felt a thing. "The sleepy passengers could not realize their wonderful escape until they had gotten out and seen the ruins of houses and heard the cries of fright. Then they thanked God for their deliverance."[39] They had good reason to be thankful. By a freak coincidence of geology, the rails seven miles behind them had been twisted into grotesque shapes by the quake, causing a fatal train wreck just minutes after they passed the site while traveling blissfully towards home.

As some of the former train passengers and their friends and families made their way south down Concord Street to the Mt. Pleasant Ferry Company's wharf on East Bay Street to be transported across the Cooper River, they found that the relative good luck they had experienced on the train and in the depot had not extended far.

> Through the darkened streets (the street lamps had been extinguished) cries of pain and the sickening groans told plainly that destruction of life and limb had been wrought. Stretchers and chairs had been brought on the sidewalk, on which the dead and wounded were being placed. Mothers called their lost children, praying at the same time that ruins had not crushed them. Fathers gathered their families in what was thought to be the safest places, and together they trembled and asked Heaven's mercy. To make the scene more ghastly the heavens were

> beginning to take on a lurid glare from burning buildings in various parts of the city.... Superstitious people thought the glare was the dawning of the day of judgment, and consternation seemed of the universe for a while. At Market wharf a crowd of passengers had gathered, waiting for the departure of the boat. The quake of course, was the absorbing topic. Reports of the condition of the city were sought and discussed eagerly, each succeeding report bringing news of new casualties. Every man who had been a soldier, agreed that the horrors of war were nothing in comparison with what had just passed.[40]

When Carl McKinley finally arrived at his home on Coming Street, he found the same level of destruction he had passed everywhere along the way. Every house in the vicinity was deserted, and the alarmed residents huddled in the streets, "trembling and fearful, awaiting the end, whatever it might be. Invalids had been brought out on mattresses and deposited in the roadway, and, together with the aged and the infant, were cared for as tenderly as possible. No thought was given anywhere to treasures left behind in the effort to save the priceless treasure of life itself—suddenly become so precious in the eyes of all who were threatened to be bereft of it."

The terror-filled night seemed endless. "The long, anxious watch between midnight and day was not less trying than the shock itself," he wrote. "The suspense was indescribably painful, and had no relief for a moment, save when it gave place to recognition of the approach and presence of renewed danger. That passed, the breathless vigil began again, and the moments seemed as hours and the hours as moments until the next dread visitor had come and gone. Four severe shocks occurred before midnight. Three others followed at about 2, 4, and 8.30 o'clock, a.m., and every shock after the first caused even more alarm, naturally, than the first itself."

With no knowledge of where the shocks came from, and with the possibility that the worst was yet to come, everyone feared that greater disasters lay ahead. "Whether the blow had come from the sea or from the land none could say," McKinley reported. "At any moment another might be felt that would rend the earth asunder, or burst the bounds that held the waiting ocean in check and drive its waters sweeping in an overwhelming wave over all the low-lying peninsula where so many thousands were collected together without hope of escape. Night and distance shut out all the world. No word could be heard from beyond the confines of the stricken city; no human hand could be stretched to save a single soul, whatever fate was impending. The silence, save when broken by piteous cries, was oppressive in the extreme. In the late hours of the night even such cries would have been a relief to senses that were strained to so great tension to catch the first footstep of coming danger, the first low moan of the earth in the throes of convulsion."

"The air itself was strangely still," he continued. "In [my] garden an unprotected lamp burned until 4 o'clock, or later, with a flame that did not once waver. All nature seemed to be waiting in breathless suspense for the issue of the hour, of the next minute, the next moment. And then! always with startling suddenness, the great fearful Power rushed out of the darkness upon the city, shaking the ground with his tread, sending terror before him, and leaving trembling thousands panting in dismay as he passed." "It seemed," McKinley wrote, "that God had laid His hand in anger on His creation."[41]

Charlestonians were now cut off from the world. They had no way of knowing that the deadly forces of the night had struck with even greater fury twenty-two miles up the South Carolina Railway tracks in Summerville.

3

Panic in the Pines

At every attempt we made to reach the door, we were hurled backward and forward and from side to side, as if we had been in the gangway of an ocean steamer in a heavy cross sea. —Thomas Turner

August 31, 1886

At sunset on that fateful Tuesday night, the people of Summerville were preparing for their evening rest. The day had been unusually hot. Scorching sun had beaten down on the village, and scarcely a breath of wind ruffled the pines. By early evening, the dust had settled on Main Street, and most of the villagers were home with their families. By 9:51 p.m., children were already asleep in bed, and their parents were ready to retire. Oil lamps glowed through lace-curtained windows. Sleeping hammocks, shrouded by mosquito nets, were strung on porches to catch the meager evening breezes. The strange booms and shakes, which had spooked the local people during the previous week, had largely been forgotten. The essence of the evening's serenity was captured by a Summerville resident, who wrote, "except for an occasional soft voiced conversation, the town had yawned and stretched and prepared to curl up until morning. If a town crier had been present, he would have inspected his surroundings and confidently intoned an 'all's well.'"[1]

For a hundred years before the 1886 earthquake, Summerville was known chiefly as a quiet, healthy, restful place. Located on a ridge of relatively high ground covered by pine forests, it became a haven for local planters, who started visiting the village to escape the biting insects, yellow fever, and malaria that infested their swampy Ashley River rice plantations from April through November.

The nineteenth-century village was the successor to a much earlier settlement, located five miles to the west, and known as the village of Dorchester. In 1696, a brave group of Congregationalist emigrants from Dorchester, Massachusetts (whose ancestors had emigrated from Dorchester, England), moved to the South Carolina Indian frontier and founded a religious community twenty-two miles north of Charleston on the Ashley River. They built their house of worship, the Independent Congregational Church, on the Dorchester road, which ran northwest from Charleston to Dorchester, and then continued on as a trail to Indian trading posts further inland. It was known locally as "the White Church" for two reasons. First, it had been named in honor of The Reverend John White of England, sponsor of the Dorchester, Massachusetts, settlement in 1630. In addition, its builders whitewashed the brick structure to indelibly differentiate it from its nearby religious rival, the Anglican parish church of St. George, Dorchester, in the village of Dorchester.

The Church of England, or Anglican Church, became the official state church of the South Carolina colony in 1706. As Dissenters—those who refused to worship according to the rites and rituals of the Anglican Church—the Congregationalist residents of Dorchester faced overt religious and political opposition. By 1756, after struggling for three generations, they abandoned the village and moved their entire community to Medway, Georgia.

With none of the original settlers left, what little remained of the village was soon deserted and fell into disrepair. By the time Summerville was first settled in 1785, scavengers had been dismantling the old parish church and other abandoned Dorchester buildings for years. They reused the brick on their plantations and to build houses in the new village of Summerville, which was incorporated in 1847. By 1820, the parish church of St. George, Dorchester, was in ruins. By 1858, only the bell tower remained.

The only remaining structure of any size was Fort Dorchester, built in 1775 overlooking the Ashley River in Dorchester village. The fort was constructed using a concrete made of tabby—oyster shells embedded in a lime mortar obtained by burning oyster shells. The result was a material as strong as newly cut granite. By 1886, its walls stood as firm as ever, although the wooden and brick structures inside had long ago been dismantled and the materials carried off.

Unhampered by religious constraints, Summerville grew. It got its first railroad connection with Charleston in 1830, and by 1886, the village was home to about 1,840 people. Catering as it did to Charleston's business elite, the village had a post office offering three daily mail deliveries, "express and telegraph offices, churches of all denominations, good schools, both public and private, for both sexes, good shops of all sorts, and very able physicians."[2] For its residents and numerous visitors, Summerville was almost idyllic.

On the night of August 31, 1886, it was life as usual for the village that had become a suburb of Charleston. Trains rolled out of Summerville five times a day for the forty-minute trip to the city. Thomas Turner, president of the Charleston Gas-Light Company, was enjoying his bungalow residence in Summerville. That morning, before leaving for work at his Meeting Street office in Charleston, he noticed that his cow, which was turned out each day to graze, was reluctant to leave and repeatedly returned to the barn. When his busy day at the office ended, Turner returned to the village via the South Carolina Railway commuter train about 7:00 p.m. After dining and spending a pleasant evening with his family, he left the house to water his horse and see that all the gates were closed. The night was "unusually sultry but clear," he later wrote, "and beautiful starlight."[3] His family members had already gone to bed. He would recall what happened next for the rest of his life.

> I had been out in the garden and admired the beauty of the evening, and was entering the door of the Hall when, without any rumble or premonitory symtoms [*sic*], just as I was stepping in at the door, for a single instant the floor seemed to sink from under me. I seized the door jambs to steady myself, when the door seemed to go down in front of me at an angle of about 25 to 30 degrees. It was so sudden and unexpected, that I was thrown forward into the Hall about ten feet—and as quickly thrown backward, ere I could fall onto the piazza, and again thrown forward into the house.
>
> At this moment I observed my Sister-in-law crawling on all fours, she having been thrown from her room on the Southwest which she was just entering into the middle of the sitting room, and, amidst the rolling and rocking of the building, she managed to get into the Hall, but unable to regain her feet. At this instant, we observed the upper part of a kerosene lamp, which had been jerked off its stand, fall on the floor, and bursting, the oil took fire, leaving the base of the lamp intact on the bureau, and, amidst the roaring and violent motion of the house, we managed to extinguish the fire with pieces of carpet and rugs from the doors, and immediately received another shock which threw us from side to side of the Hall.
>
> Having got the members of my family together, and supporting my niece who was in a fainting condition, we endeavored to leave the

house, amidst the crush of falling chimneys and plaster—but, at every attempt we made to reach the door, we were hurled backward and forward and from side to side, as if we had been in the gangway of an ocean steamer in a heavy cross sea. After some little delay, we succeeded in reaching the garden. I then went back to get some wraps and chairs for the ladies, and again experienced severe shocks and rumblings—these were repeated at intervals of several minutes during the whole night but not of as violent a nature, although the earth wave was quite perceptible, and at times upsetting a small stable lantern placed on the ground. Walking about the lot during the night I was conscious of a sulphry smell, very strong at times, which I attributed to gases escaping from the fissures in the earth.

On examination of the house, we found that the hearths and fireplaces had been shaken out, and gone down under the building, carrying fenders [fireplace screens], fire irons etc. with them. The chimney to two bedrooms on the West, was thrown to the Southwest and the one to the East room, was thrown to the Northeast and the house was moved and the piles [pilings] carried over 13 inches in a Northeasterly direction. The main chimney communicating with the dining & sitting room, seemed to have crumbled at its base, the main body sliding down and driving the mantles into the middle of the floor. As we considered the house too dangerous to occupy under the circumstances, my family camped out in the garden for several nights, and, procuring some bedding from the house, they made themselves as comfortable as circumstances would permit in a portion of the

> stable, from which they were compelled to fly in the night several times during the week they remained until I could get them to town.[4]

The final federal report on the earthquake would vividly describe Summerville's plight.

> The main shock had struck the village without any warning. It broke upon the village not as a *crescendo,* but in full force. At the first impulse, before people had time to realize what was happening, they were tossed from side to side, or flung to the ground from which they could not rise. For a few moments it was impossible to stand or walk. Most of the people were within doors. First one corner or side of the house was violently raised or knocked up, then another corner or side. The structure seemed to dance up and down from the effects of rapid and powerful blows beneath it. The floors and ceilings were warped and twisted; the timbers groaned and crackled; the chimneys, crushed at their bases, sank downward, carrying fireplaces, mantels, and hearthstones through the floors to the ground below.[5]

In both Charleston and Summerville, tall, heavy wardrobes filled with clothes proved to be major hazards when they toppled over. In Summerville, a woman was reading in her home when the earthquake hit. A wardrobe fell next to her chair, pinning the skirt of her dress to the floor. Knowing that her baby was sleeping upstairs, she tore off the trapped portion of her clothes, grabbed the infant from its crib, and fled the house. An ailing woman was nearly killed when a wardrobe fell and crushed the headboard of her bed.[6]

It was only fitting that John B. Gadsden, the former skeptic, had one of the most memorable experiences during the earthquake. Gadsden had worked late in Charleston on August 31 and was unable to reach home until about 7:00 p.m. A friend persuaded him and his father, the principal of Summerville High School, to attend a political meeting at the school that night.

The elder Gadsden had been elected president of the organization at the meeting, and in his acceptance speech, he remarked, "The last time I was elected president of a political club, and in fact the only time in my life that I had anything to do with politics, was in [18]61. I trust that nothing as serious as that which then called for my service [i.e., the Civil War] is now at hand."[7] His son would later recall those words, noting, "Many have since said, 'Should we have to choose between the horrors of war and those of an earthquake, the former would quickly be our choice.'"[8] Six weeks after the chaos that followed, the younger Gadsden would record his experiences for posterity.

> After all business before the meeting had been transacted, and just as a vote was about to adjourn, the shock which has proved so fatal to this part of the country, and which has been felt over such a vast space of territory, came upon us with no warning, like a strike of lightning, exactly at 9:51 p.m. I was sitting near a large double-door, and was on the point of rising to my feet when it came. I was seized by an irresistible force which seemed to carry me, whither it would. The floor, ceiling, walls, and window-frames seemed to be dancing, struggling, and doubling, as if to break themselves, and the whole house appeared to be (no other word can express it) galloping down hill. The terrible roaring and rushing noise, which filled the air, did not so much frighten as awe me, and I trust that never again may the feelings that I experienced during the shaking of that building come to me or mine.

> While the disturbance was at its height, no human voice or cry could be heard, and falling chimneys and walls were not noticed until afterwards. Standing as I was in the doorway of the hall, I was taken up into the air and thrown by the struggles of the mass of men behind me down the steps which led from the hall; but, strange to say, was not in the slightest degree hurt. And right here let me note a fact of which I will speak further on,— that is, that to my own knowledge and that of others the force of gravity was considerably lessened; for I seemed to fall slowly to the ground and not feel the slightest resistance when I reached it. I seemed to myself to fall as a feather or any other light substance would, and yet I fell a distance of from six to seven feet.[9]

Gadsden's fortuitous landing was not the end of his peril; the building immediately caught fire.

> Just before I was thrown from the hall, I saw that the lights had been turned over and that the oil had set the floor of the building on fire; so as soon as possible, though gasping for breath through my violent exertions, I made my way into the hall again, and with the help of others extinguished the flames, and not until I left the hall for a second time did the rumbling and shaking of the building cease.... I then anxiously sought for my father, who was by reason of his position one of the last to leave the building, and together we started for home, thinking of the dear ones we had left there, and ignorant of their fate. I reached home, running the whole way, some time before my father did, and found the family collected in the yard unhurt. We

> began at once to make them comfortable as possible under the thick branches of a large hickory tree, and there we spent the night.[10]

Summerville was in chaos. Everyone was severely disoriented and terrified of what might happen next. Secure in knowing that his immediate family was unhurt, young Gadsden set off to find his eleven-year-old sister, who was spending the night at an uncle's home about a mile away.

> Anxious for her [his sister's] safety, and wishing to have the family together, I saddled my horse and rode over for her. The sights and sounds I saw and heard during that ride will cling to me through life like a horrid dream, made real by the groans and terrified screams of the people I passed on the road; by the terror of my horse as he splashed through the water which poured from the fissures and covered the road in some places to the depth of two inches; while screams and howlings from terrified animals sounded in my ears, and, to crown all, successive shocks shook the earth every few minutes beneath me. I found my sister unhurt, but very much unnerved, for she had retired for the night, and was almost asleep when the shock occurred. Taking her before me on the horse, I made my way back as fast as possible.[11]

The earthquake had radically altered life for Gadsden's family and his neighbors. "All that night [August 31] and for three days afterwards the shocks continued, ten to fifteen minutes being the average length of time between each shock, and from that time up to this writing [October 11, 1886]," he recalled, "not one day or night has passed but we have been reminded of the terrors of that memorable night by numerous shocks, some slight and others again so severe as to cause us to leave our houses. In fact we did not return to our house for two weeks, although the building itself was but slightly injured."[12]

Colonel John Averill, Summerville's mayor, who was reading in his house when the great shock arrived on Tuesday night, was also rudely shaken out of his evening routine. "There came a crash, and the house, a two-story building, seemed to be lifted up, set down, tilted, shaken and twisted. The lamp and bookcase were overturned, a table seemed to be dancing on the floor, and the pictures on the wall appeared to be falling," he said. "Three distinct shocks were felt, and the utmost alarm and confusion prevailed throughout the night. Heavy detonations or shocks were heard or felt at intervals of fifteen minutes, while, as if to add terror to the scene, water was spouting at many points from the fissures [cracks in the earth's surface]."[13]

Realizing that he was responsible for the safety of the village, Colonel Averill joined his constituents outdoors to take stock of the situation. As superintendent of transportation of the South Carolina Railway, he had an additional responsibility—ensuring that trains could travel safely along their routes. Saddling his horse, Averill began a reconnaissance of the town and the surrounding countryside, concentrating on the railroad tracks

The fatal northbound derailment near Ten-Mile Hill

as he rode. What he found just a mile outside the village frightened him more than anything he had ever encountered.

At 9:35 p.m., the South Carolina Railway night passenger train left Charleston bound for Columbia via Summerville. As the train steamed northwest up the tracks near Ten-Mile Hill, the ground suddenly dropped away. As the quake hit, the train began to ride the earth's waves like a boat in heavy seas, plunging down into the trough of each wave and then rising up on the crest. A final wrenching motion to the right and left hurled the train down the railbed embankment, throwing the locomotive and its tender off the tracks and into a ditch. The black fireman, Mr. Arnold, who was leaning out of the cab, was thrown to the ground by the powerful lurch of the first shock and was gravely injured, as was the engineer, Mr. Burns.

Edward Holman, a black porter, was aboard the train when the shock derailed it. Although injured, he assisted the stunned and distraught passengers to safety and helped extricate the crewmembers from the wreck of the engine. Then, in the midst of aftershocks, and ignoring the pain of his injuries, he began to walk back towards Charleston to warn succeeding trains of the danger. Holman was unaware that his efforts would come too late to stop another endangered train arriving from the north and headed toward Summerville.

There had been great anxiety concerning an overdue southbound South Carolina Railway train from Columbia. It was scheduled to arrive in Charleston at 10:00 p.m., but because of the downed telegraph lines, the train had not been heard from, an ominous sign to those waiting for news. It was the last of the special excursion trains to run that summer, bringing hundreds of tired, happy vacationers home from the cool resort villages in the Blue Ridge Mountains. The train pulled out of Union Station in Columbia a little after its 4:30 p.m. scheduled departure time on Tuesday, but everyone was in a good mood, and few minded the delay.[14] Those extra few minutes were to make a dramatic difference in the lives of everyone aboard.

Despite the slight delay, the southbound train was running at its normal speed. Just after it passed Jedburg, about two miles northwest of Summerville, the earthquake hit, throwing the train into the air. As quickly as the train went up, it came down, falling at nearly a forty-five-degree angle to the east. Gregg Chisolm, a homeward-bound Charlestonian, reported that the entire train, traveling forty miles an hour, was lifted completely off the tracks. "Then there was a reflex action, and the train righted and was hurled with a roar as of a discharge of artillery over to the west, and finally subsided on the track and took a plunge downward evidently—the descending wave," Chisholm said. "The engineer, Mr. Keyes, put down the brakes tight, but so great was the original and added momentum that the train kept right ahead. It is said on trustworthy authority that the train actually galloped along the track, the front and rear trucks of the coaches rising and falling alternately. The utmost confusion prevailed. Women and children shrieked with dismay and the bravest heart quailed in momentary expectation of a more terrible catastrophe."[15] Miraculously, no one was injured, and the train was not derailed.

After behaving like a demented roller coaster, the train stopped, and in the darkness, its terrified passengers heard the moans and hymns, pierced by frightening screams, of black residents who lived near the railroad tracks. As the engineer began to back his train up, the work of the earthquake became evident. In its leaps and gallops, the train had actually passed over earthquake-deformed tracks that had been twisted into an S-curve. Cautiously, the engineer reversed the train and proceeded towards Summerville.

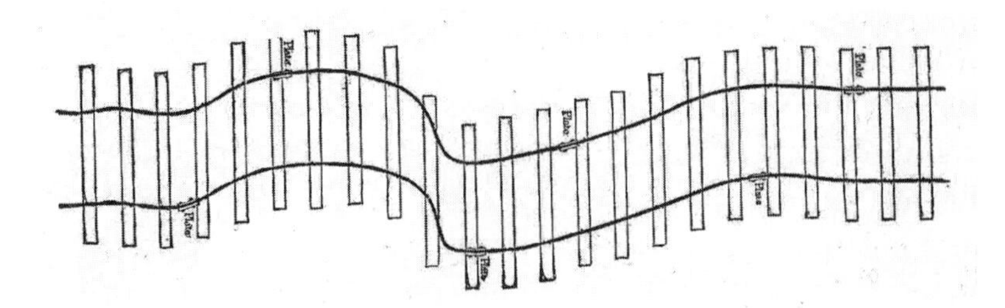

Rails bent in an S-curve

Repairing the track near Summerville

When the train was within a mile of the village, a frantic cry boomed out from the night. "For God's sake, stop!" At almost the same moment, a sharp explosion cracked through the air. It was the detonation of a railroad torpedo, an emergency warning device that Colonel Averill had placed on the track after his horrifying discovery. Severely damaged rails lay ahead. Halted by Averill's commanding voice and the warning torpedo, the train finally crept into the Summerville station using a side track.

Caught up in the excitement, J. R. Boylston, of Allendale, Clarendon County, South Carolina, a young mail agent aboard the train, found the first shock—and the resulting excitement—more fascinating than terrifying. When the train first stopped after hitting the deformed rails, Boylston took a seat on one of the crossties to take in the surrealistic scene. "Bonfires lit up the horizon on every side, and their eerie glow made frightened faces paler and eyes wider and brighter," he recalled. "Everyone spoke in whispers." Then the second shock hit, and it grabbed his full attention. "Another tremor came and the tie I sat on moved completely from under me. I got fright in a double dose that time!"[16]

The stranded train remained in Summerville for the night, while repair crews labored to fix the deformed track. According to young Boylston, the sunrise was glorious the next morning and did much to calm the passengers who had been panicked the night before. While the train was waiting in Summerville, a messenger arrived with tales of destruction in Charleston. "One lady of a prominent family, who was a passenger, received the news of the death of her son calmly," Boylston recalled. "Nothing seemed real."[17]

While the repair work was under way, a special train arrived from Columbia, carrying railroad officials and The Reverend Ellison S. Capers. A former Confederate general turned Episcopal priest (and later Bishop of the Episcopal Diocese of South Carolina), Reverend Capers was the rector of a church in Greenville, South Carolina, at the time. He brought with him a flask of whiskey, which he passed around to help calm the nerves of the terrified passengers. Mercifully, the flask never seemed to run dry. Fifty years later, J. R. Boylston recalled. "I wanted a sip so bad, but as it came to me the bishop intervened with 'you're too young, son.'"[18]

The train stayed in Summerville until 7:20 p.m. on Wednesday, when it departed for Charleston. Its passengers arrived safely in the city about 9:30 p.m. They were stunned by their surroundings. Young Boylston walked to his father's home at the corner of Smith and Calhoun streets. There he found the family's flower garden "given over to tents for those afraid to stay indoors," he said. "On that one lot six babies were born during the twelve hours after the first shock. All our food had gone to those families camping in our yard. My first bite for more than forty hours was the next morning before I left for Columbia. An old Negro mammy had saved me a cup of coffee and her last bun. That was the best food I ever ate!"[19]

As their friends and relatives in Charleston had done, the people of Summerville spent the remainder of the night of August 31 and the morning of September 1 in the open air. The great shock at 9:51 was followed in eight or ten minutes by another "of great power but inferior in force to the first. Throughout the night many minor ones were felt.... The number of these was never reported."[20] Colonel Averill, keeping watch over his town, wrote:

> The night passed slowly; the camp fires burned brightly on all sides, but few, if any, closed their eyes. The day dawned, and with it was seen the condition of the town. Not a house was occupied. The inhabitants were in the streets or their

> yards. Here a house had been shaken off its foundations and was flat on the ground, a wreck. Few, if any, chimneys were to be seen; many had gone down entirely; others had lost their tops only. Next could be seen a house with its piazza broken off and crushed or a roof crushed by a falling chimney. Go inside of some of the houses with me. Here is a cottage of about four rooms. Externally it appears to be uninjured. We enter and you see a complete wreck. Chimneys have gone down through the center of the house, carrying with them mantels, furniture, and everything in their way; ceilings were broken; floors badly sprung; timbers shattered, and everything covered with plaster from the falling walls. Truly it was a picture never to be forgotten.[21]

As dawn came, and the badly frightened villagers could see their surroundings clearly, the extent of the devastation became apparent. Matthew F. Tighe found his calling that day. The son of Irish immigrants who had settled in Summerville to farm, thirty-one-year-old Tighe was a clerk who dabbled in writing.[22] He eventually found the hamlet, and even nearby Charleston, too confining for his aspirations, which were fulfilled when he became a newspaper correspondent in Washington, D.C. He never completely lost his southern roots, however; he named his youngest daughter Dixie. That morning, Tighe practiced his prose by describing the fate of the village where he grew up. "The ruin and devastation were found to be complete," he said. "There was not a home that had not been made desolate in greater or less degree. All the chimneys had disappeared. Walls were rent in twain, ceilings had fallen, and in numerous cases the houses that rested on wooden blocks or masonry were levelled to the ground. Other houses were split from top to bottom, showing yawning chasms in their sides."[23]

The shocks continued unabated through the day, and with each passing jolt, more people decided to abandon Summerville for the Upcountry, where they hoped they would be safer. At 2:00 p.m. on Wednesday, the South Carolina Railway brought in a train of five passenger coaches to evacuate terrified villagers to Columbia. The train was quickly crowded to near- suffocation, and the crew added three boxcars to accommodate the refugees. By Sunday, September 5, an estimated 1,500 people had evacuated Summerville for Columbia and other points inland, leaving the village nearly deserted. The few who remained met in prayer under the stars.[24]

4

Chaos in the Palmetto State

You know the earthquake scared you and don't you say it didn't. Your neighbor will not believe it.
—The *Pickens Sentinel*

September 1886

As telegraph lines were restored, and news was exchanged between Charleston and the rest of the world, it became apparent that the earthquake was not just a Lowcountry event. It had not only shaken the entire state, it had also sent a seismic wake-up call throughout the eastern half of the nation and beyond. Ultimately, earthquake intensity reports and information were gleaned from as far north as Toronto and New York; south to Havana, Cuba; west to Omaha, Nebraska; and east to Long Island and Bermuda.[1] The editors of *Harper's Weekly* put the totally unexpected event into perspective: "That so vigorous an earthquake centre would be established on the Atlantic coast as that which last week wrought havoc and destruction in South Carolina was not anticipated or believed to be within the bounds of probability."[2]

In the days following the earthquake, the traumatic event generated story after story throughout the state. Modern seismologists took detailed accounts of damage written in 1886 by people and newspapers representing more than 205 communities in South Carolina.[3] These reports were then converted to intensity values and rated in Roman numerals from I to XII. On this scale, an intensity of I signifies a seismic event not felt, or felt only under especially favorable circumstances. An intensity of XII means that practically all man-made works were greatly damaged or completely destroyed.

In the Summerville area, and near Langley in Aiken County, the intensity level measured X. The ground cracked and fissured. Well-built wooden structures were severely damaged, and most masonry and frame structures were destroyed. Railroad rails were bent, and dams, dikes, and embankments were severely damaged.

In Charleston, the intensity level measured IX. General panic ensued. Well-built masonry structures were considerably damaged, and walls or entire buildings collapsed. Frame buildings were shifted off their foundations, and underground pipes cracked.

Almost the entire coastal area of South Carolina registered VIII. There was general fright and alarm approaching panic. Trees shook strongly. Sand and mud were ejected from the ground. Brick buildings suffered considerable damage, and wooden houses partially collapsed or had thrown out panel walls. Chimneys, columns, and monuments were twisted or overturned. Heavy furniture was moved conspicuously or overturned.

In the northwestern third of the state, most places experienced intensity VII. There was general alarm, and everyone ran outdoors. Some or many found it difficult to stand. Trees and bushes were shaken moderately to strongly. Large church bells rang. Damage was slight to moderate in ordinary buildings. Chimneys cracked and fell, and damage to plaster was considerable. Windows and furniture were broken.

Once all the damage reports were converted to intensity values, lines were drawn to connect areas of equal intensity.

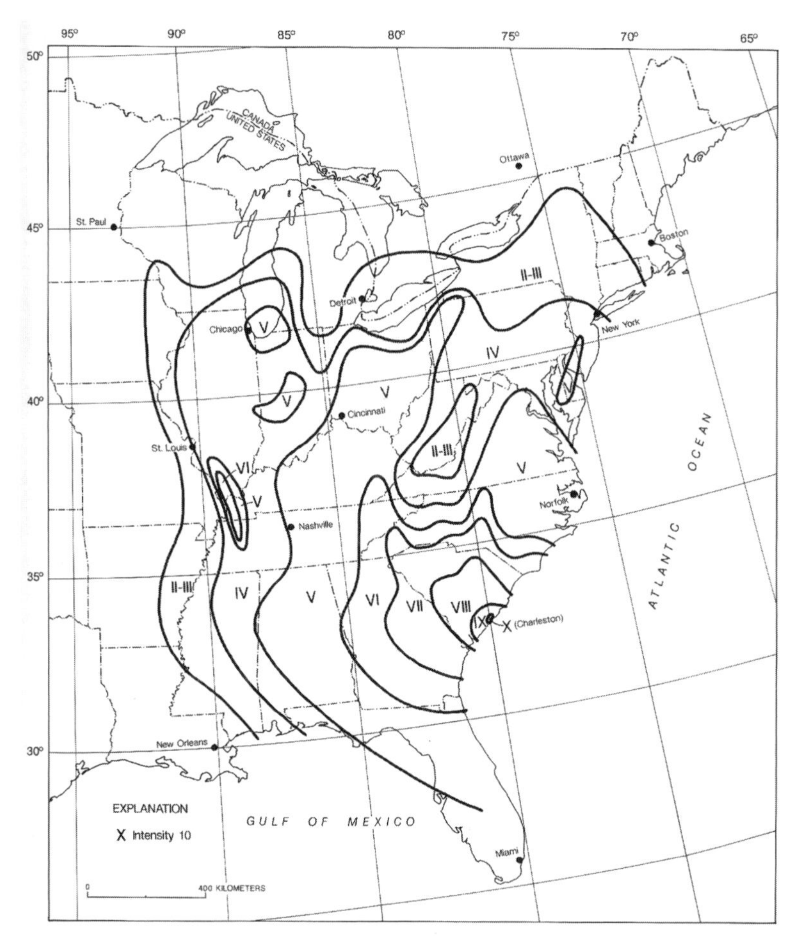

A modern isoseismal damage map of the 1886 earthquake

The result was an isoseismal map that showed the intensity of damage experienced in each part of the state, the region, and the nation.

Although the most extensive damage took place in the Charleston-Summerville area and along the coast, where population was densest, the earthquake spread a dry tidal wave of destruction throughout 2.5 million square miles, affecting all of South Carolina and most of the East Coast. It was the largest earthquake ever felt in the United States since the New Madrid, Missouri, earthquakes of 1811-1812, and by far the most lethal.

North of Charleston, across the Cooper River, is Mt. Pleasant, at that time a quiet village of 742 souls, 362 blacks and 380 whites, most of whom were involved in fishing, farming, or lumbering.[4] Its neighbors to the north, Sullivan's Island and Long Island (later renamed the Isle of Palms), were narrow, lightly populated barrier islands that separated the mainland from the Atlantic Ocean. Two steam-powered ferries, the *Sappho*, operated by the Mount Pleasant and Sullivan's Island Ferry Company, and *The Planter*, owned by the South Carolina Steamboat Company, provided the only way to travel from Charleston to Mt. Pleasant and Sullivan's Island and back.

Captain Cherry was at the wheel of the *Sappho* on August 31, 1886, as she made her 9:30 p.m. run from Sullivan's Island to her dock in Charleston. About twenty minutes after leaving the pier in Mt. Pleasant, the boat had just passed Buoy No. 4 in the Hog Island Channel in Charleston Harbor. The water was calm, the tide had started to ebb, and the captain was lulled by the peace of the evening. Suddenly, he was jolted out of his reverie by a sensation that the boat had run afoul of a thousand logs that seemed intent on pounding her bottom out.[5] The captain looked at his compass and his watch. It was ten minutes to ten, and he was right on course—but something was very wrong. It seemed as though some demonic hand had ahold of the vessel and was jerking it violently back and forth. Then everything returned to normal.

About eight minutes later, the captain recalled, "it felt as though the engine below had grown tired of its enforced servitude and was about making a dash for liberty."[6] Mystified, he signaled the engine room to find out what had happened. His engineer, Mr. Dogan, answered, "All right here, sir; guess we're on bottom." The ferry quivered and pitched vertically, but the captain and crew could find nothing wrong with it, and there was no sign that they had grounded. Then the strange activities ceased, leaving the baffled seamen to continue their journey. Moments later, a small boat approached, its occupants loudly hailing the *Sappho*. "Is the ferry intact?" they wanted to know. "Has anyone been injured in the earthquake?"[7]

When the first seismic wave hit *The Planter*, it was moored at Charleston's Accommodation Wharf on the Cooper River. Thomas Ferguson, the first mate, was sitting on the aft hatchway when he felt a tremor, and the boat moved forward. Then the ferry violently pitched backward then forward again and finally rose up nearly level with the top of the wharf. Almost immediately, it dropped down again. Ferguson's first thought was that something had collided with his boat, and it was sinking. He got up to rush forward and drop an anchor to keep the vessel from drifting away. However, to his surprise, he found it next to impossible to stand up.

An incredible roaring and the sounds of commotion on the streets convinced Ferguson that something extraordinary was happening. His curiosity piqued, he managed to get off the boat and began to make his way along the street. Startled by a loud crash and huge cloud of dust that appeared in front of him, he looked up to see a shower of bricks and mortar raining down, aiming for his head. Now convinced that he was in the midst of an earthquake, Ferguson raced back to *The Planter* in record time. No damage had been done to the boat or the wharf, and passengers from the train that had just arrived from Columbia soon filled the ferry, anxious to get across the river to Mt. Pleasant and learn the fate of their families.

When they reached the village, the officers and crew of the ferry were relieved to learn that their families were uninjured, though frightened, and were anxious to hear about relatives and neighbors in Charleston. "[The Mt. Pleasant residents] appeared more calm than the people in the city," wrote newspaperman Louis Beaty, "probably because a brick-and-mortar rain and the terror of falling walls had been spared them."[8]

In Mt. Pleasant, the first short shock was preceded by a low rumbling, which seemed to come from the east and got louder as the force of the earthquake grew stronger. It was over quickly but was followed immediately by the second awful convulsion. Houses tottered, reeled, and vibrated on their foundations like a child's top just before it falls. Trees swayed and bent, sweeping the ground with their lower branches. The shock was so severe that many people were thrown violently on the floor or against the walls and had to crawl on their hands and knees to open air. Everyone felt nauseated.

Outside, the air was filled with shrieks, groans, and prayers from frightened people. In the darkness, deafened by a hideous roar that seemed to come from beneath their feet and the noise of cracking timbers and crashing bricks, even the bravest souls were nearly immobilized. A strong sulfurous odor was perceptible throughout the night. When it seemed that things could not possibly get worse, the earth opened and belched up huge sluices of stinking mud and water in the lower part of the village and along the beach. Now it felt as if the ground was sinking, and the frenzy to escape increased. After about two minutes, the main shock, the destruction dealer, had passed.

Slight tremors and rumblings followed in short intervals during the remainder of the night. Village residents went sleepless, and many watched in horror as the aftereffects of the earthquake became apparent in Charleston just across the river. Newsman Beaty wrote:

> None except infant eyes were closed in Mount Pleasant during Tuesday night. Each successive tremor of earth brought those who had ventured indoors flying again to open air. Knots of people gathered in the streets and on vacant lots, and discussed the phenomenon, speculating on its origin and the probable links of its continuance. Families camped in their yards beyond the reach of housetops and timbers in case another quake should throw them down. Not a breath of air was stirring.... It was an unnatural night, and the vigil-keepers prayed for light, while they watched the tongues of flame that shot up in four different quarters of Charleston, showing that destruction threatened the city in and another and scarcely less terrible form than the earthquake.[9]

Mt. Pleasant was not completely unscathed. Campbell Gadsden's home sustained considerable damage when its supporting pillars and chimneys fell. The oven at Schulz's bakery was badly cracked. Marble and china items were broken in the house of Thomas Ferguson, the first mate on *The Planter*, and many similar instances were reported.[10] Public buildings were mostly intact. Each corner of the Berkeley County courthouse in Mt. Pleasant was cracked from top to bottom, but town officials believed that the damage could be repaired with the addition of iron girders. The county jail was intact, and no prisoners escaped.[11]

The villagers were largely spared the horrors that engulfed Charleston. Although chimneys had been shattered, no houses collapsed, and there were no fatalities. The landscape, however, had been transformed. Along the beach and throughout the village, hundreds of "sand blow" mounds had been formed when sand and fresh water were ejected from below. The water in nearly every well in Mt. Pleasant turned milky white and was unfit for use. However, when the mud was cleaned out of the public well in front of St. Paul's Lutheran Church on Pitt Street and the clogged pump was repaired, the water returned to normal. A twenty-foot-deep public well in the village had filled with soft mud that was forced upward with so much energy it blew off the well cover and overflowed into the street for some distance. In many cases, water and sand spouted four to six feet high. People poked tree branches and cane fishing poles down the crater holes, but no one hit bottom. A large depression in the village that had been dry was now filled with fresh water.

Countless cracks or fissures opened up, ranging from a few inches wide and a few feet long to one on the Boyd Brothers' plantation, near Shem Creek, which was "large enough to contain the body of a man."[12] The largest fissures were in Hilliardsville, a hamlet nearby. "One on Capt. W. M. Hale's farm would contain four men. From this opening must have been thrown thousands of gallons of water and a dozen cartloads of mud and sand."[13] Throughout the area near

Charleston, the ejected mud was blue and was often mixed with something that looked like a combination of mica and ashes when it dried.[14] The Mt. Pleasant health authorities took the precaution of scattering lime over the mud in case it was a health hazard.[15]

In the hamlet of Cordesville, about thirty miles north of Charleston, the lumbermen who fashioned railroad crossties quit the woods after Tuesday because the shocks were so severe. "The chickens made an unusual noise," a *Berkeley Gazette* correspondent reported, "and the cattle and dogs seemed alarmed. There was also quite a severe shock at ten minutes to one [in the morning on] September 1. All the shocks were accompanied by and preceded by a rumbling noise like distant thunder. It has left every one with an uncomfortable and nervous feeling. A great many families left their houses and spent most of the night in the yard, as in several cases the chimneys were flung down, and the houses shook as if they would fall."[16] A number of light-hearted young men "remained extra long when they went to enquire after the young ladies' welfare in the evening," the Mt. Pleasant newspaper later recounted, adding that one of them "staid 'till breakfast."[17]

Sullivan's Island lies adjacent to, and north of, Mt. Pleasant. Edgar Allen Poe, who was stationed there as a soldier from 1827 to 1828, used the island as the setting for "The Gold Bug," his famous story of buried pirate treasure. When the earthquake hit, those who had personal connections with the low-lying barrier island were terrified by thoughts that a tidal wave might swamp it. Rumors quickly circulated throughout the state that the ocean had indeed rushed in and washed away the island and all its inhabitants.

Despite the quake, Captain Cherry on the *Sappho* was able to get to Sullivan's Island and make a return trip at 12:30 that night with the news that no one there had been seriously injured. One woman on the island had the presence of mind to send a note back with the captain, reassuring a friend in Charleston, "All safe and sound, though somewhat shaken. All the neighborhood has camped with us in the farthest corner of our yard....We shall spend the night on Keenan's piazza, because it is low down—that is near to terra—I had almost said terra *firma,* but I refrain."

While Sullivan's Island was violently shaken, homes were rocked from their foundations, chimneys were thrown down, and china and furniture were ruined, no houses were destroyed or lives lost. There was one group for whom the event would be especially memorable. Amid happy shrieks and much laughter, a number of young people and their parents were dancing at a hop at the Moultrie House Hotel on the island when, above the strains of the band, came a low rumbling noise. Five seconds later, the building rocked so hard that it seemed to be on the verge of collapse. As the timbers creaked and groaned, laughter turned to screams as families rushed for the doors. A second shock came, and then a third, and the partiers joined the throng that was gathering in the streets. When the shocks were over, the only reported injury was to a Miss O'Connor, whose ankle was badly sprained.

Sullivan's Island's New Brighton Hotel was uninjured, and the venerable walls of Fort Moultrie, constructed to defend Charleston from the British during the Revolutionary War, were left intact, but the chimney was thrown from the lamp of the rear beacon of the Sullivan's Island Range Light, which was erected in 1848 to guide ships over the Charleston bar.[18]

The survival instinct caused many islanders to take the next ferry out, but within ten days after the earthquake, emigration off the island to higher ground had ceased. Indeed, by September 10, the *News and Courier* stated with confidence that "there is quite enough of this season of delightful weather yet in store, and there is no place where it may be spent more pleasantly than on Sullivan's Island."

Sullivan's Island's closest neighbor, the Isle of Palms, known then as Long Island, had no permanent settlers until Nicholas Sottile built a summer home there in 1898, so the earthquake passed unnoticed. At the Bull's Bay Light Station on Bull's Island, twenty-five miles northeast of Charleston, the lighthouse shook so much that it knocked the lens off its pedestal.

The keeper recorded the first shock at 9:45 p.m., and others came "every five minutes up to 2 o'clock [a.m.], then every half hour up to 10 o'clock the next day."[19]

In McClellanville, a small fishing village thirty-five miles northeast of Charleston, the unexpected first shock threw the residents and their animals into a state of excitement bordering on panic, and several houses were shifted from their places by several inches. Everyone rushed outdoors and, in most cases, stayed outside until daylight. Most people had no idea what had happened until the shocks passed. "All kinds of things were imagined, for instance, that a mob was trying to destroy the house; others thought that the last day had come, and still others felt sure that it was a special visitation of providence."[20] The church bells rang, and families flocked to places of worship. Throughout the village, prayers, hymns, and pleas for deliverance could be heard. On September 7, a black woman who had died of fright was buried.[21]

At the 150-foot-tall Cape Romain Light Station on Raccoon Key, about forty-one miles northeast of Charleston, the lighthouse keeper was in his house when the shock hit. He reported "a gradually increasing rumbling, sounding something like a battery of artillery or a troop of cavalry crossing a long bridge." The assistant keeper was thrown to the floor as the light station rocked.[22] "It seemed a miracle that the tower and dwellings were left standing," the keeper reported, but "little or no damage was done to either, with the exception of a chimney thrown down and others cracked." Nearly one thousand cranes nested on the key during the summer months, and they all flew about, making a "fearful noise" during the earthquake.[23]

The port city of Georgetown, fifty miles northeast of Charleston, was established in 1729 and had evolved as the prosperous center of South Carolina's fertile rice-producing region. The advent of emancipation in 1865 dealt a death blow to rice production because the labor involved was so odious and grueling that few workers could be enticed back into the fields without the threat of the whip. By 1882, the charming small city, with its handsome houses and once-busy harbor, was home to 1,811 blacks and 746 formerly wealthy planters and local merchants.

Georgetown's desirability in the nineteenth century was hampered by its competition with Charleston and by its proximity to the rice plantations. Surrounded by tens of thousands of acres of standing-water rice fields, it was an unhealthy place from spring until the first frost in December killed the mosquitoes that carried malaria and yellow fever. The earthquake hit in late summer, when the wealthier residents had long since fled the area for cosmopolitan Charleston, the beaches, the mountains, the health-and-pleasure spas of New England, or Europe. In the rice fields, only a scattering of white overseers and a dwindling number of share-cropping black field hands remained.

When the earthquake struck Georgetown, Walter Hazard, editor of the *Georgetown Enquirer*, noted that the afternoon and evening had been "intensely close." The thermometer read eighty-four degrees even by 7:00 p.m. At 9:50, "suddenly, a faint vibration of the houses was felt, accompanied almost simultaneously by a low rumbling sound, like the angry utterings of distant thunder. To the writer, who was seated on a piazza facing the Southwest, the sound seemed to come from that quarter. After the first shock of stupefied amazement had passed, the real nature of the phenomenon became all too apparent. The roar increased in volume and the vibrations grew more rapid and violent, until the houses swayed and rocked like ships in a storm. It was as though they had been rudely clutched by some invisible hand and shaken furiously to and fro. The timbers creaked and groaned, bells rang, and lamps tottered wildly on their pedestals, while the glittering pendants of hanging lights, clashing against each other, emitted fantastic music." Georgetown was besieged by aftershocks. The *Enquirer* noted ten between 9:50 p.m. and midnight and a dozen more the next day.[24]

Dr. M. S. Iseman reported that brick buildings undulated, chimneys fell, brick parapets were dislodged, and few structures survived without some effects. Virtually all of the residents fled their houses.[25] The chancel, frescoes, and tower of the Episcopal parish church of Prince George, Winyah, were badly damaged. Several craterlets were discovered in the village, one about fifteen feet in diameter and five feet deep and two others about six feet in diameter and about two feet deep. No injuries were reported. The Georgetown Lighthouse on North Island, which marked the entrance to Georgetown Harbor, was pounded by about eight shocks, "preceded by a rumbling noise as of thunder." The lamp was overturned, and the chimneys of the keeper's house were cracked, but no one was injured.[26]

A week after the initial quake, Georgetown residents were still experiencing two or three shocks every night. Despite strong recent rainstorms that had decimated the rice crop and substantial unrepaired damage and near starvation caused by the effects of the 1885 cyclone in the rural areas, Georgetown dug deep into its impoverished pockets and formed a relief committee that immediately raised $1,100 ($22,100 today) to help the Charleston earthquake survivors.

The people of Pawley's Island, twelve miles north of Georgetown, believed they had received odd warnings of the impending earthquake. A female resident reported a "very stillness of the ocean, more quiet than [she] had ever seen it" before the shock and also noted that the water was a peculiar deep purple. Another islander, Mr. LaBruce, remarked to those near him that the ocean made a peculiar sound. Later, residents would tell a *News and Courier* correspondent, "Cattle and poultry became crazed. The cattle would low and bellow and paw the earth, and run confusedly around the lot; the poultry in houses would put up a mournful cackle and crowing. Those not housed and roosting in trees would seek shelter in the dwellings of the inhabitants." Pawley's Island experienced shocks and cratering similar to those in Mt. Pleasant. The Pawley's Island Hotel, the only one on the island, was immediately closed. Many residents evacuated, fearing a tidal wave or further earthquakes. Between Pawley's Island and the North Carolina state line, human habitations were few and far between.

Horry County, then a sparsely populated rural area whose residents were primarily farmers and lumbermen, lies further northeast of Charleston. In Conway, the county seat, the first two shocks were so severe that most of the substantial buildings were in danger of being shaken to pieces, and the brick jail trembled on its foundation. Chimneys collapsed, clocks stopped, crockery rattled, and vases were thrown from bureaus and books from tables.[27] Many people sought refuge on the shallow-draft steamer *Maggie*, which plied the Waccamaw River between Georgetown and Conway. The vessel was soon full, but the shocks were felt as plainly on board as on shore.

On the day of the quake, a young college student named Noah W. Cooper was teaching at Pineville School in Horry County, between Conway and Aynor. He lived nearby with the family of Mr. Mayberry Mishoe. That night, he had retired early but was awakened by a terrible noise. "The house was cracking and twisting and moving up and down," Cooper recalled. "Everyone in the house jumped out of bed and ran onto the porch. It came to me instantly that this was an earthquake. Another great roar like thunder underground and a movement of the earth made us dumb with fear." The huge old oaks in the yard swayed back and forth like pampas grass, and the puffy white cotton bolls in the fields rose and fell like waves. Then, with a thundering roar, the chimney collapsed.

The Mishoes must have been frightened that their house would sink into the ground because Cooper felt called upon to reassure them. He told the Mishoes that he had read about earthquakes in Matthew F. Maury's *The Physical Geography of the Sea: And Its Meteorology*, an 1860s classic of American scientific literature. He assured them that earthquakes rarely swallowed things up in sandy areas like theirs.

Surrounded by the noise of his wailing and shouting neighbors, Cooper went to try to save his friend, and the Mishoes' kinsman, Bill from being frightened to death. The tremors were so strong that he had to grab a bush to avoid being thrown to the ground. He found Bill on his knees near the front gate of the house where he lived with his wife and mother-in-law.

With sweat pouring from his face, Bill was praying loudly, promising God that if he were spared, he would sin no more. Cooper saw Bill's mother-in-law sitting in the middle of the yard in a rocking chair, praying aloud without fear. "Oh, Mr. Cooper," she cried. "God is disgusted with the wickedness of the people. I am not afraid, I am ready to go, and I expect the world to be ended by daylight. I'm going home. Glory to God! Hallelujah!"

"Seats in Horry County churches were not easy to find the Sunday after 'de quake,'" Cooper wrote, and the earthquake produced a religious revival of immense proportions—for a short while, at least. Whiskey stills were dismantled and hundreds gave up the devil's brew entirely. There was also a profound effect on Horry County's winemakers. Fearing that that the seismic event was "God's message to a sinful world," they immediately started tearing up their vines in repentance.[28] As one local historian observed, "Mother Nature had done in a single evening what preachers had tried to do for years: bring a great revival to Horry County."[29]

A fissure near the Ashley River

Southwest of Charleston, the prosperous plantations of St. Andrew's Parish, especially those west of the Ashley River, took a severe beating from the earthquake because they were located between Charleston and Summerville, close to the fault line that was later found to have triggered the earthquake. An area ten miles long and several miles wide was sliced open by countless small fissures and sand-blow holes up to ten feet in diameter.[30]

The Ashley River, one of the most beautiful waterways in the state, was a rich producer of agricultural wealth and a main avenue of commerce from the seventeenth century through the end of the nineteenth century. With its lofty—for the Lowcountry—bluffs and wide, sweeping curves, the river produced panoramic vistas that enticed wealthy early settlers to build magnificent plantations, such as Middleton Place, Drayton Hall, and Magnolia Gardens, on its banks.

Built in the early eighteenth century, the plantation that became Middleton Place was acquired by Henry Middleton in 1741 through his marriage to Mary Williams, whose father, John, owned 2,200 acres in the area and built the first home on the site. Middleton built his beautiful, three-story brick mansion facing a bend in the river. Two symmetrical flanking buildings were constructed on the horizontal axis of the house. Between 1741 and 1750, Henry Middleton, later president of the first Continental Congress, had magnificent gardens, terraces, and ornamental ponds and watercourses laid out at Middleton Place by an English landscape gardener.

A century later, during the Civil War, this piece of paradise was ravaged by a detachment of the 56th New York Regiment, under the command of Gen. Quincy A. Gilmore. The soldiers looted and burned the main house and the two flankers, seizing or destroying many books and furnishings. The north flanker and the main house were reduced to partial, blackened shells. The south flanker, also burned, was partially rebuilt by Williams Middleton, Henry's great-grandson, but by 1875, the economic depression during Reconstruction left Middleton Place a shadow of its former self. *Harper's New Monthly Magazine* said of it, "The old oaks, the hedges, the elaborate terraces and ponds still remain, but the place is deserted, and the spirit of melancholy broods over it."[31]

When the earthquake hit the plantation, large fissures appeared in the earth, ripping open the garden's terraces. The landscape rolled like a flag waving horizontally, and the Butterfly

The burned ruins of Middleton Place prior to the earthquake

Lakes at the bottom of the terraces were sucked dry.[32] When geologist Earle Sloan visited Middleton Place on September 19, 1886, he found that the ceilings in the remaining south flanker were severely cracked, and the chimneys were badly injured. The power of the earthquake was evident, and the shock damage had come vertically from below, destroying what little was left of the burnt-out walls of the main house. Sloan found further indications of "increased violence" on the grounds. The earth was severely disturbed, and innumerable craterlets had appeared. Some were eight feet across, and several were still actively spewing water on the day of Sloan's visit.[33]

A few miles south of Middleton Place lies Magnolia Plantation with its beautiful gardens. It is the site of the Drayton family tomb, which was constructed prior to 1700 by Thomas Drayton, Jr., and became the resting place of subsequent owners and their families until 1891. The marble plaque on the face of the tomb was decorated with cherubs, but it was vandalized by the bayonets and rifle balls of Union soldiers, who burned the plantation house in 1865. The earthquake added a large crack to the plaque.[34]

Immediately south of Magnolia Plantation lies Drayton Hall, the home of the Drayton family since the early eighteenth century. Its Georgian-Palladian architecture represents the oldest surviving example of its kind in the American South. It was also extremely fortunate because it was the only plantation house on the Ashley River to escape the depredations of the Union army. Its charmed life continued, when its handsome main house was spared any significant damage by the earthquake. Of its two smaller flanking buildings, however, the northern one, used after the Civil War as a caretaker's house, was severely damaged by the earthquake and had to be torn down. The southern flanker, a kitchen, was probably also damaged by the earthquake. It survived, but fell victim to an 1893 hurricane and was then torn down.[35]

Nearby St. Andrew's Parish Church, built in 1706 and the oldest surviving church building in South Carolina, was severely shaken and badly shattered by the quake, which collapsed three gable ends of the church, opened a wide seam from top to bottom, and created numerous other cracks. The transept and windows of the church are still slanted to the north.[36]

South of Charleston at the mouth of the harbor, the 150-foot-tall Morris Island Lighthouse vibrated heavily. The keeper, who was standing at the door of the tower, heard a rumbling noise and felt the earth tremble. The trembling increased until it had "a strong tearing and jerking motion." After a second shock, the keeper went up into the lantern compartment of the tower. A third shock was so strong that he could hardly stand. The lens of the lamp swung from southeast to northwest, back and forth, about three or four times a second and was thrown out of position. The tower sustained two considerable cracks and some smaller ones. Fissures that measured two to four inches wide and from ten to one hundred feet in length were found in the ground nearby.[37]

All of James Island, south across the Ashley River from Charleston, was violently shaken, and many people were unable to flee their homes until after the first shock had passed. In hundreds of places, the earth opened in long cracks, many of which gushed cold water mixed with sand and blue mud. Everyone on the island slept outdoors or on their piazzas and suffered from the nausea that affected the majority of the earthquake victims.

On Wadmalaw Island, then a lightly populated, isolated farming community thirty miles south of Charleston, The Reverend P. S. Jenkins, an Episcopal priest, pleaded for help. "The chimneys of three respectable white widow ladies have been completely demolished in my immediate vicinity. In addition I would say that there are at least half a dozen homes so badly shaken as to be dangerous in any high wind, such as we have to expect at this season of the year. Now I know for a certainty that these ladies are not able to rebuild, for should they use any moneys that may be in their possession, they would soon be in want of food and clothing."

Displaced tracks of the Charleston and Savannah Railroad near Rantowles

The enormous forces that shook, shoved, twisted, and pulled the land were reflected in the lateral displacement of the tracks of the Charleston and Savannah Railroad. The railroad's Ashley River drawbridge was jammed shut by the quake, which shoved both river banks toward the center. Near Rantowles station, about twelve miles due west of Charleston, the tracks were flexed into an S-curve, and for the next twenty miles, the roadbed was alternately elevated, depressed, and distorted to the north and south as much as fifty inches.[38]

In Walterboro, forty-five miles northwest of Charleston and twenty miles from the Atlantic coast, the inhabitants braced themselves as an estimated twenty shocks battered the small town. The local newspaper correspondent wrote that few houses escaped injury. Both the courthouse and jail were damaged, and Mr. Jasper Rice and Mr. L. Bellinger had their houses set on fire by overturned oil lamps. Both were able to extinguish the blazes.

From Osborne, twenty-two miles west of Charleston, a newspaper correspondent noted that by 1:00 p.m. on September

1, he had felt twenty-six shocks, which moved his house and the earth below it back and forth three feet from northeast and southwest, pushing it up and down, in an oscillating motion. Closing his report, he wrote, "The writer starts for Charleston, which is said to be destroyed."

In 1886, Beaufort, about sixty miles southwest of Charleston, was a small seaport. At the start of the Civil War it was quickly occupied by the Union army and thereby escaped the destruction later spread by General Sherman. By 1882 it was home to 1,274 blacks and 465 white residents. On August 31, it was shaken by a severe shock at 9:50 p.m., with numerous others in the following forty-eight hours. Chimneys fell, clocks stopped, pictures and mirrors crashed to the floor, and church bells throughout the town tolled of their own accord.[39] Newspaper reports to the contrary, St. Helena's Episcopal Church, built in 1724 on a foundation of rock-hard tabby, was spared, but other large masonry buildings in the city suffered severe cracks.[40]

Nowhere was the physical damage more obvious than at the Beaufort Arsenal on Craven Street, which had numerous cracks in its massive walls. As was the case throughout the state, some buildings were severely affected, but others close by were not. The handsome neoclassical building housing Beaufort College, chartered in 1795 as a preparatory school and junior college for the local planter families, was spared, but the nearby Beaufort County courthouse suffered severe cracks.[41] Weir's Pond, on Duke Street, was fed by an underground artesian well, but the earthquake choked off the water supply and the pond went dry.[42]

Everyone in Beaufort raced to the street in nightclothes and stayed outdoors all night. The prisoners in the old brick jail prayed, sang, and begged to be let out. Many citizens thought the world was coming to an end and rushed to their churches, but E. B. Rogers, a local resident, stated that the white churches did not open. The Reverend Arthur Waddell of the First African Baptist Church also refused to open his church, stating that he had held a prayer meeting a few hours earlier and only

Earthquake bolts now reinforce the Beaufort Arsenal

thirteen brothers and sisters came, and if this was the end of the world, it was too late to pray as they had had time to come to the church sooner.[43] Although Waddell may have frowned upon lukewarm believers who sought last-minute divine intervention, he took up a collection the week after the earthquake and raised $31.31 ($629 today) from his financially strapped congregation to help the Charleston sufferers. The Port Royal newspaper noted, "the generous impulse of our colored people in proportion to their means, [which] shows the influence for charity's sake that their pious men exerted upon their respective flocks."[44]

The Reverend R. F. Bythewood, a black minister of the Tabernacle Baptist Church, did throw open the doors immediately after the earthquake and was amazed to find white worshippers flooding in with his own congregation.[45] Bythewood later served as one of the members of a bi-racial Beaufort charity committee, which successfully raised $250 ($5,025 today) to aid Charleston.[46]

Ladies' Island in Beaufort County holds the unique distinction of having the only person whose life was *saved* by the earthquake. "A woman who was momentarily expected to die, being surrounded by a crowd of sympathizing friends praying for her, and trying to smooth her way into the next world, was impressed with new life and strength by the first shock of the earthquake, for it caused her to arise from the bed, dash out of the house, and run a race in which she distanced all of her pursuers in getting to open ground. Her condition has improved ever since."[47]

In Port Royal, a small coastal village five miles southwest of Beaufort, the shock moved houses on their foundations, destroyed chimneys, and threw people to the ground. The editor of *The Palmetto Post*, S. H. Rodgers, wrote, "The houses felt like some huge monster had struck them three violent blows and then grasped them and endeavored to shake them to pieces. Goods were thrown from the shelves, pictures fell from the walls, and hanging lamps swayed to and fro like the pendulum of a clock as the buildings were rocked from side to side. Great consternation was felt, and we were not surprised to hear the shrieks and screams of terrified women and children, for the man who dares to say that he was without fear is surely devoid of feeling."[48] At the Hunting Island Light Station, sixty-five miles southwest of Charleston and ten miles from Beaufort, Mr. M. B. Trevett, the keeper of the light, reported shocks that were strong enough to stop his clock.[49]

The huge seismic event quickly made its way into the South Carolina Midlands. "Mom" Agnes James, an eighty-year-old ex-slave living in Claussen, Florence County, had no problem remembering "de quake" when interviewed in 1937. She had been on her way to a prayer meeting with her infant in her arms when the roaring and shaking hit. Disoriented, she thought that her baby had been torn from her grasp. Screaming for help, she began to search the ditches frantically, trying to recover the child. Amid the confusion, her husband reached her side, hollering, "Agnes, what IS the matter with you?"

"My baby is lost. Dear Lord, where is my baby?" she cried.

Looking at her in bewilderment, her husband replied, "Agnes, you must be crazy. There's the baby in your arms!" It was a long time before Mom James recovered from being nearly scared to death.[50]

Though the people of Florence, the county seat, were terrified by the quake, which cracked walls and threw down chimneys there, they expressed their love for the people of Charleston by quickly offering free board and lodging for five hundred refugees.[51]

Margie Daniel Epps was seventeen years old when the earthquake struck her home in rural Greelyville in Williamsburg County, sixty miles north of Charleston. "The first shock at about 11 p.m. waked us up," she said. "It seemed that the second shock must have been as great as the first. It was as though a giant caught the west corner of the house and jerked the room where I was sleeping. The poultry and hogs made frightening noises. Chickens wandered about in the yard, clucking. Leafy Graham, a colored woman, prayed and prayed so loud that her voice could be distinguished a mile and a half away."[52]

In the town of Marion, one hundred miles northeast of Charleston, forcible shocks were accompanied by the dreaded rumbling and roaring that terrified everyone across the state and threw Marionites into an awestricken uproar. The first tremor sent crockery flying and loose furniture tumbling. Brick walls cracked and chimneys were broken. The next day, a resident wrote with great understatement, "There will be but little sleep here tonight."

Columbia, the state capital, about 120 miles northwest of Charleston, had experienced the Summerville-centered tremors of August 29, but the residents were in no higher state of earthquake awareness than anyone else in South Carolina. There was no subtlety to the earthquake's arrival in Columbia on August 31. The city was badly shaken, producing widespread panic. The walls of the courthouse were severely cracked when the first shock hit, and the cracks grew larger as aftershocks led to some settling of the building. By September 9, cracks in the portico were so apparent that people assembling at the courthouse to attend a meeting for the relief of Charleston were afraid to enter the building.

At the South Carolina College (now the University of South Carolina) in Columbia, the residence of Professor Joynes on the campus was "well nigh destroyed." Built in 1802, the two-and-a-half-story structure was the oldest on the campus. Its massive walls were eighteen to twenty-two inches thick. A reporter noted, "The coping fronting the campus is almost ready to topple, the brick having been broken clear across like so much glass. The massive chimneys were wrenched on their foundations and toppled off the roof. There is not a room in the house but was damaged, and daylight can be seen between the broken walls dividing the L from the main building."

The reporter continued, "The next most injured building is the DeSaussure section of the College. The immense chimneys are wrenched off and the gables of both ends have fallen. The north wall is also sprung where the crack was made by the [New Madrid, Missouri] earthquake of 1811. The eastern gable is also sprung. The rooms in most all of the buildings suffered by having the plastering cracked and thrown down." Even today, the pervasive presence of iron earthquake bolts used to save and strengthen buildings on the Horseshoe of the main campus demonstrates the extent and severity of the shocks.

A few blocks away, the statehouse also suffered. Some of the granite blocks in the southwest gable were separated at the joints. A pilaster to the right of the portico was also "jarred out of harmony with its pedestal." In the state penitentiary, the convicts knew they had no chance to escape injury or death if the building fell. The prisoners were wracked with fear, and their howlings were awful, the newspaper reported.

Not everyone in Columbia was panic-stricken by the earthquake. Indeed, some took it all in with great equanimity. Sarah Ciples Niles Goodwyn (known as Sallie), a fifty-five-year-old Camden native and member of the Episcopal Church, felt that the frenzied reaction had been excessive and unnecessary. She

Sarah Ciples Niles Goodwyn

wrote from Columbia on September 13, 1886, "The panic from the earthquake has just subsided. You would be surprised to know the number of people who slept in their clothes for nights. We did not do any such foolish thing but went to bed at 12 o'clock the night of the first shock & altho we had several after we went up stairs we staid in bed until the next morning. Some people staid in their carriage houses 3 or 4 nights & we hear that Maggie Dunlap & her children slept down stairs in their clothes for several nights. I can't understand such a panic in Christians for death is sure to come to all & it will come in God's own way whether in earthquake storm or by disease & we cannot escape it."[53]

Everyone in Barnwell, a rural town about 110 miles northwest of Charleston and a two-hour ride on horseback from the Savannah River, was frightened when the first shockwave hit. Mr. W. E. Collier, a local resident, reported that houses rocked and shook as if about to fall, furniture was moved, walls cracked, and plaster fell in buildings throughout the village.[54] The residents felt six shocks on Tuesday night before midnight, six more on Wednesday, and lesser ones in the days following.

George H. Bates, a local attorney and the Barnwell-area correspondent for the *News and Courier*, wrote, "The prisoners in our jail were alarmed to such a degree that their piteous cries could be heard over the entire town begging for help and praying to be released." The town quickly offered an initial $250 donation ($5,025 today) for the Charleston sufferers, along with an offer to provide "good and comfortable shelter among us" for two or three hundred refugees—an enormous outpouring of charity for such a small town.[55]

In Bamberg, about ninety-one miles northwest of Charleston, several brick storehouses and the handsome residence of Gen. F. M. Bamberg were badly damaged when walls cracked and plaster fell. Residents who normally slept in brick houses moved into wooden buildings for safety. Many other people piled their bedding out in the open air, trading the possibility of contracting malaria and other diseases for the risk of being crushed inside their houses.

When the earthquake hit Midway, a small village just southeast of Bamberg, one man thought dynamite had been placed under his house. He ran outside with a gun, and seeing a man running and yelling, took quick aim and pulled the trigger. Fortunately, the gun misfired and the innocent bystander escaped.

The earthquake created chaos on the western edge of South Carolina, especially in Aiken and across the river in Augusta, Georgia. In 1886, Aiken, whose basic income came from nearby cotton mills, was a delightful small town, already famous as a flower-filled haven, health spa, and winter resort for rich, horse-loving northerners. It was situated on the South Carolina Railway's westbound route from Charleston, and trains passed through Aiken en route to Augusta. Although the area had felt small foreshocks in the preceding week, nothing had prepared the residents for what hit at 9:51 p.m. on August 31. The city's two newspapers left detailed reports of the event. The night staff of one paper, the *Journal and Review*, had personal experience with the earthquake as their office swayed from side to side while the floor seemed to heave upward and the building threatened to fall.

Joining the victims throughout the rest of the state, Aikenites fled to the streets for safety, fearing that the worst was yet to come. By eight o'clock Wednesday morning, the city had felt fourteen shocks, each preceded by a "violent rushing sound." A number of substantial brick buildings were damaged. One drugstore lost many of its medicines when bottles crashed to the floor. Chimney tops were broken off, and chimneys toppled onto roofs. No one was killed.[56] A few miles away in McCormick, the Calhoun Mill, with its adjacent general store and post office, was a popular spot for political gatherings. The well-built cotton mill was wrenched by violent shocks but survived with only minor damage.[57]

Havoc also rode the rails in the western part of the state on Tuesday night. The September 1 edition of the *Atlanta Constitution* reported wrecks and considerable loss of life on the South Carolina Railway between Augusta, Georgia, and Charleston. The initial report told of four wrecks: the two near Summerville and two others near the South Carolina-Georgia state line. It offered two theories for the derailments that caused the wrecks: tracks warped and twisted by the stresses of the earthquake and tracks submerged by water gushing out of broken dams on cotton mill ponds. By the next day, it became clear that both theories were correct.

The train wreck at Langley Mill Pond

At Langley Mill Pond, near the cotton mill village of Graniteville, South Carolina, a dam broke as a result of the first shock, sending a four-foot wall of water across the railroad tracks, washing out the railbed and forming a small pond just beyond where the rails had been. Instantly, the tracks were twisted in the shape of an "S," their iron bars bending as if they had been heated. Some of the bars were turned up, the ends lifted high in the air. The whole track was zigzagged. The night was coal black, and a short curve hid the damaged, submerged rails.

Completely unaware of its fate, the night express train from Augusta to Charleston thundered out of the darkness and into chaos. "The night passenger [train] running into the misplaced tracks was wrecked.... The engine followed the dislocated track and plunged into the smaller pond," a reporter wrote. "The fireman [Henry Ivy or Ivie] jumped and was thrown with tremendous force into the water and instantly drowned. The engineer [Mr. Reynolds], stuck to his post, and though the engine was completely submerged, managed to escape with two broken legs."[58] Miraculously, none of the passengers was reported to have been hurt.

Moments later, a similar tragedy played out, this time near Horse Creek in Aiken County. There, another break in a mill dam sent an eight-foot flood into the path of another South Carolina Railway train. The locomotive, its tender, and two cars loaded with livestock were washed off the tracks and completely submerged. The engineer, Mr. Brisenden, survived, but his black fireman, Jack (also cited as Adam) Simmons, was killed, and the horses in the stock cars drowned. Two black residents near the creek also reportedly drowned, and a mounted rider reported finding the bodies of an entire family nearby.[59]

A seismic wave from the great quake washed across the western border of South Carolina into Georgia. Just north of Augusta, on the west bank of the Savannah River, the U.S. government had built an arsenal in the second decade of the nineteenth century. The original location was found to be unhealthy, and the arsenal was moved to higher ground, now the site of Augusta State University. The Augusta Arsenal was seized by Georgia at the start of the Civil War and used to produce powder and munitions for the Confederacy. After the war, it was reoccupied by federal troops.

When the 1886 earthquake struck Augusta, it did more damage to the arsenal than wars ever had. The Benet House, which served as the commandant's residence, and Rains Hall, used to house junior officers, were especially hard hit. On September 2, Maj. J. M. Reilly, the commandant, filed his damage report with the U.S. Army Chief of Ordnance in Washington.

> I was seated in a hammock on my porch, my wife was putting the children to sleep upstairs

> when I heard a rumbling noise and felt a shaking of the house. I thought that gas had exploded in the cellar. I hurried into the hall, the noise and shaking increasing until the house swayed and rocked like a ship on a rough sea. My wife called me to come upstairs and get the children out. When I reached the room the rocking was as great as to render it difficult to stand. The plaster was falling all around. I got the smaller child from his bed and told my wife to hurry the elder ones downstairs. We got out amidst falling plaster and cornice, a noise as of rolling thunder and a creaking of mortar and bricks indescribable. When we reached the front piazza of heavy granite columns and body, it was swaying as a tree in a strong breeze. My wife hesitated to cross it but I told her to go on and we got out safely.[60]

The two brick structures at the arsenal were the only ones in the vicinity that were seriously damaged. Major Reilly noted, "Wooden houses vibrated, but did not sustain serious injury. To withstand earthquakes, houses must not be rigid as are the officers quarters here or rather were our quarters; for I consider them as hardly habitable hereafter, certainly not now. They are racked into fissures all over. The parapets fell from them and the exposed timbers are rotten.... I consider our escape as miraculous, a little more trembling of the earth would have buried us all in the ruins of our houses."

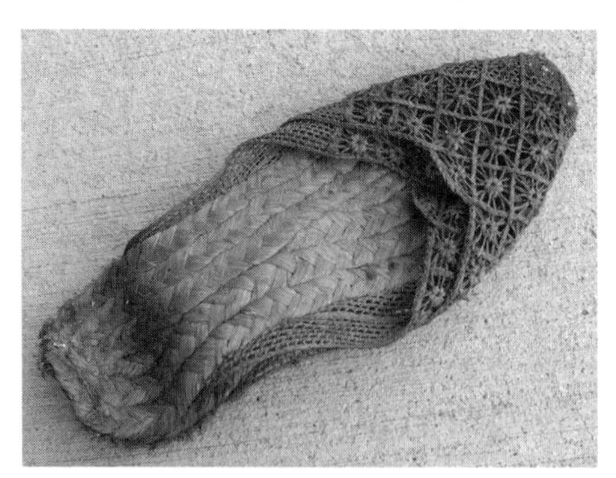

A concealment shoe found in Rains Hall

Over the course of the next year, both masonry houses were rebuilt, one with an added article of protection for good luck. During a 2006 restoration of Rains Hall, a worker found an immaculately preserved pair of colorful lady's slippers between the outer brick wall and the plaster-and-lath inner wall. Research determined that the elegant blue-and-pink slippers were "concealment shoes," good-luck charms often placed in walls near doors, windows, or chimneys during construction or repair to keep out evil spirits. An analysis of the shoes determined that they dated from the 1870-1880 period and were probably placed in the wall during work done after the earthquake.

Even though the effects of the great quake lessened somewhat as they reached upstate South Carolina, they were still memorable. Throughout the northern tier of counties in the South Carolina Piedmont, the shocks were severe enough to cause general fright. In Pickens, a small town between Greenville and the North Carolina state line, six distinct tremors were felt, each accompanied by a low, sullen, rumbling. The local newspaper reported the universal themes of fear and disorientation familiar to millions of people throughout the Southeast after August 31. "The first tremendous throb of the earth threw the whole animal kingdom into confusion. All was dark, except what little light came from the melancholy stars. Affrighted neighbor called to affrighted neighbor through the lonesome darkness. Cows lowed, chickens cackled and dogs howled. These things heightened the terror of the second shock, which came in about ten minutes and continued about one minute as [had] the first."[61] Guilty consciences may have plagued a few people who stood in the earthquake's path. The *Pickens Sentinel* reported that two men, probably moonshine whiskey distillers, were in their tent high atop a mountain when the earthquake struck. "They heard the roar two minutes before the shock came. They verily thought the end had come."[62]

August 31 was the first day of the cotton-ginning season in McColl, a small Marlboro County town in the extreme northeastern part of the state. That night, Thomas Gibson and three friends were playing poker in the second-floor storage room of Gibson's general store. John Piplin, a county commissioner, recalled that his congregation had been admonished not to play "the devil's calling cards" any longer, and he was a bit nervous as the deck was passed to him. No sooner had Gibson reached for a card than the men heard a deep groaning, all the more eerie because the store's supply of caskets—a normal part of a country store's wares—stood stacked against the nearby wall. When the full force of the earthquake hit, the groans were replaced by shaking, and the caskets started to rattle and dance. That was enough for the men, who dropped their cards on the table and beat a hasty exit. Within minutes, everyone in the village had fled into the streets. The next day, word of the rattling caskets started making the gossip rounds, and the second floor of Gibson's general store became the town's chief tourist attraction for several days.[63]

Near the North Carolina state line, the people of Spartanburg shared the common experience of their brethren in other parts of the state: unnerving shocks and aftershocks, cracked buildings, shattered chimneys, falling plaster and dire, unconfirmed rumors. "The earthquake came, and terror, in some cases amounting to wild frenzy, took possession of many of our people," the newspaper reported. The first warning came at 9:46 p.m. The rolling, rumbling sound preceding the vibrations seemed to come from the west or northwest. The first tremor, said to have lasted three minutes, was the most severe. Spartanburg suffered a total of ten shocks by nine o'clock the next morning. Some of the old people in the city recalled that the shocks were equal to those they felt from the New Madrid quakes of 1812, which they referred to as "severe."

Sam Lewis, a Spartanburg resident and former slave described as "gnarled as a fire-blackened old pine stump," scratched his head in 1939 when interviewed by a writer working for the South Carolina Writers' Project, a branch of the Depression-era Work Projects Administration. Harris, like most people of his generation, had no problem remembering the fateful night that the earth roared and moved. He was a youth at the time, and the Harts, his former owners in York County, had just bought him new shoes. Shoes in those days had brass around the toes, and he was proud to own them. He and others walked seven miles from his home to York to see what the earthquake had done there. Lewis remembered seeing two stores with the roofs caved in. When the earthquake struck his home, he ran outside and got under the house. Later, Lewis realized that what he had done was crazy, but he had been too scared to realize that the house could have fallen on him.[64]

Mary Davis Brown

Mary Davis Brown, a devoutly religious Presbyterian woman who lived in the Beersheba community, about five miles west of York, South Carolina, was profoundly influenced by the earthquake. On the morning after the event, the seventy-one-year-old woman wrote in her journal:

> Wensday morning Sept 1 a day & night long to be remembered. The great earthquak of 1886, the like has never been felt in our country. The houses shook till the inmates in manny places left theire home, horry sticken, looking everry moment to see the earth open & all be swallowed up alive. The darkness was awfully frightening, all gone home but I have a guilty consience this night that the children came to [their] Farthers house for comfert & advice in such a trying time & received no better advice & encouragement than they have got this day. My Farther in Heaven, I pray the thou would poor thy holy spirrit out on each one of [them] from this long to be remmerable day that by thy grace & by thy strenth we will live more to thy Glory & Honor & for the good of oure own dying.... The excitement of the earthquake is not got over yet. It is heard yet but not so bad. Charlston & Summervill is the worst. Theire is

> a hunderd thousand inhabitants in Charlston, their houses is nearly all shook down. They have fled to other places & living in tents.[65]

Within the first few days of September, the lives of everyone within 250 miles of Charleston were changed forever by the earthquake. People who lived through it would always mark the day of "de quake" as a watershed; the day that separated the first part of their lives from whatever happened after the event. As later generations would always remember where they were at the moment they learned about the attack on Pearl Harbor or the assassination of President John F. Kennedy, the "earthquake generation" would always know exactly where they were and how they felt at 9:51 p.m. on the sultry, still day that ended the month of August 1886.

5

BETWEEN HELL AND DAYLIGHT

A bad earthquake at once destroys our oldest associations: the earth, the very emblem of solidity, has moved beneath our feet like a thin crust over a fluid.
—Charles Darwin, *The Voyage of the Beagle*

Before dawn
Wednesday, September 1, 1886

The first two shockwaves had roared through Charleston only eight minutes apart on Tuesday night, sending the terrified citizens fleeing into the streets. Then came two more before midnight. Four others struck before daybreak on Wednesday, followed by four more before the end of the day. Every shock created more panic than the previous one, since no one knew when the tremors would end. When the shocks continued, it seemed that Doomsday was imminent, and the next jolt could snuff out all life on earth. "The sidewalks and the streets were dotted with mothers, wives and daughters, with their protectors, awaiting in anxious expectation still another agony," a reporter wrote. Then, as if violent shocks, falling buildings, and dying neighbors all around them were not enough, Charlestonians lost their ability to light the darkness with gas or electricity or communicate by telephone or telegraph. Within sixty seconds of the first vibrations, the city went dark and mute. And then came the fires.

At the time of the earthquake, Charleston was fully equipped with the most modern communications facilities. It was served by two telegraph agencies: the Western Union Company, with offices throughout the business district, and the Southern Telegraph Company, with offices in the Pavilion Hotel at 11 Broad Street. Alexander Graham Bell received the patent for his telephone in 1876, and in 1879, The Southern Bell Company opened the state's first telephone exchange for a small but growing number of progressive Charleston residents and businesses. Service was strictly local, and the lines did not connect with any telephones outside the city limits.

Charleston had the distinction of being the first city in the state to have electric lights. The privately owned U.S. Electric Illuminating Company of Charleston was incorporated late in 1881, when it installed a generator and several lights.[1] The fad did not catch on, however. The company failed to interest a sufficient number of customers, lost money, and the promoters abandoned their work. In 1886, the company was restarted and installed a new generator on Queen Street. As a demonstration of civic pride, and in an effort to project a progressive portrait of his city, Mayor Courtenay ordered electric lights installed to illuminate the market halls on Market Street and brighten Marion Square, the city's largest open space.[2]

In 1886, stylish gaslights were still the city's chief source of illumination for its streets. Homes and businesses were also supplied with gas by the Charleston Gas-Light Company, which had been incorporated in 1846 and began illuminating Charleston two years later. Its handsome, Palladian-style headquarters at 141 Meeting Street was completed in 1878 and is used today by its successor, the South Carolina Electric and Gas Company. The gas-production plant was located on Charlotte Street, in the northeast part of the city adjacent to the

Cooper River. The first shock shattered its brick walls, and the walls of its coal storage sheds burst, bringing down the roofs. The huge gas storage tank, some ninety feet in diameter, held tight, although it was shoved eight inches to the southeast and rocked back again. The gas mains underneath the streets were so badly broken that leakage increased more than three hundred percent.[3] Fortunately, divine providence stepped in, and there were few explosions. The company quickly shut off its distribution lines until repairs could be made, and the city remained without gas until after September 6.

While Charleston was spared the horror of large-scale gas explosions, it did not escape a more common hazard. Because the earthquake took place at night, and many residents were awake at the time, the oil lamps that illuminated bedrooms and sitting rooms became the main source of fires. Numerous lamps were thrown over by the shocks, and many blazes resulted. A reporter wrote, "The air was thick with horrid rumors, and the lurid glare of the fires but heightened the general ruin."

Charleston volunteer firemen in 1855

Despite his absence at sea when the earthquake hit, Mayor Courtenay's farsighted leadership had already contributed to the effectiveness of the disaster response. During his first term in office, and over the strong objections of generations-old, tradition-bound volunteer fire companies, Courtenay had combined all the companies and their equipment into a single, salaried, city-managed firefighting force that would serve the needs of all Charlestonians. The opposition to consolidation was strong because the original firefighting companies functioned not only as public safety providers but also as social and benevolent organizations. In addition, there were both friendly and professional

A Charleston fire insurance company plaque

rivalries between them. Owners who had their buildings protected by fire insurance typically displayed the plaques of their insurance companies on their most prominent exterior walls.

Prior to Courtenay's modernization, the city had furnished buildings for independent firefighting companies, bought fire engines, and paid the salary of the engineer (company fire chief) for each company. The rest of the officers and men were true volunteers, unpaid. Courtenay's measures brought consistent standards to the firefighting force and turned them into a professional organization that could respond as a unified team. Under Courtenay's direction, the city council appointed a Board of Fire Masters on October 17, 1881, and the city-paid fire department was organized on January 1, 1882.[4] Firefighting equipment was deployed throughout Charleston at strategic locations. Fire hydrants, fed by the city's two artesian wells, were installed citywide and were augmented by a system of fire wells, which provided a backup supply of saltwater for the fire engines to pump.

When the earthquake struck, Charleston's fire department consisted of 103 officers and men, mostly of Irish descent. The department had eight fire engine or truck houses, nine steam-powered pump engines, seven hose carriages, two hook-and-ladder trucks, twenty-eight horses, two fuel wagons, three alarm bells and one alarm tower, and a modern fire alarm telegraph system to serve a population of 60,000.

In 1877, Charleston became one of the first cities in the United States to install the state-of-the-art Gamewell Fire Alarm System. Powered by 250 batteries, it consisted of 97 signal boxes connected by 30 miles of wire stretched over 450 telegraph poles to the alarm telegraph office.[5] The elaborate system had its central location in Fire Station (Engine House) No. 2, the department headquarters, on Queen Street. When a fire was spotted, a citizen could go to the nearest alarm box, pull the lever, and sound an alarm at the stations, identifying the location of the box and, therefore, the fire.[6] The alarm system had proven itself invaluable and extremely efficient in numerous instances. However, the first earthquake shock snapped the wires, rendering the fire department deaf to emergency calls. This would be the first of many setbacks for the firefighters.

As flames from multiple fires began to appear, the lack of the alarm system meant that calls for help had to be delivered in person. Getting a message to the fire department entailed fighting through crowds of hysterical people and climbing over heaps of rubble. It was no wonder that firemen were delayed in responding, and when eight fires broke out at once, the force had to be divided to take care of all them, causing a greater than normal loss of property.[7]

When firefighters finally did arrive, there was no water pressure in the hydrants for the first hour after the fires broke out. The nipple from the reservoir to the standpipe of the city water plant had been broken, and all the water in the pipe drained out. Two city engineers rapidly fashioned a new nipple, and the firemen were able to pump water from the mains again.[8]

To add to the growing calamity, debris from fallen cornices blocked the doors to several of the firehouses, including Engine House No. 2, on Queen Street, where the front parapet wall was damaged beyond repair; Engine House No. 3, on Anson Street, whose rear wall was also beyond repair; and Truck House No. 2, on Meeting Street. Engine House No. 5, on Archdale Street, was shaken into a dangerous ruin. The fire engines and hose trucks from these stations were initially unable to reach the streets. In addition, at Engine House No. 2, all of the fire horses stampeded, escaped the firehouse stable, and "ran in wildest affright through the upper part of the city, snorting and neighing, to the terror and alarm of all they passed."[9] They were not recovered until the following morning, when they were located on a farm in the northern part of Charleston Neck. In all, only four of the city's nine steam pumper engines were immediately capable of being deployed to serve the entire population of Charleston on the night of August 31.

The fire department had an experienced leader. Chief Francis L. O'Neill, an Irishman by heritage and a Charlestonian by birth, was forty-three years old and in the prime of his life when the disaster struck. A trim, handsome, dark-haired man, he sported a long, meticulously waxed Prince William mustache. The eldest son of John F. O'Neill, a grocer, and his wife, Mary, O'Neill was educated in the Charleston schools and graduated from Georgetown College in Washington, D.C. In 1859, just days after he turned seventeen, he became a member of the Vigilant Fire Engine Company and helped work the brakes on the hand-pumped fire apparatus, "The Little Giant."

Fire Chief Francis L. O'Neill

When the Civil War broke out, O'Neill was nineteen, and he quickly enlisted, serving as an orderly sergeant in the Irish Volunteers, which later became Company H of the 27th Regiment, South Carolina Volunteers. He became a member of the Confederate signal corps and served in the thick of the fighting at Forts Sumter and Johnston. He also carried out one of the most dangerous jobs in the signal corps: serving as an artillery spotter from the steeple of St. Michael's Church—which was the chief aiming point when the Union artillery fired on the city.

When he returned to civilian life, O'Neill entered the wholesale and retail grocery business. He was appointed third assistant fire chief in 1869 and served as such until 1874, when he was promoted to chief. He was known as a "faithful and efficient officer, of excellent executive ability, and deservedly popular with the men under his command."[10] Under O'Neill's direction, the response to the eight fires that threatened to engulf the city was as well-organized and efficient as possible, given the extreme lack of equipment and resources that he faced. An observer who saw him and his men in action wrote, "I saw several firemen thrown from the ladders, but none of them seriously hurt. Chief O'Neale [*sic*] drove from one outbreak to another, and succeeded in holding all in check, except the [one on] lower King Street."[11]

The first alarm was brought in by a breathless citizen on foot. A fire had started about 10:00 p.m. in a small wooden house owned by Mrs. William Henry Wagner at 37 Legaré Street, in the lower part of the peninsula. A gas lamp had exploded, and the house quickly burst into flames. Engine Company No. 1, from Chalmers Street, and Engine Company No. 2, from Queen Street, were dispatched. The men of Engine Company 2 had to clear away building debris before they could reach the scene, and the loss of their horses forced the firefighters to drag the pumper to the fire themselves.[12] Both companies arrived too late to save the house. The occupants, Miss Mary Hasell Gibbes and her sister, escaped the conflagration, but the building was a total loss.[13] Fortunately, it was insured. Immediately next to it, the residence of the Misses Smythe was reported "badly scorched" by one account and "burned down" by another.[14] Julius O. Goutvenier, the foreman of Engine Company No. 2, was overcome by a stroke brought on by the stress of fighting the Legaré Street fire and was not expected to live. He later rallied and ultimately survived.

The second fire was caused by an overturned oil lamp in a house on Blake Street, and the third fire broke out uptown, in a house at the southeast corner of George and St. Philip streets, near the College of Charleston. The house belonged to the estate of Mrs. S. Schwing and was occupied by John Cleary. It was also a near-complete loss. Two adjacent two-story houses were completely destroyed. One was owned by 2nd Lt. Frank J. Heidt of the police force.[15] Due to the lack of alarms and the shortage of equipment, the three buildings burned for nearly an hour before the firefighters arrived. Because there was very little wind blowing at the time, most of the neighbors were able to save their houses by soaking them with buckets of well water.

The fourth fire, an accumulation of several smaller conflagrations, burned a devastating swath up King Street in the heart of the city's chief retail district. It began just north of Broad Street, torched the entire block on both sides of King, and burned out at the Quaker Cemetery on the southwest corner of King and

The "Burnt District" on King Street

Queen. The *News and Courier* lamented that "there was no water and no engines, and the block bid fair to be destroyed." The blaze consumed a row of prime business establishments, along with the apartments above them. This type of building on King Street housed a great many of Charleston's retail merchants and many of its middle-class citizens.

The first blaze on King Street broke out in the attic of a two-story wooden structure four doors north of Broad Street when an oil lamp was knocked over during the first shock. The building was occupied by Cinda Fowler, a black woman. It took the fire department an hour to get to the scene, and firefighters would spend all night and half of the next day conquering the King Street inferno. The fire burned both sides of the street in an enormous swath of flame that engulfed fourteen buildings between numbers 112 and 132, most of which were total losses.

Although there were few direct casualties from this fire, the economic loss to the owners was enormous. J. H. Muller's three-story brick building at 112 King, rented by Marx Berman for his clothing store, was burned, as was the Wing & McMillan stove store next door.[16] Thomas Fickling's two-story wooden building at 115 King, which housed Mrs. Fox's bakery on the ground floor and Joseph Aarons' family on the second floor, was also destroyed. Mrs. Amanda Davis lost her wooden house at 114 and the two-story duplex at 118-120 King, both rental units. The fire left her in dire financial straits and her tenants homeless.

Bootmaker Andrew Marshall lost his shop at 116 King. At 118, Richard J. Morris, a tinsmith and roofer who also sold stoves, grates, mantels, and japanned (varnished) ware, suffered the same fate as his neighbor: his store, and the dwelling house above it, burned to the ground.[17] At 119, Edward D. Andrews sustained the total loss of his "fancy and dry goods store."[18] Across the street at 120, S. A. Prince's small building caught fire and was consumed.[19] One door to the north, Carl Karish (Karl Karesh), one of Charleston's many Jewish clothiers, saw his clothing, boot, and shoe store at 122 King consumed by the flames.[20] David Palmer, a black man, broke his leg during the fire at 123 King and was sent to the city hospital.

Number 124 King, a rental property owned by Richard J. Morris and occupied by A. J. Castion, a busy caterer, confectioner, and baker of "wedding and fancy cakes," was burned. Castion lost the site of his catering business, but his three food stalls in the city market, where he sold venison, ducks, birds, and "game of all kinds" were not affected.[21]

A three-story brick tenement house at 126-128 King owned by J. K. McCauley was totally destroyed, but insurance covered most of the damage. It had housed Michael McDevitt's bootmaking and shoe store and John E. Dohen's house-, sign-, and fresco-painting business, which also sold wallpaper and window shades.[22] Across the street at 127 King, Mrs. Mary A. Frazer's boarding house was spared.

Just up the street at 130 King, the three-story brick building owned by R. J. Morris and occupied by J. R. Tully was damaged by the earthquake and by water from the firefighting.[23] The site may also have housed Abraham Sewalsky's dry goods and clothing store.[24] Only one engine was available to fight this, the worst of the eight fires. There was a fire hydrant immediately in front of 130 King and two fire wells a hundred yards in either direction from the heart of the inferno, but there were not enough fire engines to pump the water onto the burning buildings. Nearby, a similar tenement, owned by J. L. Macaulay, and housing two shops, was also destroyed.[25] In one of the shops, Washington Joiner, a black merchant, had wisely insured his stock of boots and shoes for $1,000.[26]

Louis P. Goutvenier, undoubtedly a relative of Julius O. Goutvenier, the firefighter who suffered a stroke battling the Legaré Street fire, saw his dry goods, notions, and variety business at 132 King go up in flames.[27] Tragically, the store was located just a block from Engine Company No. 2, which was already struggling to respond to the emergencies all around it.[28]

The inferno relentlessly ate its way up King Street, coyly sparing some buildings while incinerating others. Number 134 King was occupied by James J. Carey, a plumber and gas and steam fitter, and was the first building north of the King Street fire zone to survive. Ironically, it was so badly damaged by the earthquake that it was listed as "positively unsafe" and was recommended for complete demolition.[29]

Several blocks further north up King Street, another fire broke out immediately after the first shock and burned fiercely for two hours before three engines could arrive on the scene. Lacking city water, firefighters opened a tidal drain and pumped saltwater on the fire. Chief O'Neill personally directed the efforts there, commandeering every available citizen to cut wood to burn in the steam engines' boilers and to hold onto nervous fire horses.

The blaze gutted numbers 419 and 421, which belonged to the Charleston Gas-Light Company. Both buildings were fully insured. The inferno at 419 presented a huge hazard because one of the company's main gas lines was metered there, and flames lapped at a distributor valve.[30] The first engine that responded, from Engine Company No. 6 on John Street, was dragged to the scene by the firemen because their horses had run away. "Those working here were in imminent danger all the time from the explosion of the gasometer," an observer wrote, "and their bravery can not be too highly commended."[31]

Edward S. Burnham's rented drug store at 421 King was burned, as were numbers 423, owned by M. Friedburg, and 425, owned by Joseph Mintz, which was home to William McComb's dry goods store.[32] Number 424 was severely damaged by the shocks and recommended for immediate demolition, as was number 432.[33] The buildings at 427 and 429, both owned by Mrs. L. S. Witcofsky, sustained major fire damage.[34]

Other burning King Street properties lay just a block north of the Charleston Orphan House, providing the children with a terrifying view that night. The two-story tenement at 481 King was occupied by Dr. William R. Bull, a dentist, and Mrs. E. Meyers and was destroyed by the flames.[35] The Schaladaressi Brothers' fruit store, a two-story wooden building at 487 King Street burned to the ground.[36] Before the fire engines could arrive, the fire spread to the two-story wooden building to the north, at 489 King, which was used by Joseph Mintz as a clothing and jewelry store. About a week after the fire, a suspicious man who gave his name as John Cox was arrested in St. Matthews, South Carolina, with a watch and several articles of jewelry in his possession. They were identified as stolen property from Mintz's store, and the accused thief was brought back to Charleston for trial.

There were many acts of true heroism during the King Street fires. These are exemplified by a woman who lived on the second floor of one of the burning buildings. Coughing from smoke and besieged by flames, she kissed her screaming infant child, tied it securely in a feather bed, and threw it from a window to the ground. The child was unhurt.[37] The mother's fate is unknown.

With firefighters struggling to keep up with the overwhelming disaster on King Street, the remaining two fires were left for divine intervention to extinguish. One destroyed two houses on Cooper Street behind the Enterprise Railroad Company's horse stables, but it eventually burned itself out.[38] The other was caused by an overturned oil lamp at the home of Mr. J. W. Brandt at the corner of Bull and Gadsden streets and was extinguished by the owner and his neighbors after $200 ($4,020 today) in damage was done.

Fires also threatened the Mills House Hotel. When the earthquake struck, Julian A. Selby, who had just begun printing an afternoon newspaper in Charleston, was standing on the east side of Meeting Street, opposite the hotel. He recalled:

> The first shock threw down one or two columns in front of Hibernian Hall, which created such a terrible dust, that I could not see, and instead of running over to the Mills House, where we were quartered (my wife, daughter Maggie and son Gilbert), that I got twisted [around] and ran down Meeting street. Just then came the second, and I got a portion of the debris in my face and on my body, and down I came across the railroad [streetcar] track. I could only have been laid out a few seconds, when I was lifted up by a tall gentleman, who escorted me towards the sidewalk, where we tripped over the fallen telegraph and telephone wires.... My family gathered on the open lot on the northeast corner of Queen and Meeting Streets, where several of our [newspaper] carriers soon joined us and stuck there the rest of the night. Fire had been discovered in several rooms in the Mills House, caused by lamps being upset. My son and myself ran up through the rooms and with any water or means accessible extinguished the incipient flames.[39]

One of the fire department's pioneers died as a result of the earthquake. Jeffordson C. Richardson, sixty years old, was crushed by a falling wall and fatally injured at his residence at

16 Friend (now Legaré) Street on the night of August 31. After being extracted from the ruins of his home, he was taken to the home of Mr. S. J. Pregnall, on Wharf Street, "where he received every attention that the best medical skill could give." He died on September 14 and was remembered as one of the most active and efficient volunteer firemen in the city. His obituary noted that "he served as president of the old Axe Company and later of the Pioneer Steam Fire Company. He built the first steam fire engine ever seen in Charleston. It was built at the Taylor Iron Works in Charleston, but owing to the prejudice which prevailed against innovations at that time, the use of the new apparatus was not allowed until the great fire of 1861, when it was called into service and demonstrated its ability."

Most of the fires, which started about 10:00 p.m., were contained by 1:30 in the morning, but the King Street fire near Broad continued until 3:00 a.m. As the fires were finally extinguished, the exhausted firemen kept watch from the remaining firehouses. Two engines, with their horses and men, were stationed in a park at the corner of Wentworth and Meeting streets, four blocks north of the city market, to be able to respond more quickly to future emergencies.

The bravery of the firefighters earned them the thanks of the city and the respect of Francis S. Rodgers, chairman of the Board of Fire Masters. He wrote, "Amid all these disabilities, the firemen displayed their devotion to duty, and amidst the crumbling buildings and walls, and entangling wires obstructing the streets, moved their apparatus to the several scenes of conflagration, and carrying their hose by hand over ground so obstructed, and still trembling under earthquake shocks, by their supreme efforts saved the city from further disastrous destruction by fire."[40]

During the King Street inferno, a miracle of sorts was reported at the Guard House, the main police station. "On the top of the ruins of the porch of the Main Station was seen a Cross, the emblem of Christianity, which shone with resplendent brightness," the *News and Courier* reported. "There never was a cross on the building as far as known, and the appearance of this one soon attracted attention as it stood out of the ruins. A policeman...pointed to it and said, as he uncovered his head, 'It is that that saved us.'" This "cross of fire" story soon entered the realm of Charleston urban legends. The escape of the station's off-duty policemen, however, was indeed remarkable. Thirty men were sleeping in the upper story of the building when the quake hit and fatally damaged the building. No one was hurt.

With tremors, fires, crumbling buildings, and screaming human beings all around them, the people of Charleston must have truly imagined themselves in hell. Some were luckier than others. The experiences of an unnamed woman who lived in the Fourth Ward, just south of Calhoun Street, were typical of those of thousands of Charlestonians. She wrote the following account to a friend in Springfield, Maine:

> On that fatal Tuesday night my husband and I sat by the student lamp, reading. On account of the heat there were few lights in the house. My oldest daughter was on the piazza, two younger talking with my sister about going to bed. The air was close and still, but not noticeably so. The doors and windows all through the house stood open. A quiet, serene home picture, house and inmates presented; when—just here my hand trembles—suddenly under the southeast corner of the room there is a heavy, deep rumbling, indescribably awful sound. Then the heaving of the earth. Then from around are heard the smashing of glass, a clattering, shattering sound through all the house, and, above a cracking of walls and chimney tops.
>
> This all in a few seconds, and terrified, horrified, bewildered, we spring up. Instantly my thought was 'The earthquake! Run out of the house!' And screaming 'Run out! Run out!' I rushed into the piazza. When I reached the top

of the ten steps leading into the garden, the floor, piazza, house, the whole world it seemed to me, swayed beneath me. I staggered down the steps. I was almost stifled and blinded by clouds of something, I knew not of what; it was caused by the falling of the chimneys, and two piles of bricks fell within a few feet of the steps, down which I ran. My fearful thought was 'the earth has opened and this is our end.' Ah, at that moment, the thought of God was all that kept reason on her throne. I reached the front gate, ten feet from the house, and turned for the first time to look behind me.

The house stood, our beautiful loved home. I took, as I felt, my last glance at it, but it was not the home, but the dear home folks I was looking for. I was sure they had followed me, and I found myself alone. I rushed back, screaming 'For God's sake, come out!' Then little H__'s voice sounded through the stillness. 'Oh, where is my mama? I want to die with my mama.' I called out, 'Here I am,' and she rushed into my arms. And then my husband and the others came out to me into the street. We thanked God we were together once more, and then stood fearing the next moment. 'We are all here?' But no. Our son, where is he? He had gone back to bed sick, his father and I had seen him comfortably fixed for the night. We turned again to the house, all calling out to him in piercing accents that must reach into his room, his ear, and after what seemed long minutes we saw him running to us.

Then came a second shock! But not like the first. As this was less violent, a ray of hope came to us that the worst was passed. And then we turned to look in each other's faces and ask how each had fared. My husband had rushed with me to the top of the steps, but the shaking was so violent that he had to cling to the post for support, and E___ clung to him, crying, 'O, papa, what is it?' My sister and M___ ran too, to the piazza, but, blinded by the dust, they stumbled against the railing and held on and prayed. Then the tremor ceased and they ran to where I was. Our son said he was asleep, but the noise awoke him; the bed shook under him, and the sounds of the crackling walls and falling chimneys around and above told him that the predicted earthquake, we none of us feared, had come. 'I am lost, I can never get out,' he said. But it all ceased. 'Where was everybody?' Hope returned, and springing up, he ran to where we were. An other slighter shock; and we heard in the distance the wailing and groaning of human beings. And now after the horror of hearing and feeling another horror struck the sight. Had we not had more than we could bear? But now to the right and to the left appeared bright gleams of fire.

My thoughts flew to the house. I had expected never to enter it again, but urged by the fear that the lights might start a conflagration we went back. Our parlor lamp stood in just the same place on the table, the student lamp had fallen to the floor and had gone out. The gaslight still burned in the dining-room, and up stairs a dim light flickered in the breeze. We put out the lights below; but did not venture up stairs. A third shock and we all clung to each other. Oh, when will it cease? And now we were thankful to find our servants safe; they having escaped by the Montague street gate. Then we look for our neighbors, many of whom had gone for the summer, and were glad to find all safe and standing near us, though we had not seen them. The air became cooler, and A___ went in and brought wraps, etc., for the women and children. The men brought chairs from the piazza onto the ground, and there we sat, except when tremor beneath drove us into the street, and watched the fires, pitying those who had this added sorrow to face. Passers-by told us things were in an awful state down town. A fireman said the horses belonging to his engine had escaped during the shock and the engine could not be taken to the fire. If this was the case with every engine, what was to be done?

Oh, what a sad and pitiful sight we all presented, sitting in groups outside the homes we dared not enter yet. In one group there was a sick lady who had been carried out of bed; and a paralytic old gentleman who could not stand on his feet, but his servant stood by him, a faithful helper; a mother with helpless little children drawn from their beds in their night dresses, each quiet and composed—no screaming, no crying—a hushed, despairing crowd, all alert for the slightest sound or motion.

Toward two or three o'clock all was still, the fires going down; exhausted and chilled, we felt we must go in. A wooden house being the safest, we persuaded some of our neighbors to go with us into our house; the sympathy and comfort of numbers would help to keep up our spirits. We sat in the wide hall with the door open, all ready to rush out. The gas was lighted now in the hall and we could see each other's blanched and frightened faces. The poor children! How it made our hearts ache to see the expression on their dear faces. But they bore their part bravely

> and never gave way for a moment. No one slept, but waited and watched for the day.
>
> The long-looked-for day dawned and we all went out to see what had happened. Every house had suffered, some hopelessly, some exceedingly, some slightly. Ours had lost two chimneys, one crashing through and breaking a large hole in the kitchen in the same way. A few cracks in the plastering, and that was all. Our servants prepared some hot breakfast, of which we were in sore need. We had just eaten, hurriedly, as we did everything just as if expecting something to happen, when—that distinct, deep rumbling sound again, and all dashed for the street. When this alarm was over the gentlemen went in and gathered up hats, umbrellas, etc., and we sat in the street until 1 o'clock, when the hot sun drove us into the hall, and we sat there with the door open. Gaining some confidence, A___ and I started to fix the house, and our friends went into their houses to move some things, as they were not safe to stay in. Only a few vases in the parlor fell, and some of these were not broken. Nearly everything on the top shelf of the sideboard fell and was broken. The castors fell, but not a cruet was broken. In the cupboards, nothing was broken. Our experience was exceptional. The house is strongly built of wood on a pile foundation, and it must have given with vibrations I suppose.[41]

In many areas, the gas streetlights had been felled by the great heaving of the earth. The resulting darkness only heightened the fears of the crowds who had abandoned their houses for the relative safety of the streets immediately after the first shock. As other tremors rolled through the city before daybreak, a reporter noted that by 3:00 a.m., "Every park, square or vacant lot in the city was occupied by people. It is safe to say that the whole of Charleston passed the night out of doors. In many cases shattered homes were revisited and the women and children were provided with clothing and covering.... The dead were laid in the open air, the wounded were provided with temporary pallets in the lawns or on the streets, and every one waited patiently for the coming dawn, which would at least lighten the horrors of the occasion.... In many places prayer meetings were improvised, and at many a street corner could be seen kneeling groups of all ages and conditions, supplicating the Almighty to grant them mercy and protection in the hour of danger."

As newspaperman Carl McKinley made his forays through the debris-filled streets that evening, he passed by Marion Square (also known as Citadel Square), at the intersection of Meeting and Calhoun streets. It was nearly midnight, and a frightened crowd had collected. The enormous public square—the size of four football fields—was a safe place. In addition to its role as the city's largest public park, it was used as the drilling and parade grounds for The Citadel, South Carolina's military college. Even if the buildings on all four sides of it collapsed, the people assembled there would be safe. As a result, the square was besieged by terrorized refugees.

McKinley watched, fascinated, as people poured in from every direction. Streetcars, carriages, and other vehicles lined the streets surrounding the square, while the horses that pulled them stood stock still, their heads lowered as though they were anxiously sniffing the ground. The park became a central place to take the dead and wounded, which created added grief for the already dismayed refugees. The bodies of the dead were laid on the ground and covered with shawls or sheets, while volunteer physicians and surgeons tended to the wounded.

McKinley lamented, "Exaggerated rumors as to the number of the killed spread throughout the city soon after the shock, causing needless pain to many who, though spared the sight of the scenes of suffering and death so near to them, yet feared for the safety of relatives and friends of whom no tidings could be heard."

The dead and wounded in Washington Park

As he stood on the outskirts of the gathered throng, he observed something unusual. The light from flaming buildings only a hundred yards beyond the square, illuminated the half-clad, filthy, exhausted, and terrified crowd, but no one paid any attention to the approaching fire. Everyone was in a fog of shock and focused only on the earth beneath them, waiting for it to dissolve beneath their feet. This phenomenon happened all over the city. Fire was a known quantity and could be dealt with and controlled; earthquakes were entirely unpredictable.

All open spaces became public gathering places that night. In the one-acre area known both as City Hall Square (its original name) and Washington Park, at the corner of Meeting and Chalmers streets between the Fireproof Building and city hall, people began to congregate as soon as it seemed safe to move about. As in Marion Square, no tall buildings stood nearby, putting the refugees out of range of falling walls. Many of the wounded were carried into the park, which was located in one of the most heavily damaged parts of the city. Soon, hundreds of nervous, restless men, women, and children congregated around the central flowerbed.

One of the first survivors to arrive in Washington Park was lucky to have lived through the trip at all. Edmund Lively, of Richmond, Virginia, was walking in front of the City Hospital on Queen Street when the side of a house fell on him, badly injuring his back and head. "He crawled from under the debris," the newspaper reported, "and saw some [people] at the store on the corner of Mazÿck and Queen streets, whom he supposed to have been killed, as he left them lying on the sidewalk. He staggered on as far as the City Hall Park [five blocks away], and there fell completely overcome."

One noble woman who stepped up to help that night was Mrs. P. F. Murray, of 197 Meeting Street. Near Mrs. Murray's house, a mass of falling bricks struck Rosa Lee Murray, a black woman, breaking her leg and arm and inflicting other severe injuries. She died at the City Hospital soon thereafter, unrecorded in the city's death records. Three weeks later, the victim's elderly uncle, Andrew Sumter, walked to the newspaper office to tell of Mrs. Murray's good deeds. He stated that the white woman "went promptly to the colored girl's assistance, tenderly cared for her, placed her on a cot and stayed by her during the whole of the fearful night, ministering to her wants." The newspaper printed that "Sumter's acknowledgements of Mrs. Murray's kindness and feeling were affecting in the extreme, and every word seemed to swell up from a truly grateful heart. His language was simple but meant a great deal, and his story means more than has been said or can be told."

As the night wore on, the search for the dead and wounded continued. Among the city's early heroes were the numerous brave men and women who rushed into tottering houses and brought out the casualties and the infirm. "Stretchers were improvised out of shutters, doors, and loose planks, and the dead were conveyed to open space," noted an observer. "Others rushed in for clothing and bedding and the park was soon filled with panic stricken women and children. Here and there mattresses were laid on the grass on which slumbered infants unconscious of the terrible scenes enacted around them. In one place lay an old lady very ill from typhoid fever, whose condition had been seriously aggravated by the terrors of the night."

In the midst of crowds of people paralyzed with fear, other heroes and heroines rapidly made their appearance. Sadie Gibbes, a thirteen-year-old girl visiting from Columbia, South Carolina, had run into the street in the middle of all the chaos. Then she smelled smoke and heard the screams of a terrified child. With no regard for her own safety, she rushed into a shaking house, ran up the stairs into a burning room, and rescued a stranger's baby, stopping to wrap it well before she brought it down.

Another heroine was brought to light through a letter to the editor of the newspaper about a week after the earthquake. The author, a man identified only as "A Grateful Heart," and a resident of "the East," wrote, "While the brave deeds of men are being recorded, let me recount those of a noble little woman, whom may God and all the angels forever guard and protect in the future. I speak of Mrs. Fowler, who with daughters, women worthy of such a mother, ministered to the sick and distressed on Tuesday night." An hour after the first shock, Mrs. Fowler returned to her house, not knowing if it would collapse around her and her daughters at the next moment. They collected all the medicines and clothing they could carry and set out to care for the sick and half-clothed survivors who were starting to assemble in the park.

"All night long nearly, though worn out by fatigue and the shock, these untiring women moved like angels of mercy among God's 'little ones,'" wrote the grateful man. "Many of the feeble and bruised were saved from death, I believe, by the cups of hot coffee and tea with which she supplied them. I myself, a stranger from a strange land, was a recipient of her bounty. Sick and reduced in strength, and dragged into the night air in my night-clothes, a cup of coffee came to me as a boon unspeakable."

Despite being victims of the earthquake themselves, physicians selflessly attended to the wounded and the dying. Dr. John P. Chazal, Jr., a professor of anatomy and later dean of the Medical College, told a reporter that he had treated a white man and a black woman who had been badly wounded. These unfortunates had been pulled from the wreckage of the mansion at 64 Hasell Street owned by Mrs. Lazarus, a few blocks from Chazal's home on Anson Street in Charleston's Ansonborough neighborhood.[42] In one of the city parks, Dr. John S. Buist gave his professional attention to a dozen or so wounded persons, among them a young girl whose leg was broken. Doctors Manning Simmons, Peter Gourdin deSaussure, McDowell, Ravenel, and others performed heroic service on the streets. A number of babies were born prematurely, their mothers' labors induced by fright or trauma.[43] Many of those children did not survive, but none of their names appears in the city's official death record.

Frayed tempers erupted into harsh confrontations when blacks in Washington Park began loud shouting and praying. Whites, concerned that the noise would further disturb the wounded, tried to stop the emotional outbursts. "For a time a riot was threatened," wrote a reporter, "but the two colored men who were making all the noise were removed and after that comparative quiet was restored." In the following days, blacks faced many more examples of white opposition to their emotional prayers and expressions of grief.

In a park at the corner of Meeting and Wentworth streets, as one of the first major aftershocks hit, confusion reigned, creating a near panic in the now-crowded area. "Fortunately, cool heads checked the outbursts of the people," a reporter wrote, "and prevented a panic that would necessarily have followed if the people had been allowed to give vent to their feelings." "The people" was an antebellum euphemism used by the white gentry to refer to blacks.

A more sympathetic and insightful reporter found that the earthquake had produced a unifying effect on the races. Of the crowd that had streamed into Marion Square he wrote, "The

Terrified survivors pray for salvation

colored people everywhere were loud and unceasing in their exclamation of alarm, in the singing of hymns, and in fervent appeals for God's mercy—in which appeals, God knows, many a proud heart who heard them arising in the night, and in the hour of His wondrous might, devoutly and humbly and sincerely joined. Danger brings us—all of us—to the level of the lowliest," he continued. "There were no distinctions of place or person, of pride or caste, in the assemblages that were gathered together in Charleston on Tuesday night. It was a curious spectacle to look back upon; it is a good one to remember, for white and black alike."

After having made an incredible three separate forays into the night to gather information for the newspaper, Carl McKinley finally ended his workday at three o'clock in the morning. The opportunity to cover the earthquake had awakened all of his journalistic instincts, and he wanted to remain at the heart of the event. His office had been wrecked, so he walked to Washington Park, wrapped himself in a borrowed quilt, and made a temporary bed in the grass. Between 4:00 and 5:00 a.m., he was awakened by the voices of several people, who were standing over his covered form engaged in earnest conversation. As his consciousness returned, the newspaperman realized that he was being mistaken for a corpse. The weary reporter rolled over and pulled up his quilt, and the sympathetic crowd turned their attention to those more needy.

6

THE SURVIVORS

A long line of wagons was drawn up in the center of the streets, and filled with sleeping people. Two hundred people spent the night in the grave yard.
—The *Atlanta Constitution*

Wednesday, September 1, 1886

By midnight on August 31, Charleston was in ruins, the citizens were terrorized, and nerves were raw. Throughout the city, most of the survivors experienced varying degrees of nausea, the result of the violent and unaccustomed motions to which they had been subjected. They hoped and prayed that the worst was over. Indeed, they could not imagine how it could get worse—yet it did. The forces of nature took no one's plight into consideration. Glare from the fires lit up the sky, and the smoke "hovered like a pall above the city, which seemed thus threatened with destruction of what had escaped the earthquake's violence."[1] Between midnight and daylight, four more shocks rumbled through the city, shattering glass, felling walls, raising clouds of dust, and beating the citizens into an almost catatonic state. By sunrise, desperation permeated the air. But Charlestonians were no strangers to disaster. Their experiences had made them tough, resilient, and highly experienced in recovering from an almost unimaginable list of natural and man-made catastrophes.

For the first two hundred years after its settlement by whites in 1670, disease had mercilessly stalked Charleston and coastal South Carolina with morbid efficiency. Smallpox erupted in 1697, killing between two hundred and three hundred people.[2] Malaria plagued the first settlers and all who followed until the 1920s. Despite the effectiveness of quinine, an antidote that had been used by Americans to combat malaria since the early nineteenth century, Charlestonians were still being killed by the disease at the time of the earthquake. Most people, however, were more afraid of yellow fever. When this plague hit, it was frequently fatal, and death came only after four to six days of ghastly agony. A problem for most overcrowded American cities in the eighteenth and nineteenth centuries, yellow fever epidemics devastated Charleston in numerous outbreaks starting in 1699, including one in 1854 that killed six hundred people.[3]

As if diseases weren't enough, Charleston was a highly contested battleground for both armies that invaded the South: the British and the Union. During the American Revolution, Charleston saw fierce fighting. Sir Henry Clinton's redcoats landed on Johns Island, south of Charleston, in 1780. Using two hundred cannons, the British began bombarding the city with heated cannonballs, which caused numerous fires.[4] The city held out bravely for two months before surrendering, and Charleston remained occupied until the British evacuated the city in December 1782.

While war would continue to play a pivotal role in Charleston's history, fires were the chief devilment of the densely packed city. Serious conflagrations raged in 1740 (incinerating three hundred buildings), 1778, 1796, and 1810. In the summer of 1835, the worst fire since 1810 (which burned 194 buildings) broke out in a wooden tenement on the west side of Meeting Street. By the time the volunteer fire companies extinguished the last flames, 374 structures had been destroyed.[5]

The Great Fire of April 27-28, 1838, also known as the Ansonborough Fire, started at the corner of Beresford (now Fulton) and King streets and quickly spread to the northeast. It swept north to Society Street, and after consuming the Ansonborough neighborhood—some eighteen city blocks—it burned across the city to East Bay Street. As the inferno raged out of control, gunpowder was used to blow up buildings to create firebreaks. A diarist wrote, "There are about 1,000 houses burnt, [and] distress is beyond description."[6] Charlestonians could not imagine a fire of greater ferocity, but they only had to wait about two decades to see the fires of hell once again ravage the peninsula.

On December 11, 1861, the most devastating fire in Charleston's history started in a sash-and-blind factory on Hasell Street near the Cooper River at about 8:30 p.m. Although the Civil War had been underway since January 10, 1861, when Citadel cadets on Morris Island fired cannonballs at the U.S. supply ship, *Star of the West*, the Great Fire of 1861 was of civilian, not military, origin. "The wind, which was blowing strongly from North-northeast, increased almost to a hurricane," said the official report. "Building after building caught, and became, as it were, one vast sheet of flame."[7] By the time it was brought under control, it had leveled a swath of 540 acres all the way west to the Ashley River, taking with it every scrap of wood, cloth, and foliage in its path. Losses were estimated at between $5 million and $7 million dollars, approximately $100.5 to $140.7 million dollars in today's currency.

Meeting Street and the Circular Congregational Church after the 1861 fire

The fire in 1861 only added to Charleston's troubles. War had again become an unwelcome guest. For South Carolinians, the Civil War was a shared catastrophic experience. South Carolina sacrificed a greater percentage of men to the Lost Cause than any other southern state. By the end of the war, 12,992 of the state's men in uniform had perished—twenty-three percent of the arms-bearing population of 55,046.[8]

Artillery damage to the John Rutledge House

In the eyes of many northerners, Charleston was the evil cradle of secession because the Ordinance of Secession was signed there on December 20, 1860, making South Carolina the first state to secede from the Union. In addition, the first battle of the Civil War was fought at Fort Sumter in Charleston Harbor, on April 12-14, 1861. A Union naval blockade of Charleston from April 1861 through the summer of 1863 was largely ineffective, due to the Lowcountry's numerous outlets to the sea and a shortage of Union gunboats. However the landing of Union troops on Morris Island, south of Charleston, led to a new and fearsome threat. Charleston was singled out for a unique form of punishment: a 545-day artillery bombardment by Union naval and field artillery under the command of Maj. Gen. Quincy A. Gilmore. It began at 1:30 a.m. on August 22, 1863, when a 150-pound, eight-inch incendiary shell from an enormous 16,500-pound Parrott rifled cannon known as the Swamp Angel was lobbed into the heart of the city from James Island, five miles away.[9] With the city virtually empty, save for a skeleton force of Confederate soldiers, the bombing was carried out not because the city had any significant military targets within artillery range but rather to punish the secessionist devils through the wholesale destruction of private property and the persecution, through forced evacuation, of its remaining civilian population.[10]

On February 18, 1865, the last Confederate troops to abandon doomed Charleston burned supplies of cotton to keep them out of the hands of the enemy. At the Northeastern Railroad Depot near the Cooper River, a supply of gunpowder was accidentally ignited, resulting in a huge explosion that killed about 150 people. Probably an equal number were burned.[11] The resulting fire destroyed many houses on the eastern edge of Wraggsborough, adjacent to the depot.[12] Another fire started on the western shoreline of the city, the offshoot of the deliberate burning of the Ashley River Bridge, which was carried out to slow the advance of the Union army and give the Confederate defenders more time to flee.

After Union troops occupied Charleston in the spring of 1865, the venerable Mrs. St. Julien Ravenel, descended from generations of native-born Charlestonians, returned to the city and painted a grim picture. "Everything was overgrown with rank, untrimmed vegetation. . . . Not grass merely, but bushes, grew in the streets. The gardens looked as if the Sleeping Beauty might be within. The houses were indescribable: the gable was out of one, the chimneys fallen from the next; here a roof was shattered, there a piazza half gone; not a window remained. The streets looked as if piled with diamonds, the glass lay shivered so thick upon the ground."[13]

Even before fire and warfare devastated the city, nature treated it harshly. South Carolina lies in the high-probability strike zone for Atlantic hurricanes, which often lash its coast between June and the end of November. The first hurricane in recorded history to strike Charleston was a blessing; the rest were curses. In 1686, a Spanish expeditionary force from St. Augustine attacked a settlement near present-day Beaufort, then pillaged its way north. A hurricane, "wonderfully horrid and distructive [*sic*]" ended the planned Spanish assault on Charleston.[14] In 1728, a fierce storm drove twenty-three ships ashore. A disastrous hurricane in 1752 caused considerable damage along the coast as it pummeled Charleston for twenty-four hours. It destroyed the city's eight wharves and the ships in the harbor, and caused massive damage to buildings and

Aftermath of the Great Cyclone of 1885

the rice crop along the coast.[15] Many lives were lost in a 1770 storm, and other large storms hit in 1797, 1800, 1804, 1811, and 1813, along with a violent gale in 1822. An 1873 tornado demolished the Northeastern Railroad Depot on East Bay Street (recently rebuilt after the 1865 fire), killing several people and causing considerable damage. Major gales struck the city again in 1874 and 1881, followed soon by a powerful hurricane, which came to be known as The Great Cyclone of 1885.[16]

Almost a year to the day before the earthquake, a disastrous storm formed in the West Indies, skirted the coast of Georgia, and bore down on Charleston like a scorned lover bent on revenge. The U.S. Signal Service weather report for the afternoon of August 24, 1885, said, "Up signals; fresh and strong East to North winds." The report raised no alarms because brief, intense, "pop-up" afternoon thunderstorms were common in August, and they rarely lasted more than half an hour. However, that afternoon, returning harbor pilots reported especially heavy seas outside the Charleston bar, which they regarded as a prelude to "dirty weather." After sunset, the skies turned ominous as they filled with cumulostratus clouds and lightning. By 1:30 a.m. on August 25, the storm had officially commenced. By dawn, the wind had increased to about forty knots, littering the streets with broken tree branches and roof tiles.

The concept of evacuating low-lying areas in advance of a hurricane was unknown at that time. Furthermore, the lack of any appreciable advance notice, coupled with the total lack of mass transportation, made rapid evacuation impossible. People who lived on the coast were trapped and powerless to help themselves in the face of storms—and this one was a monster.

By early evening, the winds peaked at an estimated 125 miles per hour—a Category III hurricane on the present-day Saffir-Simpson Hurricane Intensity Scale. To make things worse, the strongest winds struck at high tide: 7:30 p.m. Waves driven by the storm surge crashed over the Battery, a sturdy rock seawall, flooding the low-lying city. Water rose three to six feet deep inside the city's grandest and most opulent houses at the tip of the peninsula. The winds peeled off the tin roofs of hundreds of houses as if they were made of paper. The 224-foot-tall, stiletto-shaped spire of the Citadel Square Baptist Church smashed down onto the nearby residence of Thomas D. Dotterer. Soon afterward, the gilt ball and weather vane of St. Michael's Church were ripped off and fell 170 feet to the ground, crashing onto Broad Street.

About nine o'clock that night, the winds, which had blown in from the east, gave way to an almost perfect calm. Charlestonians breathed a sigh of relief, surveyed the damage, and started to put their lives back in order. Then as now, fall windstorms were fairly common. The city generally experienced at least one tropical storm almost every year and a strong windstorm every two to three years. However, at the time, the difference between a simple windstorm or gale, which came and went in a straight line, and a hurricane, with its circular form and counter-clockwise rotation, was not commonly understood. Charlestonians were not aware that the calm they were enjoying was the eye of a hurricane.

Within a half an hour, the eye wall moved past Charleston and continued north. Now the fierce winds roared back at full strength, this time from the west. Ships in the harbor were spun around like tops and smashed into piers. A reporter wrote that "roofs, trees, fences, wharves, and ships that had withstood the force of the storm before, owing to favored positions, yielded to the attack from a new direction, and wreck was piled on wreck, ruin heaped on ruin."[17] At the time, Charleston had about forty wharves on the Cooper and Ashley rivers, and they provided the logistical backbone for the city's shipping economy. All but one were either severely damaged or destroyed.

When the storm departed the stricken city, it left behind streets so packed with debris that vehicles could not pass. Electric and telegraph wires were snarled in hopeless confusion on the remaining poles and twisted like snakes underfoot. Dazed residents looked out through gaping holes in their houses at trees stripped of leaves and branches. The cyclone caused nearly $2 million ($40.2 million today) worth of damage to South Carolina's single greatest commercial and shipping center. Before the cleanup was over, more than 10,000 cartloads of vegetable debris were hauled off the streets. Journalist Carl McKinley, who would soon become the biographer of another mass disaster, summed up the cyclone's wake:

> The brave, beloved city which had been swept by fire, stormed at with shot and shell, and occupied by a hostile army, within a quarter of a century, had now been buffeted for hours by raging seas, and shaken to its foundations by the most fearful storm that has ever visited our coasts within the knowledge of man. It was naturally to be expected that the inhabitants of the city would be dismayed by the extent of their loss. But it was not so. With the rising of the morrow's sun the work of repair and reconstruction began hopefully and bravely. Where all had

> suffered there were none to weep over a neighbor's woes. Few complaints were uttered; despair found no place even among the ruins. Offers of assistance were promptly made by the Governor of the State and from various other authorities. The answer was promptly returned in each instance that Charleston was strong enough to help itself; and help itself it did so effectually that the catastrophe scarcely interrupted the ordinary course of business beyond the few hours necessary to repair the railroad tracks leading out of the city, and to clear the rubbish out of the way of the vehicles and pedestrians on the streets. Wrecks were raised or cleared away, roofs were patched or repaired, wharves were rebuilt, the debris was removed and at the end of a month a stranger passing through the city found but few traces of the storm. Only a few weeks later [in September 1885] a disastrous tornado destroyed the town of Washington, in Ohio. Amongst the first cities in the country to offer aid in money was Charleston, which had so far recovered as to be ready to extend help to others in misfortune.[18]

After the hurricane, Mayor Courtenay had worked ceaselessly to help the city rebuild its shattered waterfront, homes, and businesses. In the process, he set two major precedents for the City by the Sea. First, he was not willing to let Charleston relinquish its dignity or appearance of civic strength by accepting outside charity. Charleston, he proudly told one and all, could take care of itself. The second precedent was a lasting one. Courtenay's city, warmed by the widespread offers of help from others, immediately became a first responder to offer aid to other stricken areas. To this day, Charleston is always among the first municipalities in the nation to send fully prepared, self-sufficient police, fire, rescue, medical, and relief personnel to disaster zones. This commendable tradition of helping can be traced directly to Courtenay's policies.

By curious coincidence, tornado-ravaged Washington, Ohio, and hurricane-stricken Charleston would soon be doubly linked. The first link was Charleston's offer to provide aid to the Ohio tornado victims just a month after the 1885 Charleston cyclone. The second was Dr. Thomas C. Mendenhall, a professor of physics and mechanics, who was then director of the Ohio Meteorological Bureau. In 1885, Mendenhall directed that a thorough study of the Washington tornado's path be made and published. Less than a year later, as a senior scientist serving with the U.S. Signal Corps, he became one of the first federal investigators of the 1886 Charleston earthquake.

Just one week less than a year after the cyclone, the proud and resilient citizens of Charleston were faced with the aftermath of an earthquake. The magnitude and extent of the disaster visible on Wednesday morning were almost incomprehensible. Mr. Brenner, a war veteran and superintendent of Charleston's Western Union telegraph office, said, "The Lord did Charleston fifty times more harm in fifteen seconds than the war did in five years. I walked through this town by the light of the fire of [18]61, and the damage is twenty times greater now than then."[19]

Few buildings in the city had escaped damage, and the inhabitants were afraid to reoccupy, or even enter, them. Those who attempted to do so were sent running for their lives by eight earthquake shocks on Wednesday, the first full day after the horrendous first shock, and three more aftershocks that night.[20] Virtually all businesses were closed that day, even the drug stores, whose supplies were sorely needed. However, human nature being what it is, the saloons kept their doors open.[21] To overcome the helplessness they felt in the face of the earthquake, some turned to drink; others to religion.

Julian A. Selby, exhausted after helping put out fires in the Mills House Hotel on Tuesday night, told the following tale: "I went into Meyers' saloon, northeast corner of Meeting and Chalmers streets, and called for a brandy and soda, an expensive drink, which was furnished. I laid down a half dollar, when the proprietor looked at me, then at the coin, and said, in a dolorous tone of voice, 'We take no money to-night.' I nearly collapsed."[22]

Meyer was not the only warm-hearted tavern operator to offer free comfort that night. A "stout, hearty man" who went by the name of King had a saloon near Market Street. According to Selby, "He made no braggadocio about not being afraid, etc., but told the police officers that anybody who needed food or drink, and [had] no money to pay for it, were invited to call at his place and get it as long as it lasted."[23] The generous Good Samaritan must have had a nervous system weaker than the strength of his compassion, for he was found dead in his bed a few days later.

That awful Tuesday night, some men were driven not to drink, but to forsake it. One of them "had returned to the city the day of the mighty 'quake. During his absence he had forsaken liquor completely. He summoned some friends to his hotel room and was preparing to have one julep for old time sake. As he lifted his glass to his lips, so the story goes, the first jarring shock ran through the city and the building began to fall around him. He was not known to drink thereafter."[24]

Walls had peeled off buildings, leaving the interiors exposed like houses being constructed backwards—rooms first. The damage was random and comprehensive. The building belonging to Maj. J. H. Robinson was a good example. Some parts were badly wrecked; others escaped injury. On one side of a bedroom, oil paintings were thrown from the wall with such force that both the canvas and frames were crushed, while on the nearby mantelpiece, a tall, slender vase remained upright and undamaged. On another wall, two or three small, framed photographs were undisturbed, while three feet away, the plaster had been torn from the wall and smashed into dust, and the beams underneath the lathwork were torn out of place. A chaise lounge was hurled across the room and broken into pieces, while chairs a few feet away were not even overturned. At the entrance to the house, one gatepost was twisted off, while the other, four feet away, was untouched.[25]

Countless buildings were exposed like dollhouses

The family of Sgt. J. P. Dunn, a police officer, suffered as extraordinary an escape as any that happened that night. Four of his children were in bed in the back room of their three-story home at 17 Tradd Street when the shock hit. Immediately, the chimney fell in, bringing down the timbers that supported the roof and its heavy slate tiles, which covered the children's beds with rafters and bricks. Dunn rushed to the room, yanked open the door, and pulled the youngsters from the debris. One of them was bloody from seven wounds on the head and body, and the other three were also injured. A reporter who visited the scene wrote, "It is little short of miraculous that any escaped with life, judging from the condition of the bed and room."

Harvey G. Senseney, the baggage master for the South Carolina Railway train that had arrived just before the quake had a similar experience. A part of the chimney fell in his room at 605 King Street just after he got in bed after completing his run.[26] Fortunately, he was unhurt.

The fear of falling buildings was universal. A black coach driver slept all night in his vehicle. "He said he had measured the houses and knew they could not fall as far as he was. This became popular before morning," the *Atlanta Constitution* reported. "A long line of wagons was drawn up in the center of the streets, and filled with sleeping people. Two hundred people spent the night in the grave yard."[27] Edgar A. Guest wrote that on the morning of September 1, "by several successful trips into the house I brought out a few chairs, some pots and pans and china. We built a fire in the yard and there we got breakfast ready. Nobody could eat. No desire for food. The water was hard to find. All cisterns were broken and water leaked out. Five families in our immediate neighborhood combined and made a large tent composed of crossties and oilcloth and here we camped for weeks."[28]

At the *News and Courier*'s editorial office on Broad Street, the nervous staff surveyed the structure and found ominous signs. The loss of the building's granite parapet left the slate roof and a portion of the attic floor exposed. Shelves and bookcases had crashed to the floor. Plaster and dust covered every square inch of all three floors—and no one knew whether the next shock would bring down the entire building.

A few courageous reporters and editors re-entered their heavily damaged offices to try to finish the newspaper. Captain Dawson's family was out of town, and he left his beautiful house—now in ruins—and came to work shortly after the first shock. One of the printers, George Simonin, had been at home in bed when the quake hit. Realizing the importance of the event that had just happened, he hurriedly took his family to safety in Marion Square and then walked barefoot to the office, clad only in a pair of trousers.[29] He, McKinley, Dawson, and several other brave men somehow managed to print an abbreviated issue. It consisted of the seven pages that were already set in type before the earthquake hit, with the breaking news of the earthquake, written and typeset on Wednesday morning, on the back page. The September 1 issue was only the second in the newspaper's history to miss having its lead story on the front page. The first was the bombardment of Fort Sumter in 1861.[30]

"The *News and Courier* buildings are injured with the rest," the newspaper said, "and naturally, the employees, finding themselves safe, hastened to learn the fate of the members of their families. These were, of course, in some fear, and it has been difficult in the extreme to bring together printers enough to make up the *News and Courier* to-day, together with the pressmen to print it. All praise to those who were so true in such an emergency.... It is with the utmost difficulty that the paper is published at all, even by changing it, at the last moment, from twelve pages to eight pages. The confusion and trouble—far beyond anything ever before experienced in Charleston—will account for all deficiencies." The editors were being overly modest. Even though the entire newspaper staff had to flee the building to save their lives, and the building itself was seriously damaged, the September 1 coverage of the earthquake was amazingly comprehensive, and the typesetting almost totally free of errors.

Later, the heroism of the journalists was applauded nationwide. "In the midst of the city's wreck and ruin," said the *Hartford* [Connecticut] *Times*, "and ere yet the earthquake shocks have ended or the broken walls have ceased to fall, the voice of the local press emphatically proclaims one thing that even the earthquake has not broken, and that is the unconquerable will and enterprise of the Charleston people. In a display announcement, which was itself put in type by compositors who stood ready at an instant's warning to drop their 'sticks' and rush or leap from the shaking building, on the recurrence of shocks that had not yet ceased.[31] The *News and Courier* sounded the reassuring note of confidence, determination and good cheer."[32]

Newspaper carriers had difficulty delivering the paper to their regular subscribers because they were stopped in the streets by people wanting to buy copies, desperate for information about the events that had so altered their lives. Even though large numbers of extra copies had been printed, there were not enough to fill the demand.

As grim as the news was, the realities of the disaster were less frightening than the horrified speculations of most of the readers. As a result, the Wednesday edition was both sobering and comforting. Carl McKinley wrote, "The rising sun on Wednesday morning looked on empty and broken homes and

The damaged Charleston City Hospital from the rear

on streets encumbered with continuous lines or heaped masses of ruins, amidst which the weary and shelterless citizens gathered together in little groups, or picked their way from place to place, wondering at the extent of the damage inflicted everywhere, and with renewed thankfulness in view of the perils escaped. Those who flattered themselves that the morning had brought an end to their terrors and trials, however, and who timidly ventured to return within doors to commence the work of temporary repair or to provide for the wants of the day, were quickly undeceived."[33]

In any catastrophe, treatment of mass casualties becomes the highest priority, but Charleston's medical system had taken fatal blows. All three of the city's most important medical institutions were simultaneously destroyed: the City Hospital, Roper Hospital, and the Medical College of South Carolina. The Medical College, a teaching institution with no hospital rooms of its own, provided free services to the poor of both races at the City Hospital and at Roper, as well as at several city-run neighborhood dispensaries. These three major institutions were situated on part of a square bounded by Queen, Franklin, Magazine, and Mazÿck (now Logan) streets. No other block in the city suffered worse damage.

The medical facilities were especially important because they offered services to everyone, not just those who could pay. Destitute residents of the city were accepted as patients if they were referred by the city physicians to the poor. Sick seamen were accepted if they were referred by their country's consul in Charleston or by the Assistant Surgeon of the U.S. Medical Health Service office at the Custom House. Paying patients from Charleston or "from away," as Charlestonians referred to those who were not Charlestonians by birthright, could check in without a referral and receive their treatment in private rooms by the physicians of their choice.[34]

The City Hospital, a Gothic structure on the southwest corner of Mazÿck and Queen streets, next to the Medical College, was destroyed during the first hour of the earthquake. Built in 1850 as a slave jail (known to whites as the work house and to blacks as the whipping house), it was converted to serve as the city's hospital, with separate wards for whites and blacks, in 1866. Because of the hospital's original use, this Charleston medical facility was the only public general hospital in the South with iron bars on every window.

Within minutes after the earthquake hit, the City Hospital was literally jolted into action. Two patients from the black wards were killed. One was Robert Redoff (or Rodoff), a forty-three-year-old laborer, who was crushed in his hospital bed by a falling piazza. The other was Florence Rector (or Flora Reston), age sixteen, who was stoned to death in her bed by falling bricks. Both were later buried in the public cemetery. The next morning, the newspaper reported, "Ten patients severely and some mortally wounded are in the Hospital, six white and four colored. They present a terrible spectacle." The living were taken outside and spent the night under the unlucky stars. Later, they were temporarily transferred to safer one-story hospital outbuildings formerly used as washing rooms and dining rooms.

On Wednesday, when the magnitude of the damage to the buildings became apparent, the commissioners of the hospital conferred with Acting Mayor William E. Huger and the city's aldermen. They decided to move the hospital from its ruins on Queen Street to the Agricultural Hall on the west side of Meeting Street. The headquarters of the Agricultural Society of South Carolina, the handsome building was the largest meeting place of its kind in the city.[35]

In a demonstration of successful public and private cooperation in the face of an emergency, Charleston's Street Department and the privately owned Southern Express Company jointly furnished transportation for the hospital's patients, furniture, and equipment.[36] Virtually everything and everyone was moved within thirty-six hours. The majority of the patients had been moved by 1 p.m. on September 1, and all but two of the rest were relocated on September 2.[37] On September 6, a black woman, who suffered a fractured skull on the premises of Dr. Forrest on King Street when the first shock hit, was the next to last patient to be moved, due to her fragile state. Six men transported her from the City Hospital to Agricultural Hall on a litter. One man whose health made him too unstable to move remained under care at the shattered hospital.

The Agricultural Society's 80-foot-by-180-foot main hall was divided into four wards: one for white women, one for white men, one for black women, and one for black men. The newspaper reported, "The hall has been admirably arranged; there is no confusion or disagreeable intermixture of races or sexes, and too much praise cannot be awarded Mr. W. D. Hard, the superintendent, for his energy and administrative ability in this matter. The medical and surgical staff and the nurses' quarters are rather crude, and there is no 'dead house,' [morgue] but these will be perfected in time." Dr. Henry B. Horlbeck, chairman of the City Hospital's board of commissioners, commented gratefully that "it is fortunate for the city that so large, spacious and suitable place was available" and described the patients as being "very comfortable."[38]

The new hospital and its patients were attended by volunteer physicians from across the city. Within a short time, William Flynn, the City Hospital's chief carpenter, had built a long frame building adjacent to the new location for the construction of coffins and for use as living quarters for the ambulance driver. The energetic Flynn went on to build 531 coffins for burial of the indigent, a new carpenter's shop, a stable, and a morgue, as part of his work to prepare the Agricultural Society Hall for medical service.[39]

While the City Hospital was attempting to resolve its crisis, the demands were heavy on Roper Hospital, its neighbor on the northwest corner of Queen and Mazÿck. It was built between 1850 and 1852 as a result of a bequest from Col. Thomas Roper, a former mayor of Charleston. Charleston's postwar destitution forced the trustees to lease the proud, independent hospital to the city. The large, handsome, Italianate building had a central core and two wings. The central portion contained offices for physicians and administrative staff on the first floor. The library of the Medical Society was housed on the second floor, and an amphitheater for medical lectures was on the third. The three-story east and west wings each had four-story corner towers and housed wards for the patients. Much of the staff came from the adjacent Medical College of South Carolina. The hospital was in excellent condition when the earthquake hit.[40]

With the first shock, both towers were wrenched from Roper's main building, and serious cracks appeared in all of the exterior walls. In addition, the top story of one of the towers was thrown onto the piazza of the east wing, leaving that part of the building "a gaping wreck."[41] The interior was heavily damaged, as was the four-story training school for nurses behind the hospital.[42] When the full damage was assessed by the light of day on Wednesday, it was apparent that both wings of the hospital were beyond repair, and the main section was abandoned for medical use. By some miracle, only two of the

Roper Hospital after the earthquake

110 patients had been killed.[43] On Wednesday, at 1:00 p.m., Roper's staff started moving patients to the Agricultural Society Hall. The transfer was completed by six o'clock that evening.

The third part of the three-part medical complex, the Medical College of South Carolina, had been established in 1822, and its handsome home was completed about 1826. A third-story attic was later added to the original classical two-story building, which had Ionic columns supporting a portico. Located next to Roper Hospital, the edifice was used as a teaching facility. At the end of the Civil War, when the Medical College had lost all means of income, the city's physicians donated their teaching services to the college gratis, charging only for out-of-pocket expenses. This enabled medical students to continue their educations. By 1883, the Medical College was described as "a substantial building, well-adapted to its purposes, and furnished with the necessary appliances."[44]

The earthquake changed all that. The portico collapsed, toppling three of its four columns, and part of the roof fell into the main lecture hall. Widely published engravings showing the damage were reprinted worldwide, and remain among the saddest reminders of the earthquake's egalitarian fury.

The Medical College of South Carolina

The earthquake also jeopardized several other key public institutions that provided care for the poorest and most vulnerable of the population. The Charleston Alms House, a home for indigent whites, was established in 1852 on the south side of Columbus Street. In 1886, the house provided "good, wholesome food, clothing, and shelter in a comfortable manner," as well as medical attention, for 84 paupers, in addition to assisting 176 needy non-residents. No casualties ensued during the earthquake, but an inspection revealed that a damaged tower had to be taken down. The ends of the building were deemed safe for immediate occupancy, and after a few minor repairs, the entire building was habitable again.

William L. Daggett, chairman of the commissioners of the Alms House, praised the staff for the "courage, duty, and devoted performance of duty which is the reflex of the heroism and fidelity" they showed during the calamity. "The House was shook up and shattered to its very foundations, at an hour when nearly all of its occupants had retired for the night," he wrote. "When it is remembered that the inmates are composed of aged and infirm people, many of whom are cripples and imbeciles, it is indeed remarkable that they were all gotten out of the main building without the least accident or injury to any one of them, and provided temporary shelter for the night."

Daggett singled out one man for special thanks: Commissioner (and city alderman) A. Johnson, who "early appeared on the scene, and by his wise counsel and good judgment aided in restoring confidence among the excited and terror-stricken inmates, as well as providing shelter for them." Chairman

Daggett also noted that the earthquake had provided at least one benefit. Immediately after the catastrophe, twenty-three of the non-residents who relied on the house stopped coming, probably because they had found work as laborers, clearing the rubble and beginning to rebuild the city.[45]

The Ashley River Asylum, an institution for aged and infirm blacks and the insane, was established in 1867. It suffered significant damage during the 1885 cyclone and was hit hard by the earthquake. The asylum cared for sixty-eight residents, but few lived there for long. Its annual report showed forty persons admitted in 1886, matched exactly by forty deaths. The dead were buried in the adjacent potter's field, also known as the "Field of Rest" or "public cemetery," the city's final resting place for those who could not afford a private burial plot. When the tremors in the earth began, the patients in the asylum were panic-stricken. One woman escaped but was recaptured. By the end of October 1886, twenty of the earthquake's indigent black victims had been laid to rest in the potter's field.[46]

Throughout Charleston, people went out of their way to help each other. Father Patrick Duffy, a Catholic priest, told the story of one such Good Samaritan. Immediately after the first shock on Tuesday night, the Sisters of Mercy calmly gathered up their resident pupils, all orphans, and escaped their shattered quarters at the Academy of the Sisters of Charity of Our Lady of Mercy at present-day 51 Meeting Street.[47] While fleeing to their convent on Queen Street, they encountered a young man, who asked if he could be of any service. Sister Agatha, the directress, thanked him but politely declined his offer. As he turned to leave, the young man noticed that one of the orphan girls had fled without shoes. He gave her his and continued his journey a bit more gingerly in his stocking feet. Charity ran in his family. The donor was Joseph Yates, Jr., son of Col. Joseph A. Yates, who would become chairman of the Executive Relief Committee, the group that organized aid efforts in the wake of the calamity.

One of the few cases of good fortune that night was the survival of the Charleston Orphan House. An enormous, five-story, 256-foot-long, residential and educational complex at 160 Calhoun Street, the orphan house had been built between 1792 and 1793 and was enlarged and remodeled between 1853 and 1855. It was home for 224 children in 1886 and was known for the high quality of the education, nourishment, and care it provided. Indeed, in 1883, a daughter of Queen Victoria stated that it was "without doubt the most perfect institution of the kind I have ever visited."[48]

The Charleston Orphan House

Nurses were employed as substitute mothers for the children and were responsible for the cleanliness and manners of their charges from sun up to sun down. They waited on the youngsters at meals, ate with them, and slept in the same or adjoining rooms. They got the children to their morning prayers and lessons and were guardians of their morals and conduct. Corporal punishment was forbidden, and the caregivers were encouraged to use kindness, affection, and a spirit of indulgence—within reason—to enforce discipline. When all else failed, disobedient orphans could be reported to the steward or matron, who would take appropriate action.[49]

Miss Agnes K. Irving, principal of the orphan house, was abroad when the quake hit, but her nurses took immediate charge of their young wards. Even in the face of an earthquake, the bylaws, which called for staff to maintain exacting order and decorum at all times, were deeply ingrained. Sobbing children were gently hushed, and those who inexplicably slept through the first shock were awakened. Because of their rigorous training, the nurses, in the middle of a major disaster with pandemonium in the streets outside, managed to dress 224 children, line them up, and take them downstairs in a quiet, orderly manner. The sleepy, tousled youngsters and their protectors then gathered in the yard near the front gate, waiting to see what nature had in store for them.

Banker George W. Williams, one of the home's compassionate, public-spirited commissioners, left his own family and hurried to the institution, fearing that the shocks and fires in that area had destroyed it. Shortly after the first shocks subsided, he reached the premises, expecting the worst, and was overjoyed to find the children standing outside, trying to be brave, "like a little band of soldiers on duty." Fortunately, no one was hurt.

The children spent Tuesday through Saturday nights sleeping in the open air, protected only by blankets. None of them suffered from exposure, as Charleston's evening temperatures in early September are generally not oppressive, and aside from the earthquake tremors, the only real hazards were insect bites and mosquito-borne diseases. On Saturday night, Acting Mayor Huger was able to send U.S. Army hospital tents for the staff and children. Sadly, the earthquake added one more inmate to the rolls. On the night of August 31, the first shock killed the father of Wilhelmina Powell, a six-year-old girl from a Lutheran family who lived on Queen Street. A year after the earthquake, her widowed mother, Doris Powell, was no longer able to provide for her and brought her to the orphanage.[50]

The orphanage buildings themselves, though damaged, had not collapsed. Six chimneys and a large flue, which adjoined the furnace that heated the orphanage, were destroyed. The base of the statue of "Charity," which topped the cupola of the main building, had been shattered, and the statue was leaning forward at a forty-two-degree angle, threatening to fall, as it was supported only by a thick iron rod that ran through its back. After a close inspection, city engineers found that portions of the building were safe for immediate occupancy, and the children were moved back inside.[51] The survival of the Charleston Orphan House and its young charges was virtually the only good news the city would get for the entire week.

Grim as things looked, here and there a glimmer of hope survived. Even before the shocks of the first few days had ceased, humor began to resurface. The hour that the earthquake struck—bedtime for most Charlestonians—brought some levity into the midst of tragedy.

"It was an awfully hot, still night," wrote the Reverend Anthony Toomer Porter, recounting a tale told to him when he returned from North Carolina, "and nearly everyone was in the bath-tub at the time of the first shock. One young man threw on his gauze undervest, ran out of the house, jumped the fence and caught the vest on a nail. He could barely touch the ground and was trying in vain to rip the vest off the nail when a party of young ladies came by, and not recognizing his condition, said: 'Mr. ____, where, oh where shall we go?'"

"'Go!' he exclaimed, 'For Heaven's sake, go anywhere, but don't come here!'"[52]

In another incident, "a staid old gentlemen, who had married late in life but had two young children, jumped out of the bath, seized his beaver hat, put it on his head, caught up the two little ones, and rushed out into the street with one in each arm. As he was hurrying along, he knew not whither, someone met him, and said, 'Why Mr. ___, do you know you have no clothes on, save your beaver?'"

"'Oh!'" he cried, and dropping both children, ran off."[53]

The earthquake, however, always seemed to have the last, grim laugh. Numerous tremors of varying severity jarred the city throughout the day on Wednesday. The predictable result was a state of endless anxiety among the people of Charleston, accompanied by the pervasive belief among many blacks and whites alike that the Bible's long-predicted days of final judgment were now upon them. One well-known agnostic, "went into the middle of the street and sent up to the Great Controllers of Earthquakers as eloquent a prayer as any Methodist minister could utter." Another man was known to make fun of his wife's piety and her claim that she had a ticket to admit her to heaven. In his excitement, he rushed out of his house pleading to his wife, "Sallie, where's ye ticket?" He was evidently determined to ride into the next world on his wife's coattails.[54]

Each succeeding shock caused more excitement and apprehension because daylight had made the full extent of the devastation readily apparent. The majority of the buildings appeared to teeter on the edge of collapse. Word that many people had been killed and injured spread quickly. Predictions of future shocks were almost universal. Rumors began that the ground had opened all over the city, although there were only seven documented instances of craterlets or fissures on the peninsula.[55]

The majority of Charlestonians decided to avoid their houses until the disturbances had ended or at least diminished. Tents, awnings, and crude shelters were erected wherever open spaces could be found. The entire population of the city was collected in the parks and streets, except for a few families who found refuge aboard ships in the harbor. At this point, there was no lack of food, but preparing a meal was difficult because few people wanted to venture back into their kitchens. The day was spent in improvising arrangements for camping out.[56] Carl McKinley, who had again taken to the streets to cover the story of a lifetime, described the overall damage.

> The general aspect of the city is scarcely a subject for detailed description, and can more readily be conceived than put in words. It is enough to say that not more than a half dozen houses escaped injury, and that the damage to all would be represented by the demolition of one-fourth of the buildings on Charleston Neck (now North Charleston); by the leveling of the houses south of Broad street; or by the destruction of a city larger than Columbia. The ruins lay piled in the streets, yards, and gardens, and the houses from which they had fallen seemed ready to crumble of their own weight. Travel was confined to the middle of the streets and was impeded there. It is impossible to estimate, even approximately, the amount of masonry that was thrown into the streets; but it may be

> guessed at in some sort when it is said that the wreckage caused by the cyclone of the year before amounted to over ten thousand cart-loads, all of which was removed within the week following. The debris in a few streets after the earthquake would have equaled in mass all of a similar kind that was caused by the [wind]storm, and every street was obstructed more or less throughout its length...What it all meant to the people of Charleston on the morning of September 1, and the emotions to which it gave rise, can not be told. But the people were familiar with disaster, and one or two days later the writer saw a crowd of common laborers busily engaged in picking out and piling bricks from the wreck of the fallen wall of a building while the standing walls beside them were being shaken almost hourly by the recurring tremors.[57]

The destruction had buried sections of the city's horse-drawn trolley lines under tons of rubble, shutting down some parts of the system when the first shock hit on Tuesday night. The trolley cars themselves were largely undamaged, as most were already out of service for the night and parked at their terminals several blocks north of Marion Square when the quake hit. However, with their ears laid back and their eyes rolling frantically, some of the panicked horses that pulled the trolleys galloped off in every direction, scattering the crowds that had run into the open to escape collapsing buildings and adding to the general pandemonium.[58]

Considered to be ultramodern transportation for their time, the urban trolleys ran on rails placed in the center of the streets and were part of the network that held the city together. Well-to-do travelers had private carriages available to them, but the trolleys were often more convenient. At the north end of the city, the trolley system dropped off and picked up passengers at the city's railroad terminals. The trolley lines also ran along the main thoroughfares of the lower peninsula: Calhoun, Meeting, King, Market, Rutledge, Wentworth, Broad, and East Bay streets, which together formed the heart of Charleston's commercial, financial, and residential districts. The temporary interruption of full trolley service further crippled the city until the streets were cleared of rubble.

The anxiety of those who were desperate to leave the city was magnified by the total inability to send or receive information. Charleston was cut off from the outside world at 9:51 p.m. on Tuesday, when the telegraph lines went down. The first two shocks badly damaged the Western Union Telegraph Company, destroying the batteries that powered the lines as well as the telegraphic equipment itself.[59] The Southern Telegraph Company's office on Broad Street was also in ruins.

The quake toppled telegraph and telephone poles, leaving behind a snake's nest of wires, snarled throughout the rubble. Passage in any direction on the streets was treacherous. As of three o'clock on Wednesday morning, "all efforts to secure connection by telegraph with any point outside of the city proved futile," a reporter wrote, "and nothing is known of the extent of the shock."

The only sources of information were friends, neighbors, and whatever was printed in the *News and Courier*. And even then, the radius of knowledge extended only about three miles from the intersection of Broad and Meeting streets, for that was as far as anyone had ventured on the night of the earthquake or the day after. No one knew what might have happened to the rest of the state—or even to neighbors as close as James Island. Nothing was known of the condition of Summerville. The Charleston peninsula might as well have broken off and floated out into the Atlantic Ocean—which was, in fact, one of the many rumors in circulation at the time.

At the *News and Courier*, Captain Dawson knew the compelling importance of notifying the world of Charleston's plight. If aid were to reach the stricken city quickly, he had to get the word out as far and as fast as possible. Manager Harris of the Southern Telegraph Company had summoned his staff to set up a temporary office and make emergency repairs. He had a line in operation by 9:15 a.m. on Wednesday, September 1, and from that line he immediately telegraphed Dawson's dispatch—the one-page report from the *News and Courier*—to Columbia. Mr. J. O. Jeffries, manager of the Southern Telegraph's Columbia office, rushed Dawson's lengthy telegram to the *News and Courier's* Columbia bureau on the morning of September 1. There, the news was quickly edited and telegraphed to the Associated Press for distribution to the nation's newspapers.[60] By Thursday, September 2, newspapers around the country were publishing the shocking story of Charleston's tragedy. In a special extra edition that day, the *Atlanta Constitution* reprinted virtually all of the coverage from the *News and Courier*, telling its readers that three of its own reporters would immediately be sent to the stricken city on a specially chartered train.

Dawson's bold initiative was roundly cheered, as was his journalistic fortitude under fire. *The Christian Herald and Signs of Our Times* wrote, "It speaks strongly for the power of the journalistic instinct in any man that in such an agitating time, and in the midst of circumstances so distressing, he had the self-control to sit down and write so able an article."[61]

At nightfall on Wednesday, Charleston was still a portrait of destruction, uncounted casualties, and mass suffering. The world had been notified of their plight, and the shaken residents of the Lowcountry would soon learn that their experiences had been shared, to varying degrees, by nearly everyone from Omaha to Long Island and Toronto to Havana. It was time for those who would rebuild the City by the Sea to step forward.

7

THE FIRST RAYS OF HOPE

Mass Gawd, be sho an come Yoself and doan sen' Yo Son, because dis ain' no time fuh chillun.
—Prayer of a woman on the Santee River

Thursday, September 2, 1886

William Elliott Huger was a man on a mission. For two days following the earthquake, Charleston's forty-two-year-old acting mayor rode through the city on horseback to survey the damage. The debris was piled far too high in the streets for him to use a carriage. He knew the terrain intimately. In addition to being a native Charlestonian and a descendant of a French Huguenot family that emigrated to South Carolina in the seventeenth century, Huger (pronounced u-GEE in Charleston) lived on Meeting Street. He was an alderman from the city's First Ward and had been appointed interim chief executive to govern the city during Mayor Courtenay's extended vacation abroad. What Huger saw on his tours brought tears to his eyes. Every block in the city had been ravaged, and every citizen was terrified. All the familiar voices that greeted him were tinged with anxiety. The extent of the destruction was mind-numbing.

For the first seven days after the quake, Huger and the city's other officials—without their absent mayor—had their hands full dealing not only with the devastation of the city but also with their own families and neighbors. In the communication void that enveloped Charleston when the first shock wave tore out the telephone and telegraph lines, personal experience and the *News and Courier* were the only ways to learn anything about how the city and its inhabitants had fared.

After the relief of having a newspaper to read on Wednesday, September 1, the city was deeply dismayed when no issue appeared the next day. On Wednesday evening, the compositors had nervously been putting together the Thursday morning newspaper. They had already experienced numerous aftershocks in a fragile building and were understandably on edge. A few moments after eleven o'clock that night, a fairly severe shock rattled the *News and Courier*'s composing room. One of the reporters later wrote, "Its advent was followed by several noises such as is [*sic*] produced by falling buildings, but no trace could be found of any building having been thrown down. The shock occurred at a critical time, as far as the newspaper was concerned. The compositors had been at work since ten o'clock a.m., and a good deal of matter had been set up in type, but not yet been locked into the composing frames which went onto the press." When the eleven o'clock shock hit, the terrified compositors left the building *en masse*.

Fuming at the perceived cowardice of the fleeing men, Captain Dawson and a number of his editors and reporters still wanted to produce the paper. Without typesetters, however, that was impossible, and Charleston went newsless on Thursday morning. "[The compositors'] view was that the composing room would be unsafe, if there were any more severe shocks, and also that their families needed their constant attention. There was, therefore, no choice but to suspend publication." The courage to return to the city's fractured buildings was slow to evolve, but the compositors did come back to work on Thursday night, much to the relief of the publishers and the newspaper's subscribers. As he was handed his morning newspaper on Friday, a merchant said, "It is refreshing to see that there is life in old Charleston yet."

The first shock on Tuesday night had downed telegraph lines and forced the operators to flee their shattered offices in order to save their lives. People flocked to the telegraph offices on Tuesday night and Wednesday, desperate to get word out to their families and relatives, but no communication was possible. Charleston was totally cut off from the rest of the world.

The telegraph companies worked without rest to re-establish service. By 9:15 a.m. on Wednesday, the Southern Telegraph Company was able to reconnect a line to the outside world, and Captain Dawson was able to send his crucial telegram to the *News and Courier*'s Columbia bureau, which distributed it to the nation via the Associated Press.

Western Union station manager Daniel M. O'Driscoll was also quick to respond to the emergency. The forty-seven-year-old native of Ireland had settled in Charleston after marrying South Carolinian Mary Watkins. After ensuring that his wife and their five young children were safe, O'Driscoll gathered his staff, set up a temporary office on Ann Street, and had several wires in operation by 4:00 p.m. on Wednesday. By five o'clock on Thursday morning, O'Driscoll ordered his exhausted workers to go home and get some rest, as they had been rewiring the system and transmitting and receiving messages without sleep for forty-nine hours. By Thursday afternoon, two lines were also restored at Western Union's office at 135 East Bay Street, and all their wires were in operation on Friday. The company generously supported the disaster relief efforts by announcing that it would forward to Charleston "any contributions of money for the sufferers and any messages pertaining thereto free of charge."[1]

The first incoming messages on Wednesday were delivered on Thursday night and Friday morning. The telegraph messengers—Western Union alone had forty runners and one horse and buggy at work—frequently returned to the office with telegrams undelivered, as thousands of people were no longer living at their home addresses. Crowds of now-homeless people filled the telegraph offices and milled outside. Because delivery was nearly impossible, clerks would call out the names of people who received messages in hopes that those waiting for news would be there to receive it.

The managers of the Southern Bell telephone exchange also showed great energy and skill and were able to replace their wires within two days after the quake. As with the telegraph companies, they were plagued with lines being broken by falling buildings for weeks to follow, but they passed none of their $1,500 to $2,000 loss on to their customers.

By an amazing stroke of luck, Mr. J. M. Gardner of Gamewell & Co., who had originally installed the city's telegraphic fire alarm system, was in the city at the time of the great shock and remained to repair and rebuild the equipment. He started work the day after the quake, but his efforts were hindered considerably when bricks fell and snapped the new wiring during the demolition of dangerous chimneys. Fire Chief O'Neill stepped in. He made arrangements with the building contractors to notify him when chimneys would be removed so that firefighters could pull any threatened wires out of the way of the falling bricks. The system was rebuilt, tested, and back in operation on September 10. It proved itself on September 11 and 13, when fires were reported, the alarms sounded, and the fire companies were able to respond promptly. The newspaper noted that the alarm tests "will be a welcome sound, as it will be an indication that the citizens are once more safe from undue delaying the reporting of fires at the various firemens' quarters."

On Friday morning, September 3, with aftershocks continuing to punctuate the days and nights, with most of the major streets still blocked with fallen debris, and with city hall sufficiently damaged as to be unusable, Acting Mayor Huger called an emergency meeting of the city council. In normal times, the council consisted of twenty-four white, male aldermen—two elected for each of the city's twelve wards—but when the earthquake hit, many were still on summer vacation, and several posts were vacant.

Because the concept of preplanned disaster response by any level of government was unknown in 1886, Huger and the alderman knew they were on their own. Save for Clara Barton's American Red Cross, no federal, state, county, or city emergency response organizations existed. There were no stockpiles of food, water, medicine, or tents for shelter. Hurricanes, tornadoes, earthquakes, floods, forest fires, and disease epidemics were considered acts of God. Individuals, families, and communities were responsible for their own welfare.

As acting chief executive of the city, Huger knew it was his responsibility to take immediate action and organize a relief and recovery effort until Mayor Courtenay returned. The immensity of the challenge was almost incomprehensible, yet the acting mayor and the board of aldermen knew they had one great asset to work with: their resilient citizens.

A Charlestonian who had traveled sufficiently to be considered a cosmopolitan said of his neighbors, "The dogged perseverance and courage with which these poor people have gone to work, in the face of so appalling a disaster, to repair damages, is something unprecedented. I can only account for it by the fact that they have been through so many great afflictions that not even an earthquake can daunt them. They are so used to being knocked down, and have been so accustomed to getting up on their feet again that the first thing that occurred to them on the morning of the first of September, when they recovered consciousness after the heavy blow they had received, was to get up again, and they got up and have been standing on their feet ever since."

By 1886, every native-born Charlestonian over the age of twenty-five had already survived six calamities: the Great Fire of 1861, the Union artillery bombardment from 1863 to 1865, the Northeastern Railroad Depot explosion and fires in 1865, the loss of the war and the resulting Union occupation of Charleston during Reconstruction from 1865 to1876; the tornado of 1873; and the Great Cyclone of 1885. Huger, his aldermen, policemen, firemen, and most other Charlestonians had experienced every form of man-made or natural disaster and could, if nothing else, keep hard times in perspective. Now they needed to use what they had learned to recover from something they had never encountered.

The earthquake's effects on Charleston mimicked those of the catastrophic fire of 1861, which consumed hundreds of buildings in a wide swath through a major residential and business district. The fire had caused mass homelessness and enormous loss of property. Two days after the conflagration, in response to the needs created by the disaster, Charles Macbeth, then Charleston's mayor, established relief committees to provide shelter, food, and clothing for the victims. Donations were solicited, and a food-rationing system was established. Sufferers were asked to come to the Confederate Court House on Chalmers Street to apply for assistance and for tickets, which would entitle them to food. Tickets could be used at a soup kitchen and a free market that sold meat and produce. The Reverend Anthony Toomer Porter, who would be highly vocal and visible during the earthquake recovery operations, had been one of the key members of Mayor Macbeth's team in carrying out the relief effort. Now the lessons learned after the 1861 fire would be the primer for parts of the earthquake recovery.

Huger and the available aldermen met at 11:00 a.m. above the office of the Charleston Gas-Light Company at 141 Meeting Street.[2] There they conferred to set their priorities and consider what aid they could muster. They recognized that one of the greatest hazards was buildings that were in imminent danger of collapsing. The aldermen decided to make a formal survey of their wards and report unsafe buildings to the acting mayor and the chief of police. This structural inspection was an urgent priority because almost the entire population of Charleston was living in the streets, afraid to re-enter homes and offices.

On Saturday, September 4, the city council met again and created a relief committee to ensure that Charlestonians had food, medicine, shelter, and security. Acting Mayor Huger was appointed chairman. Most of those who served on the original relief committee were veteran Confederate officers: men with first-hand, battle-tested experience in leadership and logistics,

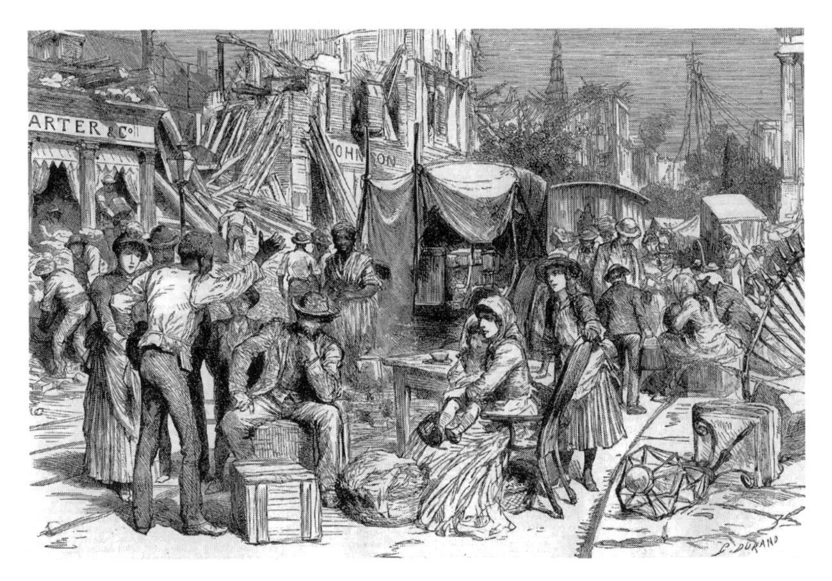

Camping out on King Street in the business district

who were accustomed to taking control, giving and carrying out orders, and getting things done without quibbling. Work was initially divided between three subcommittees. The relief subcommittee took on the duty of providing immediate, temporary, financial aid and was chaired by Col. Joseph A. Yates. Sixty-three-year-old Dr. Arthur B. Rose, a distinguished and respected physician, was appointed to chair the subcommittee on subsistence (food and medicine), and Maj. C. F. Hard chaired the subcommittee on shelter. Committee members also included the *News and Courier*'s influential editor, Captain Dawson; Gen. Thomas Abram Huguenin, the last Confederate commander of Fort Sumter, who was serving as Charleston's superintendent of streets; and William B. Guerard, a veteran civil engineer whose shattered office was located at 39 Broad Street.[3]

In scarcely more than a week after the first shock, the relief committee had formulated and implemented an effective short-term emergency response plan to meet the most critical needs: food and shelter. "There was no delay," the committee noted in a report to Mayor Courtenay when he resumed his duties on

September 8. "In what now seems to be an almost incredible short time, the committee were [*sic*] able to give shelter and food to all who required it. The work of relief is now thoroughly systemized, and so far as the duties of the undersigned are concerned, is equal to any emergency."[4] The members of the original relief committee then resigned in order let the mayor reorganize the effort as he saw fit.

On September 10, 1886, Courtenay wisely retained all the men on the original relief committee who were still available to serve. He named the new organization the Executive Relief Committee (E.R.C.), and served as its chairman. On September 14, on the motion of Captain Dawson, the reorganized E.R.C. established four standing committees to carry out its work: the Committee on Immediate Relief, the Committee on Subsistence and Shelter, the Committee on Building and Repairs, and the Committee on Finance and Accounting.[5] Each committee had three members. No blacks were invited to serve on the original relief committee, the E.R.C., or any of the subcommittees.

One of the major priorities of the relief effort was easier to resolve than the others. Food was not in short supply. The earthquake struck in late summer, when food was abundant. Virtually all of Charleston's food was grown within fifty miles of the city and was transported by horse-drawn wagons, by the railroad, or by small coastal steamships. Most of the fresh food was sold either from stalls in Market Hall, a two-block-long row of buildings between Meeting and East Bay streets, or by street vendors, who went door to door with their pushcarts six days a week. Only the terror of the first heavy shocks and the debris-filled streets interrupted Charleston's food supply, and then only briefly.

After the first few days, as the streets became safer to travel, bringing food into the city was not a problem. The shocks had produced thousands of cracks in the earth, but most were only an inch or two wide. They did not affect traffic on the Lowcountry's dirt roads, and very few fissures were reported in the city. The long-haul railroad company crews had the tracks into and out of Charleston repaired within two days. Inside the city limits, the railway and streetcar rails suffered only minor damage, but the tracks had to be cleared of debris. Parts of the lines never stopped running, and thanks to the city's efficient street department, the rails were quickly cleared, and both streetcar lines were running nearly their full routes within two days.

As for waterborne food supplies, Charleston Harbor was undamaged; the docks were sound; and the ferries and coastal steamers, which brought produce, ducks, game, and rice from the surrounding countryside, were unaffected. By September 10, even such luxury items as chocolate and fancy cakes were readily available. As a result, Charleston had access to its normal range of food supplies within two days after the first shock—except for fish and shrimp.

Although the market stalls had an adequate supply of most food, there was a limited supply and variety of fish. Charleston's fishing industry was carried out by the "mosquito fleet," a flotilla of small smacks (sailboats) owned by daring and skillful black fishermen. Their work had been crippled by the earthquake for a unique reason: most of the fishermen who supplied the market had been compelled to use the sails of their boats to shelter their families. Without sails, they could only fish in the harbor, going out only as far as they could row, and there they could catch only rough fish, chiefly mullet. The mosquito fleet went back to fishing beyond the Charleston bar by September 10, as soon as the fishermen found shelter in the new wooden refugee sheds or returned to their original homes.

Shrimp were a food staple in the Lowcountry and were harvested from tidal creeks by black shrimpers using weighted, hand-cast nets thrown from rowboats. On September 8, the same day that the city rallied to the melodious, reassuring sound of the bells of St. Michael's when they again started pealing, another noise—not as elegant but eminently familiar—rejoined Charleston's soundscape: the cry of the "raw shrimp fiend." The shrimp fiend was one of many street peddlers, including fishmongers and bakers, who sold their goods from door to door, but among his colleagues, he was special. His raucous cries of "Swimpee! Swimpee!" which rang out in the streets in the early morning, were so loud and irritating that he was described as "a source of extreme annoyance to late sleepers in Charleston." The newspaper stated, "His early and horrible cries have been the cause of more profanity than any one single thing in Charleston."

The shrimp fiend's return might have been a dubious consolation to some, but to others, it was a sign of a return to normal life. The *News and Courier* waxed ecstatic. "There was music in the sound yesterday–cheerful, hopeful, comfort-inspiring music. It was a harbinger of returning security, a reminder of the happy past, when people could retire to rest in their houses undismayed by the roar of the terrible subterranean monster that has been stirring around under our feet. The raw shrimp fiend carried hope and comfort wherever his voice was heard."

The food stalls in the row of vendor buildings between North and South Market streets were largely uninjured, although the majestic main hall at the corner of Meeting and Market streets and most of the buildings along North and South Market streets, had been severely injured. The fact that all of the market area was built on landfill that covered a former tidal creek had increased the damage to most of the area's brick buildings.

A reporter visiting Market Street eight days after the quake found the market "cheering and hopeful." The butchers were busy cutting sides of beef and pork into steaks, ribs, chops, and joints, while, as usual, flies and vultures waited for their opportunity to feast. Despite the earthquake, there was no price gouging. Beef was still selling for 10¢ to 15¢ ($2.00 to $3.00) a pound, and other meats, vegetables, and seafood were at their usual prices. The maumas (black women), with their colorful bandanas around their heads, sat in their stalls, surrounded by piles of potatoes, yams, "sibby" (Sewee) beans, onions, tomatoes, and corn. In the sweetgrass baskets so beautifully woven by these women, customers could find ample supplies of okra, peas, blueberries, blackberries, or strawberries, according to the season.

Like the food supply, the public water supply was never seriously interrupted, although the water mains from the city's two artesian wells had been broken in some places. In addition, the private water cisterns that supplied many Charleston homes had suffered some damage, leaving both these and the water mains open to the possibility of contamination from sewage runoff. On September 10, city officials directed everyone to boil all drinking or cooking water before use.

Despite the fact that sustenance and water were readily available, thousands of people had lost everything, including their sources of income, and could not afford to buy food. Dr. Rose, chairman of the subsistence subcommittee, which supervised the distribution of emergency food and medicine, had devised a plan for medicine by September 9. Those who had prescriptions from physicians certifying their inability to pay for medication would get what they needed through the E.R.C. When the needy applied to the E.R.C., a committee member would verify their eligibility and countersign their prescriptions. The patients were directed to an approved druggist, who filled their orders and billed the committee.

Dealing with food distribution was more difficult. Rose was given this responsibility on September 4, but by September 6, no food had been handed out. Embarrassed by his failure to act, Rose explained to the relief committee that a food-rationing system would be set up, but he had been unable to find a safe building to use for the purpose.[6]

During the discussion, his colleague, Maj. Hard, underscored a fundamental concept of the city's disaster relief philosophy: food and food ration cards would be issued "only to the helpless and *not to workers*."[7] Tragic misfortune alone did not entitle anyone to aid. The city government made it clear from the outset that those who had financial resources (cash, savings, insurance, or a job) were expected to provide their own food and shelter and to rebuild their homes and businesses without public assistance.

The *News and Courier* reiterated this policy in no uncertain terms. "They who are able-bodied and can have work for the asking will not, of course, receive rations from the committee." The committee would only issue enough rations to prevent starvation and only to those who were authorized to receive aid. For sufferers who needed immediate relief, the city's clergymen, who knew their neighborhoods and congregants intimately, were authorized to certify each family's need.

Ration cards were issued. To prevent fraud, the cards were only valid if signed by a clergyman, a committee member, or a person of good character vouched for by the committee, and no one would receive food without a valid ration card. The cards, which listed the number of members in each family, along with their names, addresses, and genders, could be used to obtain food at the committee's headquarters.

By September 8, one week after the earthquake, Dr. Rose was able to report that food was now being distributed to the needy. The subsistence subcommittee's headquarters, in the former Phoenix Fire Company engine house on Cumberland Street, was now in operation and was being run on "highly businesslike terms."[8] Rose's committee had purchased provisions "at whole sale as low as possible and single day rations of grist [grain], rice, bacon, and a little sugar was issued through out the day."[9] Rose noted that the rations had cost about $150 and that black applicants largely predominated.[10] He admitted that things had not gone smoothly at first, but the use of ration cards had made the distribution easier. Rose also reported that a soup kitchen would be operational in short order.

The soup kitchen opened about three o'clock in the afternoon on September 9 on a vacant lot opposite the E.R.C.'s office on Cumberland Street. It was intended to supply cooked food "to those who had no means of cooking their rations and who may be in need of nourishing food." The chief intended beneficiaries were homeless unmarried workmen and the able-bodied elderly. No ration identification card was required to receive the soup, which was prepared on four stoves by a staff of six to eight cooks. The newspaper stated proudly that "the cuisine is provided over by Augustus Harleston, an old colored man, who was many years ago a cook for Governor [William] Aiken, and who stated to the Reporter who visited the soup house that he had often prepared the soup for John

C. Calhoun." The journalist went on to write, "Bruce Howard, colored, who has charge of the distribution of the soup, invited the Reporter to try a little of Harleston's concoction. Under protest the Reporter consented and found to his surprise, that the soup was fit for a king—or an alderman." When the soup kitchen finally opened, blacks and whites stood in separate lines to receive their meals, as they did wherever food was served.

By Saturday, September 11, to increase the efficiency of food distribution, rations were issued for a week's use, rather than daily. The weekly ration for one person consisted of a small quantity of salt, three pounds of meat, one-and-one-half quarts of rice, one-and-one-half quarts of grist (specifically corn meal), one pound of sugar, one-quarter pound of tea, one-half pound of coffee, one pound of biscuits, and at the committee's option, molasses, salted fish, and one loaf of bread.

To ensure that no one in the city was overlooked or slipped through the cracks of the system, the relief committee sent clerks out to identify the especially needy, and wagons were hired to deliver food to the sick and disabled who were unable

Waiting for food rations at Cumberland Street

or unwilling to apply for aid. In the first four days the commissary was open, more than 21,000 rations were issued. The *Atlanta Constitution* quoted Acting Mayor Huger as saying that overall, the relief committee was assisting "about one-third of Charleston's entire population."[11]

Five thousand ration cards were also given to the clergy of both races for distribution, along with strict instructions defining who was eligible. This provided aid to those who might not be reached by other means, including displaced people too proud to accept charity from the city government under any circumstances. There were many of them. Dr. Rose reported that "not one in a hundred of the people who applied for rations were white. Owing to the peculiar character of the population of Charleston, this was to be expected, and the committee have adopted plans to meet the situation." Rose was saying that most former slaveowners who had fallen on hard times would not accept public charity, especially if it meant standing in lines next to their former slaves to receive it. Most of them would literally prefer to starve rather than admit to being destitute.

Major Augustine T. "Gus" Smythe, a Confederate war hero, attorney, and pillar of the community for decades, wrote to his wife on September 8, "I got $250 today for the Misses Gibbes [Mary Hasell Gibbes and her sister, who lost their house to the fire on King Street], from a private source so that their names will not go before a public Committee. I spoke to them very fully and freely, & they owned up that they would be most grateful for it. They had no money at all & did not know what to do. I will not speak of the matter publicly, tho' of course there is no harm to mention it at home."[12]

Another of Dr. Rose's chief concerns was that freeloaders from the adjacent sea islands to the south—chiefly James Island, Johns Island, and Wadmalaw Island—along with rural dwellers from St. Andrew's Parish, west of the city across the Ashley River, were flooding in to claim free rations. He feared that the ready availability of food had created a welfare-state mentality of entitlement that was corroding the work ethic of the blacks. He fumed that able-bodied black men could be seen lounging around the city during normal working hours, while their wives cooked the free rations. As he told the *News and Courier*, "Servant girls have left their employers, washerwomen have refused to wash clothes, and colored female labor is almost demoralized by the promiscuous issuance of rations."

On September 11, taking the supposed bull by the horns, the subsistence subcommittee reorganized the way rations were issued. "The committee will hereafter deliver the rations to the houses of those who need them, and for the purpose of ascertaining the needs of the people have appointed twelve canvassers, six of whom are to be named by the colored clergymen, to whom will be entrusted the duty of visiting the different wards in the city, and finding out the residences of those who need rations." From then on, wagons were hired to distribute the food throughout the city, and only those with ration cards would receive anything from the food wagons, which did not travel beyond the city limits. The committee did make exceptions for those people in extreme hardship who showed up without cards, or for aged and infirm people who came from outside the city.

The day after the food delivery program was inaugurated, the newspaper reported that William Cooper, a black policeman in the market, had overtaken a "regiment of negro women from the country. They all had crocus [cloth] bags and were on their way to town to draw rations." The newspaper continued, "As the distribution of rations by tickets [ration cards] has ceased, this army of country visitors, as well as the hosts of others from the surrounding islands will be disappointed."

Citizens who were fed still needed protection from the elements. With two-thirds of Charleston's 60,000 residents living out of doors, cool weather on the way, and the hurricane season looming over them, that would be a far harder problem to solve.

8

Tent City

We shipped to you yesterday two hundred shelter tents for the Charleston sufferers. They are goods that we have been using for making collar pads. They are not perfect goods, but they will answer the purpose of shelter until your people are again housed in their homes. —Harbison & Gathright Company
Louisville, Kentucky

Thursday, September 2, 1886

During the first week of September, prior to Courtenay's return, Acting Mayor Huger and his colleagues worked around the clock. With thousands of homes damaged or destroyed, and nearly 40,000 refugees living on the streets in fear of the next shock, providing shelter was the city council's first priority.

Many people did not wait to see if help would be provided. As soon as the railroads were running again, thousands fled Charleston and Summerville. In a mass expression of sympathy, hundreds of individuals and communities throughout South Carolina offered free or reduced-cost housing for refugees within days after the first shock. In response, the E.R.C. quickly established a Committee on Accommodations for Refugees at 1333 East Bay Street, two doors north of the post office. The committee's office, directed by S. Adger Smythe, was operational by September 9 and was open every day from 10:00 a.m. until noon. Staff members collected and posted notices of available evacuation facilities statewide and distributed applications, which they used to match applicants to available offers of shelter.

In the first days after the quake, there was a strong emphasis on removing women and children from the city to "spare them further danger and anxiety."[1] To help potential evacuees, the railroads offered free passage out of the city to those who could not afford to pay. An office was established at City Hall on September 6 to issue railroad passes to everyone who wanted to leave. The South Carolina Railway, the Northeastern Railroad, the Charleston and Savannah Railroad, and the Richmond and Danville Railroad all participated in the program. During its first day of operation, the bureau issued nearly five hundred passes to black and white families, many of which consisted of four or five members. The main refugee destinations included Columbia, Orangeburg, Greenville, and Spartanburg, South Carolina; Augusta and Atlanta, Georgia; Charlotte, North Carolina; and, Baltimore, Maryland. By September 9, the South Carolina Railway had sold about one thousand tickets, the Northeastern, about eight hundred, and the Charleston and Savannah, about six hundred. This did not include seven or eight hundred free passes distributed through the E.R.C.'s office at City Hall.

The offer of free passage out of the city inevitably produced some abuses of the system. On September 8, the South Carolina Railway notified the transportation office at City Hall that the railroad, "while being willing to assist those who are suffering from the loss of their houses, &c., can continue to furnish free transportation only to the women and children of the city where it is known they are in want or without houses or proper shelter. Those heads of families who can pay at least something should not place themselves on the kindness of the company as objects of charity." The city council responded by stating that no free-fare passes would be issued unless applicants had their applications endorsed by a physician or a clergyman. As of September 9, after evacuating thousands of people without collecting a fare, the railroads continued their aid but imposed the nominal charge of one cent per passenger mile and offered special rates "even for those who can afford to pay full price."

Despite its own earthquake damage, and the immediate need to prop up its own dangerous walls, Columbia, the state capital, took the lead in helping the Lowcountry evacuees. Columbia's reputation for compassion and hospitality was well known. Approximately two decades earlier, during the Civil War, it had taken in thousands of refugees when the people of the Lowcountry were forced to evacuate under the threat of bombardment or invasion by Union naval ships patrolling the coast. At that time, the editor of the *Confederate Baptist* wrote, "We are all one Confederate family, and it is the duty of every one of us to use his heart, head and hands to serve his country and his neighbors as himself, and particularly those who have given up their homes for the honor and welfare of the State and Confederacy."[2]

Typical of the new arrivals was the refugee train that reached Columbia at 9:00 p.m. on September 1. As the train pulled in, those waiting were deeply affected by the sight that met their eyes. Three freight and five passenger cars contained three hundred to four hundred people, whites and blacks in equal numbers. One car carried only women and small children. This particular group of refugees had spent a night of terror on the bare ground of Summerville's streets and was now jammed into train cars with other unwashed, disheveled, sleepless people, many of whom were still hysterical with terror or had given way to glassy-eyed apathy because of the emotional overload.

When they disembarked, some were met by friends and relatives and were taken to private homes. Blacks were especially quick to leave, being warmly received into the homes of others of their race. Other refugees knew no one in Columbia. If male relatives accompanied them, women and children were escorted to hotels and boardinghouses. Numerous families generously opened their homes to complete strangers.

The large number of whites who remained refused to leave the train and crowded into the passenger cars. Many were afraid to enter houses, frightened of another earthquake. Because the train could not leave the station with the holdouts still aboard, at six o'clock the following morning, Columbia's chief of police brought horse-drawn omnibuses to the station to remove those who had slept in the passenger cars and take them to quarters in the city.

Within a week after the earthquake, well over six hundred refugees had arrived in Columbia, and space had been made for many more. Fifty rooms were made available at the South Carolina College (now the University of South Carolina), and another thirty-two rooms were open at the Columbia Female College (now Columbia College). The Columbia Theological Seminary housed one hundred refugees and had compiled a list of ninety-five more private boarding rooms that could be rented at reasonable prices. "This is not the limit of Columbia's accommodations," the *News and Courier* stated. "All who come will be made as comfortable as possible."

Miss Caroline Winkler and two of her female friends, refugees from Summerville, were beneficiaries of the seminary's hospitality. "Sleeping quarters were given us for free," she said, "and we were served meals at a very small charge." A day or two after they reached Columbia, the three girls were taken ill. Dr. Talley, a prominent Columbia physician, took care of them *pro bono.* Because nearly every house in Charleston and the nearby towns needed repair, and there were not enough carpenters to get the work done any sooner, it was early December before Miss Winkler and her companions could return home.[3]

The Lowcountry refugees soon discovered that they had not outrun the earthquake's long reach. At 11:00 p.m. on Friday, September 3, buildings swayed as a vigorous twenty-second shock roared through Columbia. This fresh disturbance unhinged the refugees yet again, and they filled the streets, insisting on spending the night in the open air.[4] Despite the aftershocks, which continued to be felt in Columbia for many months, the city's hospitality was long remembered. A Summerville resident described the welcome received by the refugees:

> Everybody [in Columbia] has done more than could have been expected to make them comfortable and to relieve their wants. The good people of the Capital, in their open hospitality, have cramped themselves into small quarters, in order to give the unfortunate victims of the earthquake homes in their hour of need. Houses and rooms were freely extended to the refugees upon their arrival, and provisions and other necessities for the preservation of life, have been showered upon them continuously. They are daily visited, and every effort is made to make their stay as pleasant as it possibly can be under the distressing circumstances that make their residence necessary. The merchants in every case are extremely liberal in their dealings with the refugees and in many cases absolutely refused to accept a cent for their goods. When the refugees refused to become objects of charity, the merchants sold them the needed articles far below cost.[5]

For those who stayed behind in Charleston and lived on the streets, the first unexpected hint of good news came out of the blue—and it came from Washington. While Huger and his committee were drawing up their aid plans, the federal government was already acting to help. At the end of federally regulated Reconstruction in 1876, virtually all U.S. military troops, save for small caretaker crews at several forts, had been withdrawn from Charleston, so there were no local supplies of federal food or tents to draw upon.

Thanks to Captain Dawson, the news of the earthquake reached Washington, D.C., on September 1, and the response to Charleston's need for shelter was immediate. Without question or delay, Brig. Gen. Richard C. Drum, adjutant general of the U.S. Army, authorized one hundred large hospital tents to be shipped out by rail. They arrived on Saturday, September 4, via the Atlantic Coast Railroad and the Northeastern Railroad. It was the first relief aid from any source to reach the city. However, with minor exceptions, those tents were virtually the only kind of federal aid the city would ever see.

Indeed, many in Washington were skeptical that the damage was as great as the newspapers were reporting. The Quartermaster's Department of the U.S. Army, which had supplied the tents, also sent one of their senior officers, Colonel Batchelder, to Charleston to make an on-site inspection. A *News and Courier* reporter interviewed him at the Charleston Hotel, where Batchelder frankly discussed his assignment. "The fact is, there is a great deal of skepticism in the North as to the extent of the disaster here. People were unwilling to believe that the accounts sent from here of the havoc wrought by the earthquake were not greatly exaggerated, and it is with a view of reporting upon the correctness of these reports that I was sent here."

The reporter asked him if he believed there had been any exaggeration. "I do not think so," he responded. "I feel that the calamity is even greater than it has been represented in the press account.... I have seen enough to say that [the damage] is wide-spread. The eastern portion of the city has undoubtedly suffered most seriously, and I think there is scarcely a house but will have to be pulled down." Batchelder also explained that all the tents from the quartermaster general's stock at Philadelphia, the East Coast depot, had been shipped immediately. Other tents were stored in the West, but these had not been ordered or sent because the government assumed they would arrive too late to be useful.

Although Batchelder's official report confirmed the immensity of the destruction in Charleston, it did not convince the federal government to send any more tents. Newspaperman Louis A. Beaty wrote scornfully, "Besides neglecting Charleston in her distress the Government actually sent an army officer down to see if she were not lying about her condition!"[6] Logical or not, the government had no policies, procedures, or funds allocated for disaster relief, and the ruined city had to take whatever was offered.

Acting Mayor Huger and Col. John Averill, the mayor of Summerville, each appealed to the War Department by telegraph for more tents, but both were told that every tent that could possibly be procured had been already sent.[7] Huger's telegram indicated that if no further tents were available, his relief committee would use funds contributed from the North to erect temporary board shelters for homeless people.[8]

Ultimately, the only funds that ever flowed from Washington were individual donations from government employees, which had been personally solicited by South Carolina native William L. Trenholm, then U.S. Comptroller of the Currency, and other government officials. Trenholm also suggested that people who had old tents, awnings, or sails could send them to the Charleston or Summerville relief committees "in order that at least temporary shelter may be provided before the equinoctial gales (fall hurricanes) break over the affected people." His well-intentioned advice was welcomed, but it did not elicit the outpouring of federal relief that Huger and Averill were hoping for.

Although the first few days after the quake were dry, the nights were very uncomfortable. Julian A. Selby, a fifty-year-old Columbia newspaperman visiting Charleston, wrote, "The dew was so heavy for several nights, that the dampness went through umbrellas as if they were sieves.... The thrown-together shelters that the homeless slept in provided little better protection.... A queer sight it was, you may rest assured—babies, asleep in carriages, with umbrellas and other improvised coverings; delicate females walking around disconsolately, and men trying to be philosophical, but a muttered damn would escape occasionally, and the dew was wetting everything."[9]

Prior to the arrival of the one hundred federally provided tents, people who were desperately in need of shelter had applied for any housing that might become available. Now the city fathers had to determine who would get the first aid that had arrived. In Charleston, Gen. Thomas Huguenin, a former Confederate commander, and Maj. George W. Bell, a U.S. Army officer, took charge of the federal tents at the Northeastern Railroad Depot on September 6. The first five large hospital tents and five other wall tents (which had vertical walls, as opposed to smaller, A-shaped models designed to shelter two soldiers) were assigned to the Charleston Orphan House for use both as shelter and as a temporary infirmary for those injured in that part of the city. The Sisters of Charity of Our Lady of Mercy benefited from a gift of several tents. Three went to their convent on Queen Street; two went to the St. Francis Xavier Infirmary, which they operated; three tents were issued to the Catholic Female Orphan Asylum at 156 Queen Street; and several others were delivered to the Catholic Male Orphan Asylum at 173 Calhoun Street, all of which the Sisters operated.

Distributing the tents

Between two and six federal wall tents were erected on the vacant lot adjoining the burned-out shell of the Catholic Cathedral of St. John and St. Finbar, and they quickly became the nucleus of a refugee community. Dubbed "Camp Duffy," the camp was named for Father Patrick L. Duffy, the much-loved Catholic priest who served as chancellor and secretary of the Diocese of Charleston. It hosted a mixture of black and white refugees, but the tents were allotted chiefly to whites who no longer felt comfortable in Washington Park, which had been taken over by vocal, hymn-singing blacks. The government tents were soon surrounded by homemade shelters. A fire engine from Engine Company No. 2, whose firehouse on Queen Street had been severely damaged, was stationed at the camp, and the firemen slept in a large tent next to their horses and equipment.

The ruins of the Cathedral of St. John and St. Finbar

Other tents were erected at Hampstead Mall, a public square in the northeast part of the city. Three were placed in Wragg Square, a beautiful public green shaded by enormous live oaks across Meeting Street from Marion Square, and four were set up at the U.S. Arsenal, in the northwest part of the city, next to the Reverend Anthony Toomer Porter's Holy Communion Church Institute on Ashley Avenue.[10] Five tents were put up at the City Almshouse; two were erected at the foot of Broad Street; twelve were sent to the Battery; and about twenty fly tents were erected on Marion Square. The remaining federal tents were distributed to an unrecorded list of private individuals and "erected immediately in convenient locations."

Tents at Wragg Square

On September 6, the newspaper reported, "the desire of the committee is that, as the number of tents is limited, each tent shall be made to accommodate as many refugees as possible, preference being given always to the women and children. At least twenty persons should occupy each tent. It is estimated that quarters must be found for at least five thousand persons, and for a considerable time to come. Efforts will be made to obtain the use of the vacant houses on Sullivan's Island, and the Secretary of the Navy has been telegraphed to by Congressman Dibble, who arrived in Charleston yesterday, to ascertain if there are any Government vessels which can be sent to Charleston and used as floating places of refuge."

From the outset, public records and newspaper reports show that the large, waterproof army hospital tents, complete with wooden floors, were, with few exceptions, reserved for whites. On September 9, the *Atlanta Constitution* wrote, "Additional tents have been placed in the public squares, and for the colored people comfortable wooden shelters have been erected on Marion square."[11] Surviving photographs make it clear that some of those larger tents were used by a single wealthy family and did not provide shelter for anywhere near twenty refugees, as the relief committee had desired.

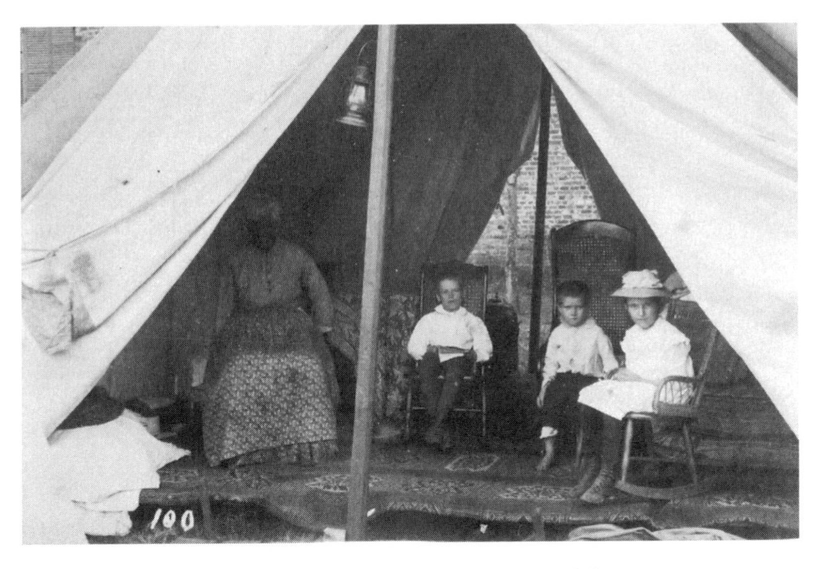

A dah (senior family nursemaid)
and her charges in a U.S. Army hospital tent

The Sunday, September 5, edition of the *News and Courier* stated, "Not one-tenth of the homeless people of Charleston have been provided for. At least five hundred tents, possibly more, will be required." The paper noted that additional tents were expected to arrive on Monday and would be distributed by the Committee on Shelter. In the meantime, people camped and took cover wherever they could find it.

Early that Sunday morning, rain had added to the misery of the refugees. About 4:00 a.m., clouds moved in, bringing light showers. A reporter noted, "The situation then really became appalling with 50,000 people exposed to the weather, having no shelter save that offered by sheets and blankets. A heavy rainfall meant sickness, disease, death."

Little could be done in the darkness, but at daybreak, people began to search the streets for pieces of wood they could use to improve their campsites. Many claimed sheets of tin from roofs that had been blown off by the cyclone the year before and were still lying in vacant lots. Refugees from Washington Park walked across Meeting Street and started to salvage wood and

Camping in misery

tin from the wreckage of the Hibernian Society Hall until a policeman arrived. He informed them that they could not remove private property without permission of the owner, and they acquiesced. "It was a well-behaved, orderly crowd," a reporter wrote, "and showed no disposition to violate the law. As soon as they were told to desist they left everything untouched."

Without access to the larger, more comfortable tents, blacks had to make do with whatever tents or shelters they were given or could make for themselves. On Monday, September 6, the relief committee and its subcommittee members met again and gave their reports. As the *News and Courier* later stated on the same day, "[Their] intention is to commence early today the erection of wooden shelters along Marion Square, with all the requisite accommodations, for the colored refugees. Quarters can easily be provided there for several hundred persons. Wooden shelters will also be erected, it is expected, [for the whites] on Hampstead Mall, and some tents will be erected on Washington Square."[12] The committee decided that in the future, it would erect tents only in clusters, and "only in exceptional cases will tents be allowed to be put up single and alone."[13] The reason was simple: tent cities made food delivery, sanitation, and police protection more efficient than having random tents scattered in back yards and private gardens.

Tents on Rutledge Avenue

The relief committee's policy on allocating emergency shelter was based on race. When Capt. J. F. Redding, the newly appointed assistant chairman of the shelter committee, turned in his report on September 7, it stated that twelve new booths (wooden sheds) capable of holding two hundred whites were erected at Rutledge Street Lake (now known as Colonial Lake). Thirty booths capable of holding five hundred whites would be erected by nightfall in Citadel Square. Forty booths capable of housing eighteen hundred whites were available in Marion Square, and forty more booths capable of holding three hundred whites and three hundred blacks were available at Hampstead Mall. Sixteen large wall tents with closure flies had been received that morning and were distributed to house about two hundred people. The race and residence of the recipients for this batch were not specified, but later photographs show that most of them went to whites or white charitable institutions.

Housing assistance poured in from other states. On Wednesday night, September 1, North Carolina Governor Alfred M. Scales sent Acting South Carolina Governor John C. Sheppard a dispatch saying, "We have news of terrible calamities in your State. How can we aid your people? Will gladly come to their relief."[14] On September 6, Scales shipped 180 large tents by special railroad car, and they arrived in Charleston the next day. Fifteen of the tents had been sent to Summerville in care of Colonel Averill. The other tents, "will be distributed to whites to day, these 165 should cover [additional] 2,000 people."[15] Sixteen tents came from the Floyd Rifles, Macon, Georgia. In an act of generosity from the entertainment industry, five circus tents were shipped in from New Orleans, lending an oddly festive air to the grim surroundings.

The New York Cotton Exchange noted that traditional summer religious camp meetings in the states of New York and New Jersey were at an end and suggested that the tents they had been using should be sent to shelter the homeless in Charleston. The Cotton Exchange also appealed to the wealthy to donate their lawn tennis tents for the same purpose. In Arkansas City, Arkansas, Sam Jones followed that advice and sent a large camp-meeting tent to Charleston.[16] In addition, a Christian publication noted, "Rev. Stephen Merritt, of New York, and Mr. W. L. Klopsch, who each had a large new tent at the recent [church] tent-meetings, promptly packed them up and sent them to Charleston, and other Christian friends have nobly given tents or money."[17]

Hart & Company, a Charleston farm implement dealer at the southeast corner of King and Market streets, received a letter from Harbison & Gathright, a manufacturing company in Louisville, Kentucky, who was probably one of their suppliers. The Kentuckians wrote, "We shipped to you yesterday two hundred shelter tents for the Charleston sufferers. They are goods that we have been using for making collar pads. They are not perfect goods, but they will answer the purpose of shelter until your people are again housed in their homes." The generous firm also sent a donation of $25 ($502 today) to aid the relief effort.

The Reverend Edward T. Horn, pastor of Charleston's St. John's Lutheran Church, was at a church meeting in Pennsylvania when the quake struck. As soon as he learned of the catastrophe from his son-in-law, Horn petitioned the adjutant general of Pennsylvania to send tents to Charleston, but he received the reply that the official had no legal authority to do so.[18]

On September 10, in addition to sending an initial $5,000 to Charleston and $1,000 in aid to Summerville, the New York Chamber of Commerce passed a resolution to call on New York

Governor David B. Hill to send tents owned by the state National Guard to Charleston, but Hill's adjutant general gave the same reply as his Pennsylvania counterpart. "The governor regrets his inability to comply with your request. The military code [of New York] absolutely forbids the loaning of military property of the state for any purpose outside of the national guard."[19]

That same day, Willard C. Fisk, the private secretary of New Jersey Governor Leon Abbett, and Capt. John McKechney, of the quartermaster general's department of New Jersey, personally escorted a delivery of three hundred fifty National Guard tents, which had been sent to Charleston at the order of Governor Abbett. The massive shipment consisted of fifty hospital tents, two hundred wall tents, and one hundred "A" tents. The first two types were large and especially prized; the "A" tents sheltered only two or three people, and then only for sleeping. The *News and Courier* noted, "Neither in New York nor New Jersey is there any law authorizing the Executive to send away any tents for any purpose outside of the State, but unlike Governor Hill, of New York, Governor Abbett did not believe this a time for red tape, having said that there was no law for earthquakes, either."

That same day or the next, 160 wall tents arrived from Ohio—and created a small furor.[20] The newspaper noted, "When the tents reached Charleston, however, they were burdened by a debt of $572.66 ($11,500 today) in express [shipping] charges due to the Adams and Southern Express companies for their transportation." According to the article, the tents were sent by Ohio Governor Joseph B. Foraker, "despite the fact that there was no law authorizing him to loan the camp equipage of the State National Guards." The E.R.C. asked S. G. Pinckney, superintendent of the Southern Express Company, to intervene, and he convinced the two shipping companies to cancel the charges. The tents were then donated to Summerville.

By September 11, the E.R.C. had erected at least 215 wooden booths, received 1,091 tents, and declared that it had enough shelter "to meet pressing needs."[21] At face value, that decision

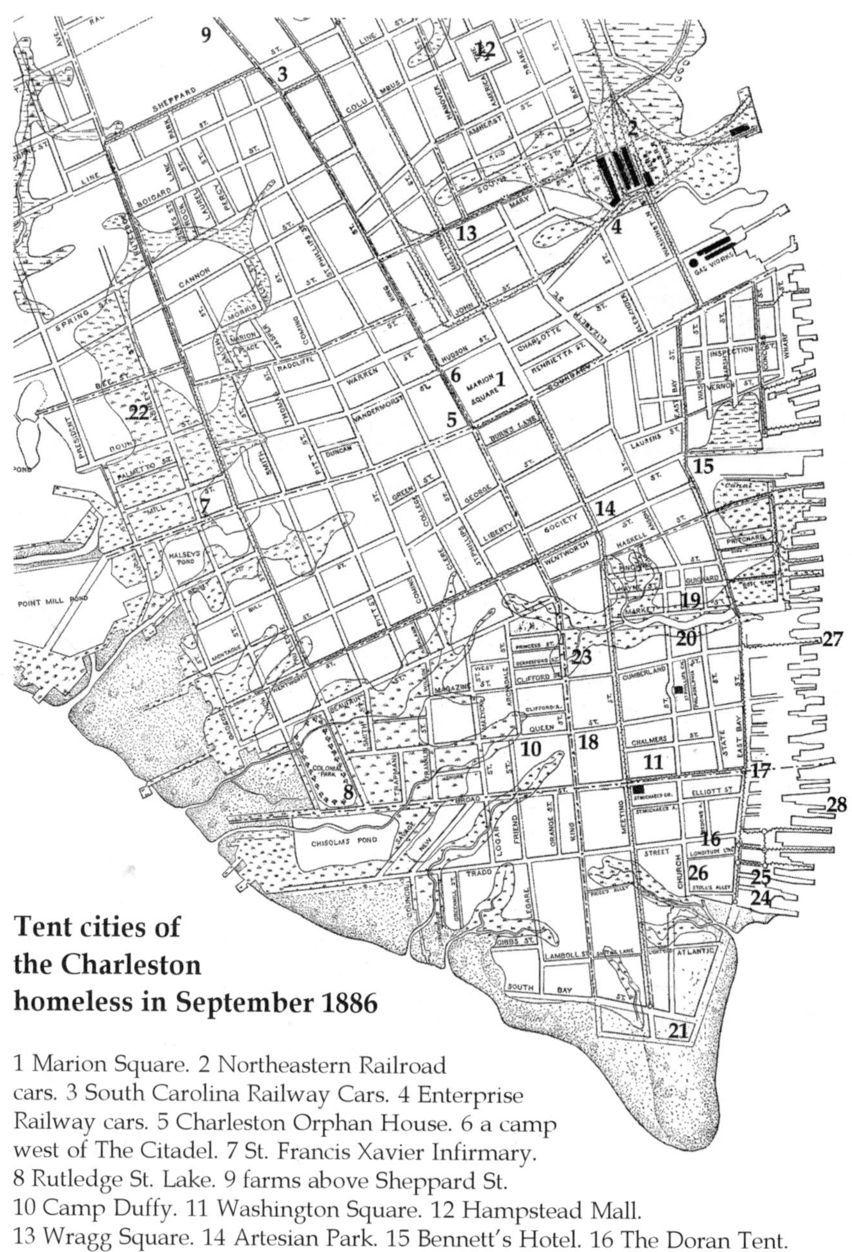

Tent cities of the Charleston homeless in September 1886

1 Marion Square. 2 Northeastern Railroad cars. 3 South Carolina Railway Cars. 4 Enterprise Railway cars. 5 Charleston Orphan House. 6 a camp west of The Citadel. 7 St. Francis Xavier Infirmary. 8 Rutledge St. Lake. 9 farms above Sheppard St. 10 Camp Duffy. 11 Washington Square. 12 Hampstead Mall. 13 Wragg Square. 14 Artesian Park. 15 Bennett's Hotel. 16 The Doran Tent. 17 at the foot of Broad St. 18 Queen's Camp. 19 stalls in The Market. 20 Linguard St. 21 The Battery. 22 U.S. Arsenal. 23 Robb's Lot. 24 Young's Camp at Southern Wharf. 25 unnamed. 26 Young's lot. 27 Steamer *Amethyst*. 28 U.S.S. *Wistaria*. Locations are approximate. Specific locations for at least eight additional known tent cities could not be identified.

would appear to have been mindless. Tens of thousands of Charlestonians, mostly poor and black, were living in unimaginably squalid conditions in flimsy homemade shelters created from bed sheets, which were incapable of protecting them from the weather. On top of that, sanitary conditions were appalling. Nevertheless, instead of putting out urgent calls for several thousand more tents, the committee chose to cut off the supply. With a huge number of Charleston's residents homeless, the decision seemed poised to inflict immense suffering on Charleston's poorest refugees, yet there was a dark, driving necessity behind what seemed to be a callous decision. The city elders were haunted by the imminent outbreak of mass disease epidemics in the rapidly growing, foul-smelling refugee camps.

By the end of the first week after the main shock, the housing patterns of Charleston had been totally transformed, and the city's social fabric had been rewoven. The fabled mansions were empty, replaced as dwellings by an immense quilt of tent cities, which filled every safe open space. The newspaper noted, "Necessity, which has been said to be the 'mother of invention,' has taught people to take care of themselves, and in the parks, public squares, and vacant lots a very great variety of tent architecture may be seen. The only material available for the construction of tents is sheeting, old rugs, shawls, carpets, matting and other miscellaneous articles of bed furniture. The people in the various sections of the city have gradually congregated together, and as the sun sets, the whole population repairs to their open-air houses."

Some of these new dwelling places were identified by their street addresses; others were given whimsical names by their residents. Because of its size, distance from buildings, and central location, Marion Square quickly became the largest tent city. "All during Thursday [September 2] the work of building additional tents progressed rapidly, and by nightfall the square, with the exception of the [Citadel] parade ground, was literally covered with tents of every conceivable style and shape. Particolored materials were used in many instances, and the scene

Erecting wooden shelters in Marion Square

was by far the most picturesque ever witnessed in the city. They were occupied by all sexes, all ages and all classes, and as the people moved to and fro in the narrow spaces between the tents, the surroundings had the appearance of the encampment of a grand army of refugees." Even before the city erected wooden sheds there, the square had already become home to about one thousand persons, mostly black.

Two rows of hastily planned, quickly built, one-story wooden sheds were erected on the Meeting Street side of Marion Square. They were constructed in rows of five or six one-room adjoining compartments. Each compartment measured twenty-five by about twenty feet and was supposedly capable of accommodating about twenty people, but not comfortably. "The sheds are roughly but substantially built and with proper ditching would afford good shelter," the *News and Courier* wrote on September 8. "These are all occupied by colored people. To-day another row will be constructed on the south lawn [adjacent to Calhoun Street] and accommodations will thus be provided for about one thousand eight hundred colored persons on this square." Compared to the primitive tents they replaced, the high-roofed sheds permitted adults to stand erect, and offered protection from the rain and winds.

The next day, the Committee on Shelter reported that fifty-two ten-foot-by-seventeen-foot sheds had been built for whites in three areas of the city. For blacks, ten sheds of the same size were built on Marion Square and sixteen on Linguard Street. Records also note that sheds were erected "at the Base Ball Park for the inhabitants of that part of the city" on September 9.[22]

The sheds had earthen floors, slightly sloping roofs, and no doors. People who wanted privacy had to hang their own sheets or blankets from the roofs to provide it. Unfortunately, the low ends of the shed roofs faced the front, which caused rain to be channeled onto the ground, creating rivers at the openings of the booths. No ditches were dug to channel the rainwater away, so the sheds were prone to flooding. Their flawed design notwithstanding, the wooden shelters filled up as rapidly as they could be built.

A week after the refugees had started crowding into Marion Square, the city installed several water hydrants for the temporary residents to use for washing and drinking. The newspaper noted that by September 8, "The committee has also taken precaution to provide sanitary closets [toilets] for the place,

Sheds and tents in Marion Square

and it is hoped that by tonight this camp will be well-policed." A sketch made in early September shows at least one of the wooden outhouses that had been erected for the thousands who had gathered in the square.

The relief committee also erected enough sheds to house five hundred white people in a vacant lot formerly occupied by the west wing of the Citadel. These shelters were intended for residents who occupied stores and houses in the upper part of King Street, and who, it was thought, would naturally want to camp close to their property. The shelters were completed on September 8, and most of them were occupied that night. Just four days later, however, the newspaper reported that the whites had quickly changed their minds. "Nobody seemed to be desirous of occupying them, and the settlement was, therefore, turned over to the colored people," the paper reported. By September 12, "excellent sanitary arrangements" had been made for the shelters to the west of The Citadel, and "about three hundred colored persons are quartered there."

A large settlement had sprung up at the tip of the peninsula at Battery Park, a flower-filled three-acre space now known as White Point Gardens. This neighborhood was an area of lavish mansions owned by some of Charleston's richest families, many of which now joined the homeless. There they had a beautiful view, in addition to some of the best tents. An *Atlanta Constitution* reporter wrote from Charleston, "The tents from the war department...were mostly put up in the park facing the battery and overlooking the confluence of the Cooper and Ashley Rivers, leading out to the bay."[23]

Two hundred improvised tents grew up around the federal tents. About six hundred people, divided almost equally between the races, slept there at night. "The nights at this place are very cool," a reporter noted, "and considerable suffering is the consequence." Camp life here mirrored the situation in most of the tent cities. During the day, some refugees went home to look after property in their shattered buildings. Some sat around in groups, recounting their startling experiences or speculating on the future. Others, mostly women, prepared food, using whatever they had at hand. Laura A. Bennett, the pregnant black woman whose restless sleep had been disturbed by the earthquake, knew about the suffering all too well. She had been injured during the initial shock, and her baby daughter had been stillborn at the Battery Park camp on the morning of September 1, after only one hour of life.[24]

In Battery Park, and probably other sites, a prayer meeting was held each evening. By the light of a storm lantern, held by one of the pastors, someone would read the scriptures. Women of both races occupied the iron benches, while the men, both white and black, stood around them. In the moonlight, under the oak trees, fervent prayers were offered. The services also disseminated practical advice. On September 7, during a sermon, a Reverend Thompson advised everyone to return home, if possible, to avoid the sickness that would surely come if the rains set in.

While Marion Square and Battery Park offered at least a modicum of comfort, Washington Park was at the other end of the spectrum. At night, the one-acre space behind city hall was home to nearly six hundred people, and no one had adequate protection from the elements. There were no outhouses or other sanitary facilities at all.

By September 7, blacks occupied all the space there. One reporter noted, "The ellipse in the centre is crowded with tents made of all kinds of material. Since the rain yesterday [September 6] many of these tents are covered with tin.... The whole square

The camp at Washington Park

is gradually being covered by tents, not excepting the lawns adjoining the walls of the Fire-proof building. Some of these are square, some round, and some triangular. The majority of them are made by driving four sticks into the ground. The rectangle thus made is covered with sheets, and the top covered with matting or carpets, or old rags of any kind that can be obtained. Others are built by driving two sticks into the ground, placing a pole over them and stretching bed quilts across them, thus forming an 'A' tent." The newspaper also noted, "in one of the shanties in Washington Square, the two sides of the *News and Courier* delivery wagon was [*sic*] used as a portion of the walls."

A reporter painted a grim picture of the camp. "The squatters on Washington Square commenced yesterday morning to repair their tents and to prepare for another night. Several of them in fact had commenced the erection of one-story wooden sheds with the evident intention of making a permanent stay. The sanitary condition of the place, however, had become so horrible that all the people residing in the vicinity were alarmed. One gentleman, who lives on Chalmers Street, had purchased disinfectants to use in the square for his own protection, and arming himself with a revolver forbade the dumping of the filth in Chalmers Street."

Smaller tent cities sprang up everywhere like mushrooms after a heavy storm. About three hundred people camped at Hampstead Mall. Another one hundred took shelter in a large garden behind the saloon operated by Bernard Heslin on upper King Street. "All the awnings in the vicinity have been taken down, sewed together and fastened to the extensive grape arbor in the garden," the newspaper reported, "thus making a large pavilion. The settlement is known as 'Camp Heslin.'"[25] About thirty families camped behind August Tamsberg's Hair Works on King Street.[26] Nearly three hundred blacks camped out at Robb's Lot on the east side of King Street, between Market Street and Horlbeck's Alley, opposite the Hotel Windsor. "The style of tents in use here are nearly all square. Prayer-meetings are held every night and the voices of the mourners may be

The camp at Artesian Park

heard all through the lower wards." The Queen's Camp, a large settlement of blacks, was located at the north end of the King Street burnt district. The brick foundations that remained after the fire provided some shelter for the two hundred people who camped there.

Artesian Park, at the corner of Meeting and Wentworth streets, named for its location near one of the city's two artesian wells, attracted about 150 white refugees. There, "the tents are apparently well made, but these constitute only a small proportion of the resting-places," a reporter noted. "There are several 'busses [horse-drawn omnibuses] there, two or three ice wagons and carts, and several carriages. These last are tied to stakes driven into the ground, and are used by white families." Two of the city's fire engines were also stationed there, and the firemen used one of the festive red-and-white circus tents sent from New Orleans as shelter.

Graham's Castle was a comparatively luxurious settlement made possible through the generosity of Robert Graham, a forty-nine-year-old native of New York, who offered his forty-stall stable at 71 Queen Street as shelter for about 150 people, many of whom were able to sleep on mattresses in the stalls.

Captain Thomas Young built a shelter on Southern Wharf, fronting on the Cooper River near South Battery. It was made from a ship's sail erected over a wooden floor covered with canvas, and it sheltered about one hundred people. The adjacent wharf shed held five hundred additional refugees. Young also had a camp of about fifty people who occupied his lot on Church Street at night. Emulating Young, James Doran, a thirty-six-year-old stevedore, who managed to get ahold of a ship's mainsail, erected the Doran Tent on the corner of Tradd Street and Bedon's Alley. When spread on a wooden frame, the sail provided covered shelter for about three hundred black and white refugees at night. Judge Patrick E. Gleason presided over Bennett's Hotel, a mammoth canvas pavilion erected in the lot at Bennett's [Rice] Mill at the east end of Society Street, which sheltered about fifty people, all white. "Comfortable quarters" for homeless blacks were erected in Linguard Street, and were quickly filled. Other tent cities sprang up at Gadsden's Green and at several farms above Sheppard Street.

The South Carolina Railway made between one hundred and one hundred fifty passenger coaches and freight cars available to the refugees. They were placed at the disposal of the relief committee and were parked in the rail yard between Anne and Line streets, just north of the railway's main terminal. "These cars can be occupied at night, and at night only, by any persons who have no other shelter," the newspaper noted. If used as intended, the cars could have accommodated about two thousand people.

The Northeastern Railroad also donated freight cars, which were parked in their rail yard at the northeast end of Mary Street. The Enterprise Railroad, which ran one of the city's trolley lines, also came to the rescue. Through the generosity of its superintendent, Theodore W. Passailaigue, Enterprise parked its enclosed trolley cars at convenient points along their tracks when they stopped running each night, which made the trolleys available to a few more of the thousands of people who were afraid to sleep in their houses. These selfless acts of charity helped, but the need was unquenchable, and even a week later, the owners of livery stables were doing a brisk business hiring out carriage buses to people who wanted a safe place to sleep while living in the streets.

Several ships belonging to the federal government provided aid and shelter. Charlestonians remembered in particular the kindness of Captain Brown, master of the U.S.S. *Wistaria*, a four-year-old, Charleston-based, steam-powered, 167-foot lighthouse tender vessel with side paddle wheels.[27] Brown "immediately sheltered and fed people to the fullest capacity of his boat, which conduct has been not only approved, but commended by the lighthouse board at Washington," the newspaper stated.

Merchant ships in port when the earthquake struck also offered temporary shelter, though they prudently kept their boiler fires banked during the earthquake scare, in case they had to evacuate the port if a tidal wave threatened.[28] The seven-year-old British steamship *Amethyst*, a large, iron-hulled vessel, was off Cape Hatteras, North Carolina, when the earthquake struck. Her first officer, Mr. Warner, indicated that his ship felt no disturbance at sea. A Charleston harbor pilot notified the captain of the earthquake when the *Amethyst* neared the mouth of the harbor. Her captain showed Charlestonians the true meaning of maritime compassion. When the vessel docked at noon on Saturday, he learned of the plight of the people and immediately set part of his crew to work cleaning one of the below-deck compartments. He had the space whitewashed and the floor covered with canvas and provided what other physical comforts he could. The *Amethyst* was able to make space for about two hundred refugees in the compartment, with room for more on the spacious main deck. Julian A. Selby was one of the ship's many guests, and a crewmember who had the night watch graciously offered him his berth. Selby was grateful for the offer, for he had spent the previous night sleeping in one of the passenger cars of the Northeastern Railroad, and he "never want[ed] such another experience on account of the terrible mosquitoes."[29]

While shelter was the overwhelming priority, during the first weeks of the relief committee's operations, the members received numerous emergency requests for small sums of money to pay for immediate needs, such as clothing and blankets. To cope with these requests efficiently, the E.R.C. authorized William M. Campbell, the city treasurer (who also served as treasurer of the committee), to appropriate $1,000 to its Committee on Immediate Relief. The chairman, Captain Dawson, was authorized to issue payments not exceeding $10 ($201 today) to any citizen deemed needy and was directed to report daily what amounts had been paid out and to whom.[30] Another $1,000 was authorized the next week, and further authorizations between $2,000 and $5,000 followed in weekly increments, the last being one for $1,750 on November 27, 1886. In all, the committee disbursed $24,750 ($497,475 today) to take care of individual emergencies.[31]

After two weeks of living in whatever shelter could be found, standing in line for food—or waiting for it to be delivered—and having their sleep interrupted almost nightly by aftershocks, the people of Charleston were exhausted. Life in the Charleston-Summerville area resembled what soldiers often report from a war zone: long periods of anxious, enforced boredom punctuated by moments of total fright. After the main shock on Tuesday night, the initial reaction was terror and the adrenalin-filled fight for self-preservation. As the days wore on, and the aftershocks continued, terror turned into a pervasive, deep-seated, lingering depression.

A diary kept by an unidentified Charlestonian for two weeks after the earthquake shows the preoccupation with the possibility of aftershocks and the ennui that sapped the strength and tried the nerves of almost everyone in the earthquake's central zone.

> Tuesday 31st Aug. 1886.— Spent this evening at S.W. Corner Ashley and Doughty and there encountered the great earthquake which brought so much destruction to Charleston. The shock struck us at 9.50 P.M. As soon as it was over the entire household repaired to the

garden and there spent the remainder of the night. We experienced several more shocks during that night but none of the same degree of violence as the first.

Wednesday 1st Sept. 1886. Returned to our home this morning. Erected a tent in our garden and there spent the night.

Thursday Sept 2d 1886. Spent this night again in our tents. Experienced during the day and night several shocks.

Friday Sept. 3d. Mr. A. H. Hay loaned me two 'tarpaulin' covers, out of which I constructed a better tent in our garden under which we slept this night. Experienced several shocks during the day and night and one <u>severe</u> about 11 P.M.

Saturday Sept. 4th. Experienced several shocks during the day and night. Spent the night in our tent.

Sunday Sept. 5th. Experienced several shocks during the day and night. Spent the night in our tent.

Monday Sept. 6th. Several shocks during day and night. Spent the night in our tent.

Tuesday Sept. 7th. Several shocks during day and night. Spent the night in our tent.

Wednesday Sept. 8th. Several shocks during day and night. Spent the night in the end kitchen room downstairs.

Thursday Sept. 9th. Several shocks during day and night. Spent the night in the kitchen room.

Friday Sept. 10th. Several shocks during the day and night. Spent the night in the kitchen room.

Saturday Sept. 11th. We heard no shocks today. Spent the night in the kitchen room.

> Sunday Sept. 12th. We heard no shocks to-day. Slept in our old quarters in the house for the first time.
>
> Monday Sept. 13th. We heard no further shocks to-day and slept in our old quarters up-stairs in the house to-night.
>
> Tuesday Sept. 14th. Went to Sumter [S.C.] to-day to see the cotton picker and returned at 9.30 P.M. Heard no shocks to-day. Slept in our old quarters to-night up stairs in the house. Strong shock heard by some 2.15 P.M.
>
> Wednesday Sept. 15th. Heard no shocks to day. Supremely happy! Slept upstairs to-night in our old quarters.
>
> Thursday Sept. 16th. We experienced no shocks to-day or night altho' others did. Slept upstairs.[32]

Within nine days after the earthquake, the committee on shelter estimated that they had housed 5,500 refugees in the donated tents and the city-built wooden sheds. Although this was a remarkable achievement in that short period of time, the shelters filled only a tiny fraction of the huge need. In its September 7 issue, the *News and Courier* estimated that 40,000 of the city's 60,000 people were sleeping in their yards, in the streets, and in public squares every night. With two-thirds of the population unable to sleep in their own houses, and overcrowded public shelters available for less than fifteen percent of the displaced people, problems were inevitable—and soon surfaced.

9

TROUBLE IN THE STREETS

Here we is going down to hell, and you won't let us even say a prayer! —A black worshipper to the Reverend Anthony Toomer Porter

September 1886

The refugees who poured by the thousands into tents and makeshift wooden shelters were emotionally overwhelmed and physically exhausted. As each passing day brought more unnerving aftershocks, their condition worsened. No one knew if the next tremor would be the final, cataclysmic event that would destroy what was left of the city—or the world. That thought alone—and such an idea was not extreme, considering that the imminent arrival of an earthquake cannot be predicted even today—haunted everyone in or near Charleston.

As the city started to take stock and rebuild, blacks faced far greater challenges than whites. Their financial resources were slim. Their housing, mostly small, wooden cottages, was rarely insured, for most were too poor to purchase insurance or were renters. In addition, their lack of education made it hard for many of them to understand that the earthquake was a random product of natural forces and not an act of punishment visited upon them by a vengeful God.

President Lincoln had emancipated the slaves on September 22, 1862, to become effective January 1, 1863, but emancipation did not take practical effect in Charleston until the Union army marched into the city on February 18, 1865. The early jubilation of the freed slaves soon disappeared, when they found that their first freedom seemed to be the freedom to starve. Fortunately, federal aid staved off starvation for both blacks and whites during the latter part of 1865.

A black policeman in Charleston during Reconstruction

After the end of the Civil War, South Carolina's government was run by the radical wing of the Republican Party. Black people in South Carolina saw a great expansion of their rights during the ensuing ten-year period known as Reconstruction. These rights included the freedom to vote, hold public office, attend state-supported schools, travel at will, congregate where and when they chose, and conduct their own businesses and church services. However, in 1876, when former Confederate general (and later governor) Wade Hampton's Red Shirts wrested control of the state government from the hands of the northern-backed Radical Republicans and their black constituents, for all practical purposes, blacks lost most of their hard-earned rights.

By 1886, a decade after the end of federally imposed Reconstruction, racial intolerance in South Carolina had reached a high pitch. With the Union army no longer present to maintain order, there was little to restrain whites from treating blacks as they had treated slaves before the Civil War. Blacks were disenfranchised of most of the freedoms they had gained since emancipation. Racial segregation was rigidly enforced. The Ku Klux Klan flourished in the rural parts of the state, forcefully reminding blacks to remain subservient. Those who refused to comply faced being beaten, burned out of their houses, whipped, shot, or lynched.

Virtually all blacks elected to public office during Reconstruction were ousted by the Bourbons, the name given to the post-Reconstruction generation of South Carolina's white upper class. Its members included the majority of the state's newspaper publishers, such as the *News and Courier*'s Francis W. Dawson; legislators; judges; municipal officials; and self-made businessmen, such as Mayor William A. Courtenay. Many prominent white clergymen, a large number of whom had themselves been slave owners, mirrored the Bourbons' entrenched beliefs in the racial inferiority of blacks and the ironclad need for segregation. Even church vestries and trustees, in their legal capacity as corporate entities entitled by law to buy and sell property, had sometimes owned slaves, who generally worked as caretakers or janitors.

Unlike the aging relics of the obsolete plantation aristocracy, who longed to return to the slave-based agrarian past and whose dreams revolved around restoring the antebellum way of life, the Bourbons were forward-looking, growth-oriented, and decidedly pro-business. At the time the earthquake struck Charleston, the Bourbons were in full control of the city.

Today, the tip of the Charleston peninsula now known as "South of Broad," is almost exclusively white-owned, and houses are quite expensive. In 1886, the area was already home to the mansions of the rich, but blacks and whites had been living close together for more than a century. Blacks lived intermingled in white neighborhoods in humble homes in alleys off the main streets, in tenements near the docks, and in the former slave quarters of the city's substantial homes.

In every other aspect of life except the location of housing, Charleston was a racially segregated city, especially when it came to education. Separate public schools existed for whites and blacks, but the amount expended for each white student in South Carolina often exceeded that for a black child by a factor of more than twenty to one. White children of elementary school age could attend Bennett's School, Craft's School, the Meeting Street School, and the Memminger School. Black children, who outnumbered whites, had only two choices: The Morris Street Colored School and the Shaw Memorial Colored School. Affluent whites could send their male children to several private college preparatory schools, including the Reverend Anthony Toomer Porter's Holy Communion Church Institute, the High School of Charleston, the Central Catholic School, or the German Academy. For education beyond elementary school, white females could attend the Charleston Female Seminary (a boarding school), the Academy of our Lady of Mercy, St. Mary's Catholic Female School, or the South Carolina Training School for Nurses, located behind Roper Hospital. Black children had three choices: the Colored Catholic School, an offshoot of St. Peter's Church; The Avery Normal Institute; and the Wallingford Academy. Few women attended college, and in 1886, all of Charleston's institutions of higher learning, including the College of Charleston, the South Carolina Military Academy (the Citadel), and the South Carolina Medical College, barred blacks entirely.

At no time was Charleston more segregated than on Sunday mornings. Prior to 1865, many slave owners brought their slaves to church with them. The whites sat in pews on the main floor, with the blacks in the slave galleries above. In many antebellum religious denominations, blacks were eligible to receive the same rites of the church as whites, including baptism, confirmation, and marriage. However, slaves attended church only if permitted to do so by their owners. Furthermore—although it did not happen in most cases—married black slaves and their children could be separated and sold to any buyer at any time, without any requirement to keep slave family units intact.

After emancipation, blacks left the white churches of their former masters in droves. The freed men and women formed their own congregations and ordained their own ministers. Some allied themselves with all-black denominations, such as the African Methodist Episcopal (A.M.E.) Church, the independent black offshoot of the white-ruled Methodist Episcopal Church. Charleston's white churches had never shared their burial grounds with blacks before the Civil War, and even those white churches that retained some of their black members after 1865 maintained segregated pews and burial places.

Outside the church doors, all public services were also provided according to race. The medical facilities that served both whites and blacks did so in separate wings or wards, and in the jails, white and black prisoners had separate cells and cellblocks. In their final years, the aged, crippled, and insane poor all lived out their lives in separate-but-unequal public facilities.

The earthquake, which could be viewed as the most democratic, egalitarian, and totally raceless event in Charleston's history, did nothing to alter the segregation of relief services provided in its aftermath. Access to food distribution was not skewed by race, for the ministers of both races were able to issue ration cards, and the soup kitchen was open to anyone with who was willing to stand in line. But when it came to emergency housing, race was the key to who got what.

The E.R.C. allotted most of the best emergency housing—the well-made, waterproof canvas tents—to whites. Blacks had access chiefly to the city-built wooden sheds or nothing at all. When it set up tents and sheds, the E.R.C. initially designated certain areas for white refugees and others for blacks. In practice, the planned segregation of the tent-and-shed cities collapsed almost immediately. The mass homelessness produced by the earthquake overwhelmed the city's resources to such an extent that it brought large-scale, unplanned public mixing of the races in the only safe places left to congregate. People of both races quickly staked out their outdoor living spaces on squatter's rights terms, with possession being ten-tenths of the law.

A prayer meeting in the camps

Through sheer force of need and numbers, the largest of the open-air parks, notably Washington Park and Marion Square, started out as racially mixed refugee camps. That twenty-four-hour-a-day, elbow-to-elbow mixing soon triggered sparks of racial tension and demagogic outbreaks of race-based anger. Much of the strife was based upon differences between white and black religious practices, which quickly produced friction among the highly stressed residents of the overcrowded tent cities.

Many white Christians in Charleston were disturbed—and even alarmed—by the intense, emotional, and exuberant worship practices of their black fellow Christians. Whites were much more restrained in their worship. Spontaneous, ecstatic experiences of the Holy Spirit or public expressions of emotions made earthquake-era whites extremely nervous. Unlike many blacks, Charleston's white Christians did not derive their religious practices from direct experience. Like their houses, plantations, former slaves, and silver plate, their beliefs and behaviors were inherited. To paraphrase psychologist William James, their religion was made for them by others;

communicated to them by tradition; learned by rote; confined to fixed, orderly forms; retained by habit; preached exclusively by men; and practiced chiefly by women.[1]

The Reverend Anthony Toomer Porter

The first *News and Courier* report of people assembling in Washington Park described the refugees in terms of gender and racial stereotypes, which included "panic-stricken" white women and children, "brave, stalwart" white men, and "quite a number of colored people assembled, as is their wont, [who] fell to praying and shouting." The perceived difference between the forty-six-percent white minority and the fifty-four-percent black majority was clearly defined by the opinion of Louis A. Beaty, editor and publisher of the *Berkeley Gazette*, a newspaper published in Mt. Pleasant.[2] Of the aftermath of the earthquake, he wrote, "The negro character, permeated with animal excitement, superstition and fetichism [fetishism], shone forth in its glory. The demeanor of the whites was calm and reassuring."[3]

On Saturday night, September 4, the Reverend Anthony Toomer Porter of Holy Communion Episcopal Church in the northwest part of the city, was exhausted. As a result of attending to the needs of his parishioners and his own family, the fifty-eight-year-old priest had not slept since a heavy earthquake shock the night before. When he lay down to try to rest that evening, he heard a whirling sound and then felt a violent tremor "as if a tremendous sledge-hammer had slammed down nearby."[4] It was yet another aftershock.

"Added to this," he said, "a large crowd of negroes had assembled just outside my wall, in the street, and they were indulging in howls and yells, screaming, praying, and singing. It was a very pandemonium. I could not stand it, so I went out to them. I soon singled out the ringleader. He was an old gray headed man, and was praying at the top of his stentorian lungs, and informing the Lord how very wicked he and all of them had been, that hell was open, and that they were all going down into its burning jaws. I let him go until, while he did not mean it, he was bordering on profanity, and was stirring the crowd round him to frenzy."[5]

Porter asked him to wind down his rhetoric, but the preacher replied that black congregations all over town were holding similar services. Porter informed him that Acting Mayor Huger had decreed that all the noise must cease at 10:00 p.m. The minister investigated, found that the order was indeed in effect, returned, and agreed to cease. Scarcely had he stopped, when a black woman came at Porter in a rage.

"Yes," she shouted angrily, "just like you buckra [white men]. Here we is going down to hell, and you won't let us even say a prayer!"[6]

Reverend Porter saw that a row was imminent. Walking up to the woman, he raised a small cane that he had in his hand and exclaimed, "Look here, I never struck a woman, but if you do not hush up this instant, I will wear this out on you!"[7]

Silence followed, and Porter walked away. He had not gone ten feet when the crowd broke out with the song, "Oh, pretty yaller gal, can't you come out tonight?" Porter spun on his heel and told them that their singing was as offensive as their praying and that it was silence he had come for. He finally got his silence, but the woman and her fellow worshippers had the last word.[8]

Two days later, under the headline, "Necessity Knows no Law," the *News and Courier* published Toomer's complaint, withholding his name and attributing it to "an influential gentleman residing in the northwestern part of the city." The "'informant,'" the newspaper said, "met a Reporter on Sunday afternoon

[September 5] and desired the statement made that the boisterous proceedings of the colored people in his neighborhood were simply intolerable. He asks that the police put up a preemptory stop to the excesses, although they may be committed under the guise of devotional exercises, for, as he said, the constant noise and tumult are doing more to unnerve the ladies and children of this vicinity than did the repeated shocks of the earthquake. The tired and harassed people are unable to find the rest so much needed after so many nights and days of patient sorrow and suffering. The speaker went so far as to say that unless the police heeded the request, that he would today organize the citizens in his neighborhood with a view of taking the law into their own hands."

A week later, in a signed editorial dated September 9, Porter laid out in no uncertain terms his philosophies on race, religion, and earthquake relief. After reminding the black clergy how much he was interested in the welfare and advancement of their race, he urged the ministers "to use your powerful influence with all your people to leave the streets and tents and go to their houses. If they cannot go to their own, there is plenty of vacant space in other houses. If it is the rent they are afraid of I will see that you are helped to provide for a month's rent for absolutely needy cases.... Break up those camps on Washington square, and even those wooden tents on Marion square."

Porter went on to claim that if northerners, who were now funding the relief efforts, were to think that their money was being wasted on food and shelter for able-bodied people who could be living in their own homes, the funds would dry up. He ignored the fact that the majority of people living in the crowded, noisy, stinking tent cities were poor, unlike Porter's congregants, and were forced to be there because their homes were thought to be uninhabitable.

He then went further, taking it upon himself to tell the blacks that they should no longer apply for food rations and that he was severely limiting the number of ration cards he distributed. "Discourage this going for rations business," he scolded them. "I know of some men getting $1.50 and $2 a day, and their wives, instead of being at the washtub or in the kitchen of their families or hired out, are crowding in for rations. This is no new thing. I went through it all after the fire of 1861," Porter continued, "and then learned the unwisdom of the system and its demoralizing effect. Up to now I have given [out] but six tickets, having shown a great many how much more respectable it was to earn their bread and dissuaded them from applying. St. Mark's [Episcopal] congregation have sufficient confidence in me to be guided by my advice, and I am sure you [black ministers] exercise the same influence." Porter made no such demands of poor whites.

About the loud outdoor preaching, he had more to say. "Do stop these repeated so-called religious scenes, singing and loud praying, and stentorian preaching," Porter demanded. "God is not deaf, and I don't suppose all the congregations are, and need not be 'hollered' at. Your educated and intelligent class do not do so; but your class-leaders and men who have been taken from the field and workshop do, and make up in sound what they lack in sense and religion. You will never elevate your people thus, and you antagonize the two races." Porter then stated, "The average white man does not make excuses and allowances for [the black refugees'] antecedents and surroundings, and looks on with contempt and says, 'What is the use to try to elevate those savages?'"

The priest ignored the fact that more than half of the city's blacks were former slaves, and the rest were the children of former slaves. He likewise discounted the fact that, prior to emancipation, they had been denied almost all of the basic human rights enjoyed by whites. Then he chided black people for not instantaneously rising above the limitations imposed upon them by their owners for two hundred years. Porter closed his epistle with the words, "This advice is offered in the most fraternal spirit to you and your people, and I hope it will be accepted as such." For the blacks he pontificated at, Porter's relentlessly patronizing words and deeds indelibly marked him as an archetype of the insufferable white bigots who opposed every freedom and avenue of progress blacks set out to achieve.

Just as the earthquake brought out the worst in people like Porter, it brought out the best in others. Some of the finest examples of thoughtful, tolerant behavior came from, of all places, the acting chief executive of the city, William Huger, and George W. Dingle, the chief executive of the Charleston police force. A number of whites—inspired and led by Porter—had taken their complaints about the noisy, black, outdoor religious services to Dingle, who was chairman of the Board of Police Commissioners. They fully expected him to order his 107-man force to sweep through the refugee camps and shut down the disruptive preaching. To Dingle's credit—and the utter astonishment of Porter and the other white complainers—Dingle chose compassionate insight over the demands of his indignant white constituents. A brave man in an era of white dominance, he summarily rejected their demands and wrote:

> There were many complaints of the excited and emotional religious services by a portion of our community, and there were some citizens who thought these meetings a proper subject of Police interference, and because of their non-action, criticized them sharply; but the Police were obeying the orders of the acting Mayor, which I, myself, would have given, had I been on the spot. It was certainly very trying to weak nerves to hear these distressful supplications and lamentations, but to the participants it afforded relief, and was certainly not intended to give offense. In a very few days these meetings were discontinued, and, as a whole, it may be truthfully said, that the colored population of Charleston behaved, through these trying scenes, with most commendable propriety.

An outdoor worship service in September

The weekend after the earthquake offered the first opportunity for formal worship. September 5, the first Sunday since the calamity, was a day of universal prayer for Charleston's Christians. "The morning broke as clear and as fresh and fair as if its rays were to illuminate and invigorate the life of a city on some less solemn and melancholy occasion," the newspaper wrote. "There was scarce a cloud to fleck the sky, and all nature was as graceful and calm as though the grief and sorrow of the previous days had been things of a very distant past. It was a day of universal prayer, a day on which the stricken people, abandoning their temporary homes of fancied security, went forth to the house of God to worship Him under the canopy of His own beautiful and all-covering dome. There was something peculiarly appropriate in this, as if God had destroyed the temples which men had builded to bring them apart from the glory of their own works to adore Him and tremble at His power in full view of its most beautiful and most terrible manifestations." This editorial introduction was followed by three columns containing extensive excerpts from sermons preached in all the city's major white churches and from the many open-air services preached by white ministers. No mention was made of sermons preached by black ministers.

That Sunday, Christian Charlestonians were united. Dispassionate scientific explanations of earthquakes as random acts of natural forces were not openly discussed. Nor did anyone publicly question why a loving God would condemn, for example, a loving mother and her innocent newborn child to die a gruesome death—and spare the life of a vicious, hardened criminal. People were looking for hope and good news, but both were often hard to find.

In the first few days after August 31, the smallest sliver of good news led to instant euphoria, most of which was quickly dashed by the arrival of more bad news. In the crowded streets, tents, and shelters, personal security was a constant concern. Everyone felt vulnerable and exposed. It was just as well that most of the survivors had not given much thought to how the earthquake had affected the men and women who were behind bars when the first shock hit.

Terrifying news soon arrived from the city police department: the earthquake had cracked the walls of the city jail. Built in 1802, and enlarged in 1855-1856, the jail sat on the southeast corner of Magazine and Franklin streets. Within seconds of the first shock, the jailer, Capt. John Kelly, one of Charleston's many Irish-American policemen, had cells full of terrified, frenzied prisoners on his hands. "When the building began to shake," a reporter wrote in the early hours of Wednesday morning, September 1, "the prisoners made a dash for the door. Capt. Kelly, however, stood at the door, a pistol in hand, and firing half a dozen shots kept the crowd back. Their shrieks could be heard for squares and many of the inmates dashed themselves madly against the bars in their attempts to escape. They were kept within doors, however, and although the building was badly shattered none of them escaped."

The comforting news that the inmates had not gotten free to savage the community was short-lived. Following the initial shock, in an act of kindness, Captain Kelly had released the caged prisoners from their cells into the exercise yard below.

The Charleston City Jail, wrecked by the earthquake

Massive brick walls three feet thick enclosed the yard, but they "yielded to the quivering earth like so much glass."[9] Gaping cracks appeared in the walls after the first shock, some of which extended from the top of the building to its foundation, and the two-story tower over the four-story rear wing was badly damaged. In addition, the horizontal shaking of the building had opened up major cracks at all four corners of many cell-block windows, making it possible, in theory at least, for the prisoners to push out the bars and escape.

No sooner had the first newspaper report been set in type than an addendum was set to follow it. Headlined "Thirty-eight Prisoners Escape," it read, "Capt. Kelly told a Representative of the News and Courier early this morning that thirty-eight of the prisoners whom he had taken down into the [exercise] yard had managed to escape. Whether any of them have been recaptured has not yet been ascertained." The number of escapees was later revised upward to forty-two.

The runaway offenders were described as having made "leg bail" during the earthquake. Among them were prisoners awaiting trial for serious crimes, including burglary, larceny, highway robbery, assault and battery, and murder. Lavinia Bradley, described as "a virago" (a powerful, manlike woman) had the distinction of being the first prisoner over the wall. An inmate colleague named Jake Singleton, "better known in city slums as 'crooked-foot Jackey'" followed her.

It was soon clear that many of the prisoners fled chiefly from the fear of being crushed to death in their cells. These inmates soon returned voluntarily to custody, knowing that it would be better to serve out their relatively short remaining terms than to be recaptured later and given much longer sentences. Such was the case of James Goff, an accused murderer, whose family asked for bail for him but had been turned down. His family promised the court that he would voluntarily surrender himself "as soon as the earthquake shocks cease." Those who remained at large consisted chiefly of hardened criminals, who faced long terms. "The cream of the gang—the prisoners charged with capital crimes, and the accomplished burglars—were among the forty-two who decamped [in the initial jailbreak] on Tuesday night," the newspaper reported.

On Monday, September 6, the *News and Courier* reported that Captain Kelly, with seven guards as escorts, had moved twelve remaining prisoners to an improvised jail in the eastern wing of the Citadel on September 4. These prisoners were locked into the watch-room, but when an aftershock struck at 9:30 p.m., they were transported to Marion Square. There, with no steel bars to restrain them, "Peter Anderson, alias Mac Serene, white, and Charles Brown, colored, escaped, leaving ten jail birds in all."

The Citadel continued to be used as a jail for several weeks, and by late September Captain Kelly had twenty-three prisoners there whom "the earthquake had left on his hands." In his words, they were mainly "volunteers," and "his favorite boarders, among whom are the now celebrated 'boy burglars,' are still at large." Since the Citadel's facilities had proven to be unsuitable, Sheriff Ferguson had a new stockade built. "He will use no rosewater treatment, but will put every new miscreant in irons who comes to the stockade. He promises to provide full accommodations for all the rogues that Charleston and the vicinity can muster," the newspaper said. The new facility was "a stout frame structure" on Franklin Street, containing sixteen cells, each with a capacity for six prisoners. When it opened for business on September 29, there were only twenty-five prisoners, most of whom were serving short sentences imposed by City Recorder William Alston Pringle for minor offenses.

For a city of 60,000 people, Charleston had a large police force: 107 men. Shortly after the first shock, the downtown contingent of the force, who patrolled the lower six wards, had been temporarily moved out of the terminally damaged Main Guard House and into its horse stables nearby. On Sunday, September 5, the force was moved again, this time to the Charleston High School gymnasium, a relatively undamaged building at the corner of Meeting and George streets. It had sufficient space for the men and their sleeping cots, which had been rescued from the ruined Guard House. The police horses were sheltered in a temporary stable erected on the north side of the grounds.

The Upper Division of the police force, which patrolled the upper six wards, north of Calhoun Street, was able to stay in its headquarters in the Upper Station, also known as the Uptown Guard House, on the west side of King Street, between Morris and Cannon streets. Its offices and holding cells were still in usable condition. Two weeks after the first shock, the Main Guard House was reoccupied, but not for long, as city inspectors declared it terminally damaged.

"Owing to the disturbed condition of the city," Charleston's police chief, Thomas Frost, Jr., put out a call for temporary policemen, hired to serve fifteen-day terms. Seven had been hired by September 5, and soon fifty of them helped regulate law, order, and sanitation in the refugee camps. They were housed in a large, colorful circus tent sent from New Orleans, which was pitched on the grounds adjoining the high school.

In addition, the E.R.C.'s Gen. Thomas Huguenin, who was also the commander of the 4th Regiment, State Volunteer Troops, called upon the Charleston Light Dragoons and the German Hussars, both of which had been antebellum volunteer military companies, to patrol the city in general and the tent cities in particular. They were also needed to alert the firemen in case a blaze broke out, as the alarm telegraph was still down. The two units were composed of proud war veterans from Charleston's blue-blooded families who had been mustered into Confederate service at the start of the Civil War. After the war, they retained their prestigious status as volunteer militia units. The Dragoons had twelve men on horseback by September 2, and by September 5, they extinguished two fires. But, as Huguenin noted on September 6, "The Dragoons, like everybody else, are now nearly worn out." He called for additional mounted volunteers to assist the Dragoons. He emphasized, "The loan of horses is urgently requested also."

The escaped prisoners—and criminals in general—seemed to have caused the city more fear than actual harm. Whether through efficient law enforcement or because criminals had more important things—such as survival—on their minds, crime was low in the weeks after the earthquake. There were no reports of looting and only scattered accounts of thievery.

On September 4, Robert Singleton and Edward Simmons, two convicts who had escaped from the City Jail on the night of the earthquake, broke into the residence of Otis Philips on George Street and stole a handsome gold watch and chain. In a few days, detectives found the thieves and their loot in one of the slums of the city, and the convicts were returned to jail.

Likewise, on September 6, Henry C. Ortmann's saloon on King Street was robbed of $100 between two shocks that night.[10] The following night, a thief made off with a gold watch and several other items from Scotsman Robert B. Dowie's house at the corner of Church Street and South Battery. The Reverend A. J. S. Thomas, who lived across from Dowie, foiled a similar attempt. At the Unitarian Church on Archdale Street, a man was caught stealing lead from the wrecked church and was arrested.

At the September 10 session of the Magistrate's Court, Judge Pringle had only three relatively minor cases to hear. He sentenced Robert Johnson to pay three dollars or serve ten days in jail for breach of the peace and being drunk and disorderly. Henry Grant, arrested for possession of stolen goods, was released, as no affidavit had been sworn out against him. Daniel Ford had been charged with assaulting Rosa Deas in Princess Street and was sentenced to pay five dollars or serve thirty days. These relatively few and petty crimes may not have given a true picture of criminal activity at the time.

Because of general disorder in the city, many crimes may have been unreported.

The newspaper noted that complaints about burglary and theft had been increasing rapidly. "A burglar broke into R. H. Ohlkers' saloon on East Bay Street through a wrecked skylight in the rear. He escaped the three policemen who tried to apprehend him," the newspaper said. "A number of unoccupied or damaged houses have been entered and robbed of bedding and ladies' and children's clothing. These cases were turned over to [police] constables and ten women were arrested and sent to jail for stealing. They are all old offenders from Princess Street and other disreputable locations, who had taken refuge on Marion Square, and were discovered with several of the stolen articles in their possession," the article continued. "It would be advisable for all other persons who have been robbed to report their losses to Trial Justice Gleason, who has determined to put a stop to the wholesale stealing now going on in the city. The constables will be ordered to search all colored encampments around town, and it is possible that before they get through there will be made several hundred arrests of people who have taken advantage of the demoralized condition of things to rob and steal."

The paranoia expressed by the *News and Courier* disappeared within a few days, and relatively few thefts or arrests ever took place. The real crimes notwithstanding, the *News and Courier* reported several times on the generally circumspect and orderly conduct of all classes and races of society. By September 23, about fifteen of the temporary special policemen were discharged, and the rest returned to civilian life as their fifteen-day enlistments expired. It was as though the earthquake had reminded the population of the value of personal integrity. In a city already known throughout the nation for its graciousness and civility, serious crime hit an all-time low.

10

THE MARCH OF SCIENCE

It ought to be understood that the whole thing is governed by natural laws and that nature works in a perfectly systematic manner.
—Professor Thomas Corwin Mendenhall

Wednesday, September 1, 1886

The morning after the earthquake, the first official news of the disaster reached Washington, D.C., via the initial telegraphic report filed by Captain Dawson of the *News and Courier*. The scientific community, realizing the unique opportunity it had to study an immense natural phenomenon first hand, immediately urged the government to form a commission to investigate the extraordinary event. This request was quickly granted, and notable scientific heavyweights were rapidly assembled to serve on the commission or advise it. They included Prof. Simon Newcomb, an astronomer and mathematician with the U.S. Naval Observatory and vice president of the prestigious National Academy of Sciences; Prof. Henry M. Paul, also a mathematician with the U.S. Naval Observatory, then serving in Tokyo; Prof. William Morris Davis, a prolific author of scientific books and articles and a distinguished

Harvard professor of geology; and Prof. Charles Greene Rockwood, a mathematician from Princeton. In addition, Ensign Everett Hayden, a scientist attached to the U.S. Naval Observatory, supervised a nationwide earthquake data and damage survey.

As he read the message handed to him by an aide on September 1, USGS director Maj. John Wesley Powell's initial reaction was probably one of dismay. As a captain in the 2nd Illinois Artillery Volunteers, Powell had been struck by a Minié ball that plowed into his right wrist as he raised his arm to give the signal to fire at the Battle of Shiloh. The wound was so severe that his arm had to be amputated below the elbow. It was at times like this that pain in the phantom limb made him clench his teeth.

What Powell didn't need on that day in September 1886 was another crisis. He and the geological survey he led had been under fire of a different kind for nearly two years during a Congressional investigation that, as he saw it, was out for his blood. The Allison Commission, headed by Senator William B. Allison of Iowa, had been formed in 1884 to rein in the proliferation of government agencies that claimed jurisdiction over science and to eliminate or consolidate those found to be wasteful or redundant. The commission also wrestled with the question of how federal scientists should interact with state scientists, academics, and other "individual investigators." Especially at issue was the continued funding level for the agencies involved, including the USGS, which Powell and his predecessor, the first director, Clarence Rivers King, felt had been under-funded for years. King, a highly regarded, charismatic, and controversial scientist and administrator, had, in fact, resigned after only two years in office when Congress granted him less than half of the annual budget he had asked for.

Powell testified exhaustively during hearings from 1884 until the commission's investigations ended on January 30, 1886, defending his proposed topographic map of the United States, which would take an estimated twenty-four years to complete given his current level of resources; explaining why his expenditures for printing scientific reports were not extravagant; and generally trying to justify the geological survey's reason for existence. In doing so, he made enemies of several prominent colleagues who felt that any government intervention in science was unwarranted. Things were especially grim for government scientists when Grover Cleveland, reputed to be no friend to science, was elected president in 1885. "In his first inaugural address, Cleveland put himself on record as in favor of strict economy, protection of the Indians and security of freedmen, and the value of civil service reform."[1]

In the end, Powell's tireless arguments paid off. When its investigations were over, the Allison Commission completely exonerated the geological survey, recommended an increase in its funding, and left it intact. It was a victory, but it came at the cost of two years of distractions and inadequate funding. Powell returned his focus to his work.

Just eight months after his battles with the Allison Commission, when he would have preferred to spend time on his activities with the Anthropological Society of Washington and the Biological Society of Washington, both of which he helped to found, and his program for ethnologic research under the Smithsonian, which he directed, he was now faced with a cataclysmic earthquake.

He had to respond; it had become part of his job. In late 1884, two small but widely felt earthquakes in the eastern states had called attention to the lack of any organized attempt to observe them in the United States. In its October 3, 1884, issue, *Science* magazine suggested that the USGS should provide instruments and observers to study earthquakes and that students of this branch of physical geography should form an earthquake club. Therefore, in November 1884, Powell, Clarence E. Dutton, and another USGS staff member conferred with some of the men who would later become members of the federal earthquake commission for Charleston—Professor Rockwood of Princeton, who had been collecting records of earthquakes; Professor Davis of Harvard; Cleveland Abbe, an astronomer and meteorologist from the U.S. Signal Service;

Charles F. Marvin, a meteorologist, also from the U.S. Signal Service; and Professor Paul of the Naval Observatory — to consider the best way to systematically observe earthquakes.

They agreed that there was only one practical scheme. First, they would rely on human observations, which would be collected by distributing circulars to be filled in after an earthquake was felt. Next, they would work toward establishing stations for instrumental observations. Marvin, who was noted for designing, building, and standardizing meteorological devices, offered to design an instrument that would be simple, inexpensive, and easy to maintain. Rockwood, Davis, and Abbe became a committee to determine the best geographic distribution of stations. The USGS would furnish instruments to earthquake observers and receive reports. The seismological investigations became part of the work of the Division of Volcanic Geology under Captain Dutton, who would play a crucial part in documenting the Charleston earthquake.[2]

Now, faced with an actual earthquake, Powell mentally ran down the list of people to send to the scene. Like the men he would choose, Powell had little formal schooling. Born in Mount Morris, New York in 1834, he was the son of a Methodist preacher whose vigorous, assertive stand against slavery alienated many townspeople and forced young Powell to stay out of school for his own safety. With the help of a neighbor who had an interest in science, Powell had educated himself, studying botany, zoology, and geology from the time he learned to read.

It was his fascination with rivers that would create national fame. In 1856, at the age of twenty-two, he supposedly rowed the length of the Mississippi River alone, and a year later, he repeated the feat by rowing the length of the Ohio River. In 1858, Powell rowed down both the Illinois River and the Des Moines River. His injury in the Civil War put a stop to all that rowing, and he became a professor of geology at Illinois Wesleyan University.

The year 1869 brought him to the river that would make his career. He developed a theory that the Grand Canyon of the Colorado, a nearly unknown region that was the source of many vague and often wild rumors, had been created by rivers, which formed the canyon as the plateau rose. On May 24, at the age of thirty-five, he and nine other men headed down the Green River in Wyoming on a journey that would cover almost one thousand miles through uncharted canyons and change the west forever. Three months later, Powell and five of the original company triumphantly emerged from the depths of the Grand Canyon at the mouth of the Virgin River. The adventurers had not been heard from since their departure and were presumed dead.

Powell had confirmed his theory. Returning as a national hero to Illinois, he promptly hit the lecture circuit to raise funds for a second expedition in 1871, which would produce what the first did not—a map and scientific publications.[3] One of the participants on the second journey was young army officer, John Karl "Jack" Hillers, who had been hired to serve primarily as an oarsman. Watching E. O. Beaman, the expedition's official photographer, Hillers became fascinated with the camera. After a disagreement, Powell fired Beaman and hired another photographer. However, other members of Powell's party noted that twenty-eight-year-old Hillers, who was the youngest and strongest member of the expedition, was willing to climb with the heavy and cumbersome camera equipment, and he had an eye for capturing images. When the second photographer became too sick to continue, Powell put Hillers in charge of photography for the remainder of the voyage.[4]

Hillers was so successful that starting in 1872, Powell directed him to focus on photographing the life and cultural aspects of Native American tribes in the Southwest. Over twelve years of shared expeditions, the two men developed a relationship of deep trust. Hillers, now supervisor of the photographic labs for the USGS and the Bureau of Ethnology, was the natural choice to photograph the aftermath of the earthquake, although he could not spare much time from his other work.

Of course, Powell had to select a leader for the earthquake effort. His ever-willing protégé, William John McGee, immediately sprang to mind. Only thirty years old when he was appointed an assistant geologist at $1,200 per year on July 1, 1883, McGee had been set to work immediately supervising one of Powell's pet projects—compiling a geologic map of the United States. This was an ambitious and formidable task, especially since McGee's prior geologic experience was confined to a study of the glacial deposits of Iowa, which he had done on his own initiative, and to a stint as a field assistant during the study of the Great Basin in the West. When not engaged in working on the map, he was supposed to study the surface deposits of the District of Columbia and its surrounding area.

Wanting to dazzle the International Congress of Geologists at their meeting in Berlin in 1884, Powell encouraged McGee to hurry and finish the project. Professor C. H. Hitchcock, whose own geologic map of the United States had been published in 1881, was hired to assist him. The huge, new, hand-colored map was completed in January 1884, largely, McGee confessed, because of Hitchcock's energy, experience, skill in geologic cartography, and extensive knowledge of American terrain and geologic literature. In other words, Hitchcock did most of the work, and McGee learned from it.

Powell immediately ordered McGee to take the huge map and recompile the data in a smaller format suitable for publication. The inexperienced young geologist probably spent many long days struggling with this task, but it was completed on time for the Berlin meeting.[5] In an ironic twist, the meeting was postponed because of a cholera outbreak in Europe. McGee had served ably and loyally ever since and had increased his geologic knowledge immensely. He was in line for a special assignment.

Completing the team from the survey would be Clarence E. Dutton, a captain in the U.S. Army Ordnance Corps, who had been detailed to the USGS in 1880. Dutton, who had worked with Powell on expeditions since 1875, shared his director's love of the canyon country and also had a long-standing interest in volcanoes and earthquakes. He was considered to be "one of the clearest, most impressive, and entertaining of writers."[6]

Because earthquakes now fell under his division, Dutton would have been the natural choice to lead the Charleston effort. However, he was somewhere in the Cascade Mountains at the time, and quick action was needed. Powell decided that the highly respected Dutton would compile the final report on the findings of the USGS team. Finally, to ensure that other relevant areas of science were represented, Powell called upon Thomas Corwin Mendenhall, an eminent physicist attached to the U.S. Signal Service, to join the onsite researchers.

Still smarting from his recent experience with the Allison Commission, Powell knew that his expenditures of both time and money would be closely scrutinized. His predecessor and first director of the survey, Clarence King, who knew how closely Congress would watch for signs of success from his fledgling agency, had directed his troops to produce immediate results of practical value. "Realizing very fully," King wrote, "the natural desire of Congress and Administration to see actual results and apply the test of a critical examination to the fruits of the new Bureau, I have called upon the members of the corps for an energy and intensity of labor which should not be greatly prolonged, and which [does not depend on more funding to be completed]."[7] Taking his cue from King, Powell encouraged his team to do a thorough job—as quickly as possible.

On September 2, McGee dashed to Washington's Union Station and paid $7 for a sleeper train ticket on the Atlantic Coast Line railroad. He arrived in Charleston on September 3. Both Mendenhall and Hillers left for Charleston on September 3, arriving on September 4. On September 5, USGS Director Powell implemented one of the two steps agreed upon in 1884. He ordered a form to be distributed nationwide, "calling for detailed information upon the subject [the earthquake], from whomsoever may have information, even of an apparent trivial character to impart." Those who were given the forms were asked numerous questions about where they were at the time the earthquake was felt, what they felt, when they felt it, and the type of ground in their localities.

Within six months, thousands of earthquake questionnaires from more than 1,600 locations had been returned to Washington, where they were tabulated by Ensign Hayden. Hayden then provided the data to the other earthquake commission officials, who were working on the larger picture: determining where in the United States the earthquake was felt, how fast the shockwaves traveled, what time the quake was felt at each place, how long it continued, whether it was accompanied by sounds, and how many shocks occurred, along with a measurement of their intensity and effects. Ultimately, the data collected and analyzed and the reports written by all the team members were compiled and edited by Captain Dutton. The result, formally titled "The Charleston Earthquake of August 31, 1886," and now known to scientists worldwide as "The Dutton Report," was published by the USGS in 1888.

William John McGee

The principal investigator of the 1886 earthquake, USGS geologist William John McGee, then thirty-three years old, had experienced it personally—but not in Charleston. On the evening of August 31, 1886, McGee felt the first shock of the quake while in bed in his room at 1424 Corcoran Street, a three-story brick structure in Washington, D. C. A dedicated scientific researcher to his core, and a man well prepared to take care of himself in the wilds, McGee carefully and dispassionately recorded his impressions of the shock waves, muted as they were by their five-hundred-mile trip from South Carolina to the nation's capital.

The tremor was observed and correctly interpreted shortly after it commenced in Washington. The room occupied during the shock was the upper [third-story] front, facing north; the bed (occupied at the time the shock commenced) heads east, the head being two or three inches from the party-wall. A wash stand with its accessories occupies the north-east corner of the room within reach from the bed; an office table stands near the north-west corner. On the table was a student lamp and above a gas fixture with globe. The bed is of average height, but the headboard quite high—8 ½ feet or more. This head-board is rather slender but carries a heavy moulding with a gable top in the centre. Over this was thrown a heavy leathern belt from which suspended a heavy (.45 caliber, 7½ inch barrel) pistol and a hunting knife of which the blade is 9 inches in length. The tips of pistol and knife rest lightly against the head-board.

The sensation conveyed by the first tremor produced the instinctive impression that some one had his hand on the foot of the bed and was shaking it to and fro; but knowing that the house had no other occupant I very quickly inferred the true cause of the disturbance. Meantime the violence of the shaking increased, and the points of pistol and knife began to rap against the head-board and at the same time the movements of the bed (which has a spring mattress) were such as to indicate a quick eastward and upward impulse—as if the mattress were kicked or struck in an upward and eastward direction. At the same time, too, rattling of the student lamp, of the globe on the gas fixture and basin on the wash stand, and of the mirror on the dressing case, together with some creaking of the bed and

> other articles of furniture was perceived. I rose immediately, lighted the gas, and noted the time as 9.54 1-2 (corrected to standard meridian). As soon as possible thereafter the washstand was drawn toward the centre of the room free of the walls, and a tumbler was nearly filled with water and placed upon it; and while waiting for the water to become still I gave attention to the swinging of the headboard of the bed and of the knife and six-shooter suspended from it. The head-board swung quite freely east and west; with each oscillation the suspended arms [gun and knife] swung free from the headboard and then struck violently against it; and although the manner in which they were suspended permitted of free movement north and south there was no tendency to swing in that direction.[8]

His meticulous notes on this tremor were a typical representation of his lifelong fascination with his chosen field. McGee was born in a log cabin near Farley, Iowa, in 1853. His father, a native of Antrim, Ireland, had left the Emerald Isle for America only two years before his son's birth. McGee's Kentucky-born mother descended from Scots-Irish and English ancestors who allied themselves with the patriots during the Revolution and later settled in Kentucky, Indiana, and Iowa. The future scientist's highly regarded skills in natural history, geology, and anthropology were self-taught because of the limited access to higher education on America's prairie frontier. He ultimately authored more than 250 scientific papers and studies, nearly 100 of which discussed geology, one of his chief passions. Although he never attended college, his colleagues deferentially referred to him as "Professor McGee," and he taught at the university level long before he was awarded an honorary doctorate by Cornell College in 1901.

As his train traveled through the pinelands of North Carolina, the scientist had a long list of things on his mind. Because of his work with the USGS, he was well acquainted with the geology of the southeast. However, despite his familiarity with the area he was to study, McGee knew that earthquake research was largely uncharted territory in 1886. Theories abounded, but no one knew what triggered earthquakes. There were no textbooks on the subject, no scientific protocols, and no tested guidelines for investigating such events. He had no idea where the epicenter of the Charleston event was located, how large the zone of heaviest destruction was, or even any assurance that food, water, or shelter would be available for himself and his colleagues. He would soon find out.

When McGee reached Florence, South Carolina, he changed trains and headed south on the Northeastern Railroad for the final leg of his journey. As his train reached a point about ten miles northeast of Charleston, he observed a number of craterlets on the eastern (seaward) side of the track. McGee soon learned from local geologists that a wide area around Charleston had extensive pockets of quicksand below the surface, and it was this water-saturated material that had been forced to the surface by the sudden, massive upward pressure of the shock waves from below.

"They [the sand-blow craters] are formed of light yellow sand which is sometimes 6 inches deep, and is spread over areas sometimes reaching 100 or 150 square feet," McGee wrote. "Strong streams have evidently issued from all of them; some, as indicated from the channels now dry, having been 1½ feet in width and 2 or 3 inches deep. A dozen or two were seen within a quarter of a mile; and as indicated by the mud stains on shrubbery and embankments the waters from them flooded the surface to a depth of 6 inches or more."[9]

Thomas Corwin Mendenhall, the next scientist to arrive, brought a wealth of first-hand earthquake experience to the team. A native of Hanoverton, Ohio, Mendenhall was born in 1841 and received his basic education in the public schools. Like Powell and McGee, he was a self-made scholar who charted his own higher education, took it upon himself to qualify as a physicist, and gained nationwide acclaim for his scientific

Thomas Corwin Mendenhall

studies, receiving honorary degrees from several respected institutions. As a professor of physics and mechanics, he served on the faculties of Ohio State University, the University of Michigan, the Western Reserve University, and the Imperial University of Japan. His work in the seismically active nation of Japan brought him into intimate contact with the geology of earthquakes. He was a professor of electrical science with the U.S. Signal Corps from 1884 to 1886, and he established several stations to systematically collect data on earthquakes. At the time he visited Charleston, he was a forty-five-year-old solidly built man with a high forehead and thin, graying hair. With his wire-rimmed glasses; full, meticulously trimmed mustache; and natty bow tie, he looked like the stereotype of an eminent, Victorian-era scholar.

Guests flee the Charleston Hotel

The first newspaper report of Mendenhall's visit appeared in the *Atlanta Constitution* on September 6. Unlike the *News and Courier*, whose reporting was generally straightforward, sober, and direct, the *Constitution*'s pages were often peppered with wry insights and openly humorous comments on serious subjects. Undamaged living quarters were scarce in Charleston when the federal scientists got there. With some misgivings, McGee, who was the first to arrive, had set up the team's headquarters in the damaged Charleston Hotel. With this as a backdrop, the *Constitution* took the opportunity to deflate the ego of the most-experienced member of the team.

> Professor Mendenhall, who studied the cause and effect of earthquakes in Japan for several years for the purpose of perfecting an instrument to foretell them, created some amusement in the Charleston Hotel last night. Since he had felt the effects of hundreds of earthquakes it was presumed he could not be induced to abandon his room at the hostelry. The building is four stories high, stucco and brick, and of antiquated architecture with very high ceilings throughout. He was proceeding upstairs to his room when the last shocks of the night came. The porter, who was showing him up, stopped on the faint approach of the sound, and when he decided what it was, dropped the pitcher of water, went leap-frog fashion over the professor and dashed down the stairs. Professor Mendenhall did not stand upon dignity, but rushed down and into the street. Later he called a carriage, and was driven to the battery and slept on the seat of the vehicle until daylight.[10]

On the same day the *Constitution* printed its jibe, Mendenhall granted a formal interview to the *News and Courier*, his usual serious demeanor intact. He noted that his role was that of a physicist, not a geologist, and that he was there "to study the physical causes of the earthquake, and particularly, to study the direction and force of the motion, amplitude of the vibrations, and, in general, questions of a physical rather than a geological character." To do this, he told the reporter, "it was desirable to get the movements and displacements that have taken place close to the surface of the earth, where masses of matter have been projected from columns, such as shafts in burying grounds, masses of marble, [etc.], in which twists have taken place." Mendenhall made many of his observations in the churchyards of St. John's Lutheran Church, the Unitarian Church, and the Second Presbyterian Church and at E. R. White's Marble Yard on Meeting Street.

He went on to express his unambiguous conviction that the earthquake was a natural event. He was distressed by many of the supernatural explanations published in the popular press. "Such agitation and instruction are calculated to unnerve and unsettle any class of people and to prevent them from rising to such an emergency as this. It ought to be understood that the whole thing is governed by natural laws and that nature works in a perfectly systematic manner," he said. "The effect of pronouncing the earthquake 'a punishment of God,'" he continued, "has been to terrorize the ignorant country people." Then Mendenhall added, "I was, however, surprised when I found that the same thing had been preached from the pulpit of this city."

The *News and Courier* reported on September 8 that Mendenhall "very largely increased his data in respect to the angles of displacement of monuments and their ornaments. Such observations are of the utmost importance from the fact that the only process of reasoning now is from the results of the cause." After four days of carefully observing and measuring the effects of the shocks on funeral monuments, brick pillars, and other manmade objects, Mendenhall returned to Washington.

Mendenhall arrived with Jack Hillers, Powell's trusted former oarsman, now an outstanding master photographer. A native of Germany, the hardy forty-three-year-old stood just under six feet tall and had blue eyes, sandy-red hair, a mustache, and a short goatee. Hillers had an "affable disposition and often ribald wit" and had honed his photographic craft in the Indian Territories of the West.[11] There he created hundreds of carefully composed and often awe-inspiring photographs of the natural wonders and of countless Native Americans. The Charleston-Summerville earthquake photographs he took are unadorned, photos-of-record views, as opposed to the artistic masterpieces he created in the virgin West.[12]

John Karl "Jack" Hillers

McGee and Mendenhall had a number of specific scientific objectives. Their primary mission was to collect and record data based on personal observations and measurements. They wanted to determine the exact locations of the earthquake's epicenter and hypocenter, the intensity of the shocks, and the severity and distribution of damage. Both men were physical scientists and did not set out to study the effect of the earthquake on the population or the operation of the city. Their reports did not mention the human aspects of the event, except for noting the recorded death toll as expressed by the city's official "Return of Deaths."

The federal investigators were traveling light. They had their clothes, money to pay for their travel and living expenses, and a large supply of notepads, pens, and pencils. The only specialized tools they brought were surveying equipment, a compass, and, for Hillers, a camera, a tripod, and a supply of pre-coated glass photographic plates. Hillers' work was evidently hampered by a lack of equipment or supplies. In the *News and Courier* of September 8, McGee complained, "[Hillers] had been much disappointed in his failure to receive from Washington certain photographic apparatus which he had proposed to employ." Nevertheless, Hillers was able to make a large number of valuable photographs to document the main geological features of the earthquake.

Because the scientific investigation of earthquakes was a new field, there were no detecting or recording instruments available for the scientists to bring or use. In a preliminary report published in the journal *Science* nine months after the quake, the authors, Clarence E. Dutton, who would write the final Charleston report, and Everett Hayden, who supervised the nationwide earthquake data and damage survey, expressed their frustration with the lack of technology. "To estimate the force of a shock," they wrote, "we have no better means than by examining its effects upon buildings, upon the soil, upon all kinds of loose objects, and upon the fears, actions, and sensations of people who feel it. In view of the precise methods which modern science brings to bear upon other lines of physical research, all this seems crude and barbarous to the last degree."[13]

The federal team quickly found that their work would be made easier by Charleston's treasure trove of documents and research aids. The Fireproof Building (now the headquarters for the South Carolina Historical Society) had survived with only superficial damage. It was then home to the County of Charleston's Register of Mesne Conveyances Office (the Register of Deeds office). At that time, South Carolina had the most comprehensive and best-maintained collection of colonial-era land records of all the original thirteen colonies. This put the land records of the city and the region from 1670 to 1886 at the scientists' fingertips and enabled the team to identify the exact location and land history of any damaged building.

Using these records, the team could see how both natural and manmade forces had enabled the earthquake to take such a violent hold on Charleston. The city they came to study is built on a peninsula that lies between the Ashley and Cooper rivers, which come together to form Charleston Harbor and then flow into the Atlantic Ocean. The peninsula is flat, with no part rising more than fifteen feet above sea level. The surface soils are largely sandy, and they rest on marl, a crumbly combination of limestone and clay.

A plan of Charles Towne based on a survey by Edward Crisp in 1704 shows the settlement flanked by the two rivers and by marshes. The peninsula itself was infiltrated from three sides by creeks, two of which nearly met each other from opposite sides. Only about a hundred yards of high land separated the creeks and kept the original city from being an island.[14] As the population of the city grew, much of the marsh to the south and west was filled in with sand to create new land. By 1739, the inhabited space on the peninsula had more than doubled. By 1886, about forty percent of Charleston's structures sat on landfill.[15] This distinction—made land (landfill) vs. solid ground (high land)—would be crucial in understanding why some buildings were heavily damaged or destroyed by the earthquake and why others—sometimes only a few feet away—suffered little or no significant damage.

Another invaluable research aid was the 1884 Sanborn Fire Insurance Company maps. Bound in a series of oversize volumes weighing more than thirty pounds each, these large-scale, highly detailed maps showed every city block in each of the city's twelve wards. They also indicated every masonry or wooden structure, showed all fences and walls, and provided the location of every pressurized fire hydrant, fire well, and firehouse, along with the number and type of fire engines in each firehouse. The maps were prepared with great care, and maintained at great cost, by The Sanborn Fire Insurance Company. Created for each of the largest cities in the nation, they were used to determine the insurance rates for any given structure based on its construction materials and proximity to high-risk (wooden) and lower risk (masonry) buildings.

With their research work cut out for them, the federal team had to find a way to offset the limited amount of time and staff available. McGee immediately recruited an array of local experts—most of whom were unpaid volunteers. Two well-respected Charlestonians, Earle Sloan, a geologist and skilled chemist, and Dr. Gabriel E. Manigault, an educator, a physician, and an eminent zoologist, were among the first to be chosen. In addition to providing his own extensive insights, Dr. Manigault sought out other highly regarded local informants. They included Frank R. Fisher, a cashier at the South Carolina Railway Company and an amateur astronomer and inventor; Edward Laight Wells, a cotton broker and prolific author; John Grimball, a former naval officer; Charles F. Panknin, a chemist, pharmacist, and professor of pharmacy at the College of Charleston;

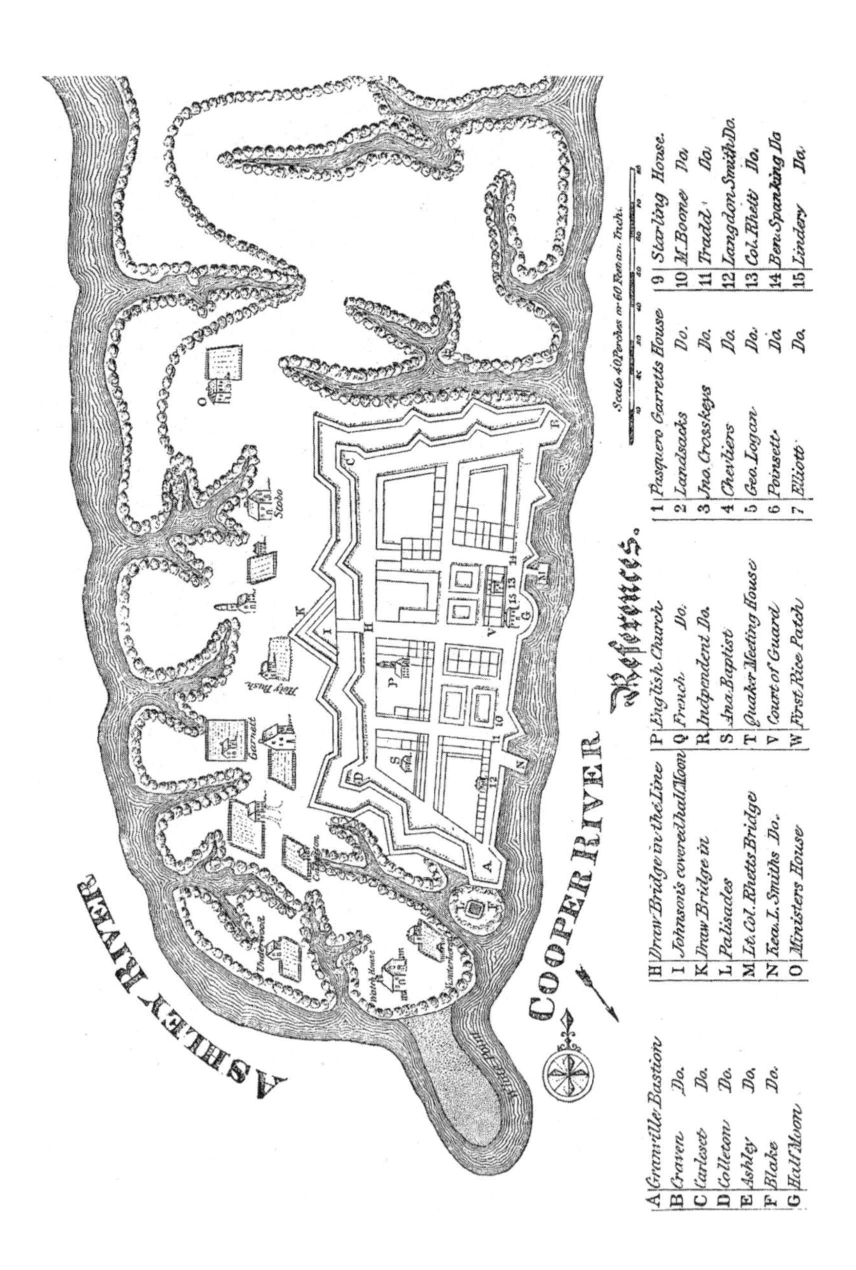

A Plan of Charles Towne in 1704

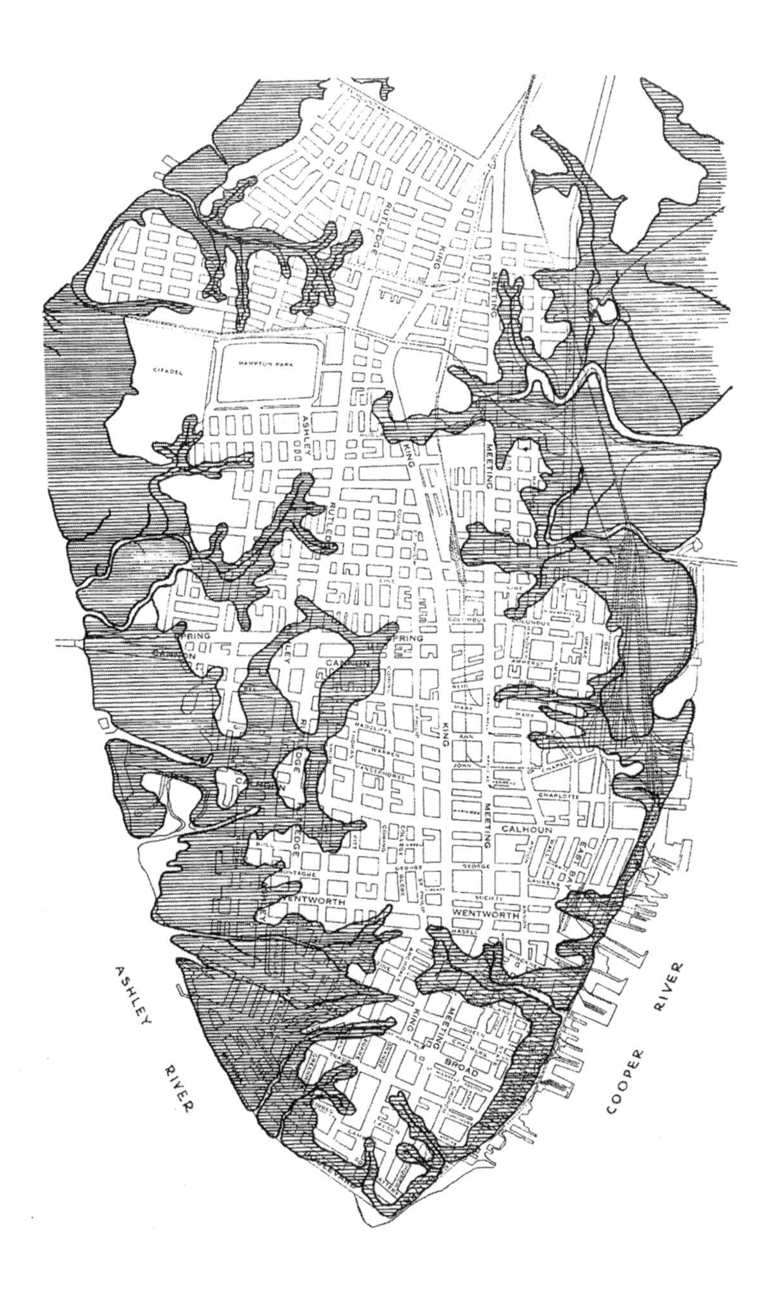

Charleston in 1886, showing the original firm ground (light) and later landfill areas (dark)

Thomas Turner, president of the Charleston Gas-Light Company; and John B. Gadsden, the Summerville resident who had extensive personal experience with the earthquake and its foreshocks. Both McGee and Mendenhall also consulted with Dr. Charles Upham Shepard, a planter with an interest in science who established Pinehurst, South Carolina's first commercial tea plantation, in Summerville. This quickly assembled group of two professional scientists and a veteran photographer, along with knowledgeable local scientists and informants and Carl McKinley, a gifted journalist, became the most highly skilled team ever to investigate a geological mass disaster in the United States.

In his week-long reconnaissance of the Charleston-Summerville area, lead investigator McGee filled his field notebooks with data and sketches, which became known as his "Itinerary Notes on the Earthquake of August-September 1886." In his brief but intense study, he was able to compile extensive data. He also won considerable admiration from the citizenry for providing the *News and Courier*, the *Atlanta Constitution*, and other newspapers with information—sometimes daily—about the scientific causes of earthquakes.[16]

In a city continually rattled by aftershocks, the ability to get information directly from a nationally recognized professional geologist was an enormous help in alleviating public panic. Until September 3, when McGee arrived and newspapers from outside the city again started arriving by train, Charlestonians could only read the local newspaper, draw their own personal conclusions about what had happened to their city, and guess what fearful things might lie ahead. One major cause for anxiety was that few people in Charleston—or the rest of the country—had any reliable information about the causes of earthquakes or what to expect after they occurred.

In 1886, serious readers among the general population did have some access to scientific data about earthquakes, but it was not widely available. In January 1886, *The Manufacturer and Builder*, a monthly national magazine of practical science, provided its readers with a brief, but accurate, description.

> The only settled facts about earthquakes are, that they are the result of some shock imparted to the rocks at a considerable distance beneath the surface, and that this shock reaches the surface in a series of concentric rings, all points on the circumference of each ring receiving the shock at the same moment, even though they may be hundreds of miles apart. In other words, all points at equal distances from the center of the earthquake receive the shock at the same moment. Although this is theoretically the case, according to well-known physical laws, still, in practice, the facts are somewhat different, for the shock is retarded or accelerated according as the rock opposes or favors the passage of the wave. The severity of the shock in a given place is dependent upon a variety of causes. These are, first, the strength of the original shock; second, the distance from the earthquake center; and third, the kind of rock on which one is standing.[17]

In Charleston and elsewhere, there was enormous speculation about the earthquake's origins, and what to expect in the future. The discussion came at a time when science was vying with fundamentalist Christian biblical teachings to be the dominant intellectual guiding force. An illustration from *Puck* magazine, published in New York, portrayed scientists as stalwart, learned progressive-thinking men defending the intellectual high ground. Their weapons were the searchlight of enlightened thinking and a Gatling gun loaded with bullets representing historical facts, archaeological facts, geological facts, evolution, and rational religion. They fought under the flag, "Think or Be Damned." From a castle labeled "Medieval Dogmatism" streamed their enemy: wild-eyed, fanatical religious zealots, carrying banners representing icons of biblical truth: Noah's ark, Jonah and the whale, the devil, and the fires of hell. The clerics rushed out the gate toward the scientists

The Last Stand: Science vs. Superstition

under the banner, "Believe or Be Damned." This pitched battle between science and religion for the minds of Americans continues unabated to this day.

Immediately following the earthquake, secular newspapers all over the country offered explanations by scientists and self-professed experts on the cause of the earthquake, but all of these *cognoscenti* were hampered by the fact that none of them had ever experienced anything like the massive geological event that had struck South Carolina. Reliable information was hard to come by. Even the world's top scientists disagreed strongly about the origin of the Charleston earthquake—and earthquakes in general. The *New York Star* quipped, "Theories, explanations and prognostications rain down upon the ill-fated city of Charleston like ripe apples before a north wind in a new Jersey orchard. The Mother Carey's chickens of science flit hither and thither over the country, screeching their prophecies of woe and desolation."[18]

Just before the Charleston quake, by interesting coincidence, the *Atlanta Constitution* ran a story entitled "The Primary Cause of Earthquakes." It described the work of Englishman John

Milne, a professor of geology at the new Imperial College of Engineering in Tokyo. He would soon become known as the inventor of a revolutionary new seismograph and as one of the greatest individual developers of tools used to observe earthquakes. Milne was quoted as believing that geological forces caused earthquakes, not cosmological or meteorological events. He stated that the majority of earthquakes were the result of volcanic action, but he noted, "Some earthquakes are produced by the sudden fracture of rocky strata or the production of faults."[19] These conclusions all proved to be valid.

Reports spread quickly that Charleston had sunk to the bottom of the Atlantic Ocean, which now reached as far inland as Summerville. During the first five days after the earthquake, rumors circulated in Marion, South Carolina, one hundred miles north of Charleston, that "[Charleston] was under water caused by an immense tidal wave sweeping in and destroying everything in its track." Ten days after the tidal wave rumor started, at least one explanation for its origin surfaced. In the area west of Charleston and east of the Edisto River, numerous craterlets erupted with water during the darkness of the night of August 31. People were sent fleeing from their homes and were "thrown into a state of utter consternation by finding water ankle-deep at the foot of their doorsteps, hearing the rush of water in the darkness all around them, and seeing it dripping from the trees while the stars were shining quietly from a cloudless sky overhead. It was this extraordinary phenomenon that, in the absence of communication with Charleston, caused the first reports to be sent out that the city had been swept by a tidal wave. Not knowing where the water came from, the terrified people naturally thought that the ocean had rolled in a great wave over the land and submerged the seacoast."[20]

Another rumor circulated that the ground under Charleston was rapidly sinking and that the only building still visible was the Custom House, whose top could barely be seen above the water. Then it was said that the whole city had fallen into its streets, and scarcely a human being was left. In Mt. Pleasant, rumor had it that, on the night after the fourth day, a tidal wave would strike the village; on the seventh day, the big shock would be repeated, only stronger; and on September 29, "with a final convulsion, time would pass into eternity."[21]

Theories circulated that the earthquake was volcanic in origin. These were fueled by reports that stinking mud, water, smoke, and blue flames had spewed from fissures and craters in Summerville, Charleston, and the surrounding area. The volcanic theory immediately gained local credibility, especially when stories spread that the erupted water was hot. A young Charleston physician voiced the opinion that the ejections of sand and water occurred because the earth's molten core had come into contact with underground bodies of water, producing steam that vented to the surface.

"Sulphurous fumes, blue mud, and stones were cast up from an active volcano which appeared near Summerville," one newspaper reported. "Scientific men differ entirely as to the character of the recent disaster. One denies that it had a volcanic origin. Another thinks that the phenomena indicated the approaching birth of an active volcano on the American continent."[22] Over the next few days, close observations of small geysers that were still flowing showed that the water was, at best, tepid, and certainly not superheated. The steam, smoke, and flames proved to be the products of overstimulated imaginations.

Major Joseph S. Powell of the U.S. Signal Corps, described as a "student of terrestrial and atmospheric phenomena," believed that the earthquake was caused by the recent spouting of a number of dried-up geysers in the far West. "The superheated gases and boiling water could not find a sufficient outlet there," he said, "so they worked their way across the continent until they struck the Atlantic side, where the earth's crust is thin and sensitive." Maj. Powell (not to be confused with explorer John Wesley Powell) predicted that shocks would continue in Charleston and the surrounding area and on the entire eastern slope.

Colonel Henry Thomas Washington, a California seismologist, had yet another theory. After consulting with his fellow scientists in Washington, he arrived at the conclusion that electricity had caused the earthquake. The *Atlanta Constitution* quickly noted of Washington, "He is one of the few seismologists who hold to this theory."[23] This concept also proved to be without foundation.

In the weeks and months after the first shock, scientific theories and oddball predictions kept flying around like mosquitoes on a rice plantation, but no one knew which expert to believe—or whether to believe any of them. Mt. Pleasant newspaperman Louis Beatty summed up everyone's frustrations. "There are strong scientific reasons to believe there will be another, though earthquakes, like geysers and volcanoes, are a mystery yet beyond the knowledge of scientists," he wrote. "There are theories in abundance, but each one lacks a link, or several links, to make the chain complete. Each theory leaves the ultimate cause as great a mystery as ever. What is lacking is a knowledge of the cause of the initial shock."[24]

The religious community had its own ideas about the origins of the earthquake. Some fundamentalist groups proposed that the South Carolina Lowcountry had provoked God's anger. The belief that the Lowcountry was responsible for bringing God's wrath down upon itself presented its proponents with a major problem. If the Carolinians of that time—who were generally conservative, church-going Christians—had indeed brought punishment upon themselves, someone would have to single out the grievous sins that that had so offended God and name the parties who committed them. None of the fire-and-brimstone proponents were willing to take the risk of casting the first stone. Nevertheless, on the first Sunday after the quake, many preachers spoke of the disaster as the act of a vengeful deity. The people of the Lowcountry, who suffered the brunt of the carnage, justifiably took offense at the suggestion that the earthquake was their fault.

The published responses in denial of this divine punishment scenario came not from agnostic scientists but from college-educated practitioners of "rational religion," notably Charleston's Episcopal ministers. In a newspaper report of a sermon he preached in the earthquake's aftermath, the Reverend Anthony Toomer Porter acknowledged, "Charleston had been visited in [my] lifetime with pestilence, fire, the sword, riot and cyclone and earthquake, [and] nothing [is] left now to endure but famine and a tidal wave." He continued, "Since these things had been concentrated on us, some [people] more pious than wise, might say we were a specially wicked people....[but the belief that] the earthquake was the direct act of God as a visitation for sins [is] not only un-scientific, but unreasonable."

About that same time, the Reverend Robert Wilson, rector of St. Luke's Episcopal Church in Charleston, circulated a flyer that echoed Porter's line of thought. Wilson was described by the newspaper as being "a theologian of broad and liberal creed, and also an amateur scientist, who has made a special study of the geological formations underlying Charleston and the low-country of South Carolina." Wearing his scientist's hat, Wilson, who wrote for *Popular Science Monthly* and *Lippincott's Magazine*, stated in the *News and Courier* that the cause of the quake was "a convulsion of immense magnitude, resulting from the cooling and shrinking of the inner crust of the earth....The centre is far away from here, and the seismic wave [from the believed convulsion] did not directly affect Charleston and Summerville." He concluded the technical part of his discourse by stating, "The earthquake is gone for, perhaps, the next ten thousand years.... All danger is long past. There will be no more earthquake." Then donning his theological robes, he said, "The man who calls this a 'visitation of God's wrath for sin' is a fanatic who ought to be silenced. God is a loving father, and not an executioner. Such talk is narrow, dangerous and false. If we suffer from the operation of the wise laws which govern the universe for general good, the compensations will come hereafter. This earthquake teaches the littleness of our power and knowledge, and our dependence on Him."

Ultimately, the majority of Lowcountry clergymen preached that although God was omnipotent and omniscient, the death, misery, and destruction emanating from the earthquake should not be blamed on Him. God, in their view, should be praised for sparing so many from its terrible effects, and the dead and the sufferers would receive their rewards in heaven.

To some, sectionalism played a role in the perception of God's intent. A woman from Flushing, New York, wrote that her minister had taken up a Charleston relief collection and said specifically that the earthquake "could not be looked upon as a visitation [of God], for no doubt there were better people in Charleston, or just as good and pious people as could be found anywhere in the broad land." She also noted, "That rather upset the theory of one of the Pharisees [in her congregation] who had spoken to me about it, and declared that the earthquake had been sent upon Charleston because she was the first to fire on the flag."

In mid-September, after reading of the Charleston disaster, Seventh-Day Adventists in Maine were preparing for "the termination of all things terrestrial." They quoted from the Bible that "there shall be famines, and pestilences, and earthquakes, in divers places," and from this they deduced that the earthquake was the fulfillment of the prophecy and an omen of things to come.[25] Following the same line of thought, *The Christian Herald and Signs of Our Times* used the Charleston disaster to help its readers visualize the terrors to follow.

> The distressing details reported during the past few days of the desolation and suffering in Charleston, S.C., from the effects of the earthquake of August 31st, enable us to form some conception of the almost universal affliction which will be produced by the great earthquakes that will accompany the dire events of the final consummation. John, describing it (Rev. 6:12-14) says, 'There was a great earthquake; and the sun became black as sackcloth of hair, and the moon became as blood; and the stars of heaven fell unto the earth, even as a fig tree casteth her

> untimely figs, when she is shaken of a mighty wind.'...It is a dreadful picture, with all the ghastly features of the Charleston catastrophe multiplied and intensified. Yet the Word of God says that the convulsion yet to be experienced will surpass even that.[26]

In many black communities, feelings of impending doom and the need to immediately repent ran deep. The newspaper reported on September 12, "At a wake held over the body of an old colored woman in this vicinity, a few nights since, the watchers were suddenly startled by a movement or movements, of the supposed corpse, which had already been placed in the coffin. On removing the lid...the corpse sat up and announced to the relatives, friends and acquaintances that 'she had been dead, but by God's command her spirit had returned to its earthly tabernacle, for the purpose of warning her fellow mortals that, on the 15th [of September], the final end of all things will come to pass, and time be swallowed up in eternity.'" The event was said to have given fresh impetus to religious revivals among black Christians.

The federal scientists did their best to counter the divine punishment and doomsday religious predictions. Through his openness, Professor McGee helped to quash many of these speculative assumptions, crackpot tales, and lurid rumors. On the other hand, he was apt to present his own preliminary guesses as if they were studied scientific conclusions. To help clear up misconceptions, the day he arrived in Charleston, McGee told an *Atlanta Constitution* reporter his theory of what triggered the disaster. First, he dismissed the idea that the earthquake was volcanic in origin, noting that the so-called volcanic mud and sulphurous waters were simply marl and sulfur salts released by the tearing motion of the earthquake and brought to the surface through newly created fissures. His explanation for the event was simple: There had been a landslide beneath the earth's surface when lighter rocks slid over a layer of granite deep under the ground. McGee believed that the aftershocks would rapidly dwindle away because, "as it took certainly hundreds and likelier thousands of years to acquire pressure enough to cause it to slip it will likely take as long to start again." He also noted that the severity and frequency of the aftershocks were proof that the worst had passed. McGee's theory of a subterranean landslide was quickly discarded, but it was, nevertheless, a preliminary attempt to explain the quake's origins based on the evidence he had collected.

When asked about a possible link between extremely strong earthquakes that struck Greece and Italy on August 27, 1886, and the disaster in Charleston, which happened just days later, McGee gave his opinion that there was no connection between the two events. As support for his contention, however, he stated, "Earthquakes occur more frequently at night than by day and oftener in winter than in summer. It has been found, too, that a series of cold summers is followed by earthquakes, and it frequently happens simultaneously in different parts of the world." The day/night and winter/summer attributions—and any significant links between weather and earthquakes—later proved to be groundless. However, his willingness to talk to reporters allayed many fears and ultimately contributed as much to helping Charleston as did the later publication of the research data he compiled.

The calming effect of McGee's explanations was severely tested by Dr. Ezekiel Stone Wiggins, who made preposterous and seemingly never-ending predictions of colossal disasters. In 1882, Wiggins, a Canadian physician, educator, and meteorologist, predicted a cataclysmic storm "which no vessel smaller than a Cunarder" could survive. It would start in the northern Pacific on February 9, 1883; cross over Toronto on March 11; ravage all of the North American continent; submerge the American coast; and inflict catastrophic damage on England, the south of Europe, and India. The newspapers gave the story such great play that fishermen refused to go to sea—and lost their income for weeks. The storm never happened, and Wiggins blamed "invisible moons" for corrupting his prediction.

Though thoroughly ridiculed in the press, he never stopped making his outlandish claims.

Wiggins became a menace to the mental health of Charleston when he claimed he had predicted the 1886 Charleston earthquake, though he never offered any proof. He then went on to predict that Charleston, Atlanta, Mobile, and New Orleans would be totally destroyed by a far more massive earthquake on September 29, 1886, at 2:00 p.m. His previous failures notwithstanding, the dire predictions gained a great deal of space in the popular press, and thousands of Charlestonians were again thrown into panic. When St. Michael's clock struck two on the afternoon of September 29—and nothing else happened—Wiggins again became the laughingstock of the responsible press. Mark Twain lampooned him in the *New York Sun* on September 29, making a spoof prediction in Wiggins' own jargon to announce the forthcoming total destruction of earth by a giant meteorite that would crash into Ottawa on October 3, 1886. The *News and Courier* labeled Wiggins a "monumental fraud," a "crank," and a "lunatic astrologer." The *Arkansas City Republican* simply wrote, "Wild geese have been observed flying south during the past week, and cool weather, particularly cool nights, has followed. As weather prophets, they are much more reliable than Wiggins."[27]

While William McGee was busy combating rumors and specious scientific claims, a scientific colleague was busy starting them. An astounding pronouncement was made by special dispatch to the *News and Courier* by Simon Newcomb, another member of the federal earthquake commission. Known throughout the world as one of the most extraordinary astronomers and mathematicians of his time, Newcomb had been named to the commission to determine the speed of the earthquake waves by determining the exact time they struck communities throughout the United States. However, Newcomb had no special knowledge of geology, earthquakes, or South Carolina. This did not stop him from issuing a strong and chilling warning. On September 5, in a moment of scientific hubris, he stated, "People camp on Sullivan's Island. This is dangerous. They are safer in their houses. A shock strong enough to level houses would probably be followed by a tidal wave that would overwhelm Sullivan's Island."

The people of the Lowcountry, most of whom lived close to the sea, were shocked and frightened by the eminent scientist's prediction, coming as it did from an official member of the earthquake commission. At the time, few local people knew that an earthquake had to originate offshore to generate a tidal wave or tsunami. The 1886 earthquake had its epicenter about twenty-two miles inland, thereby making it impossible to generate an inbound tidal wave.

Newcomb took a solid drubbing for his hasty statements from his onsite colleagues. The day after his warning was published, both McGee and Mendenhall bluntly refuted his statements and tried to assure nervous Charlestonians that fears of a tidal wave were groundless. After disclaiming the prediction, the two scientists said they were confident that the earthquakes were substantially over because water-spouting craterlets had stopped flowing. To Mendenhall and McGee, this meant that the abnormal stress and pressure, which created the mini-geysers, had been dissipated.

The following day, the newspaper said in an editorial, "We think that the telegram of Prof. Simon Newcomb in regard to a tidal wave and its probable effect at Sullivan's Island was very ill advised, though it was doubtless kindly meant. Prof. Newcomb had no authority to speak *ex cathedra* [from the throne] on the subject, for he knows nothing personally of the condition of things in South Carolina and along the coast, and his untimely and unnecessary warning served only as the occasion of alarm to people who were alarmed enough already."

Even McGee's boss was weighing in with unsupported speculations. While McGee was beginning his investigation, the international scientific community was clamoring for details of the great exotic earthquake. By September 3, USGS Director Powell had already received about one hundred reports from observers in various parts of the country.[28] That same day, the director of the prestigious British Association for the

Advancement of Science cabled Powell to "wire the chief facts of the earthquake."[29]

As McGee had probably not yet telegraphed his boss with any onsite observations, Powell responded to his British colleague's request with speculative, unverified information. Because of this, the statements made by the nation's top geological official that day proved to be wholly inaccurate, including the pronouncements that "the principal shock, causing destruction in Charleston originated in central North Carolina," and that "in the Carolinas [the earthquake] was accompanied by land slides."[30] Both statements were groundless and incorrect. Within days, debates about the origin of the earthquake filled the nation's newspapers, often adding more confusion than clarity to the public's understanding of what had happened. This tendency by noted scientists across the nation to offer hasty public opinions before making a thorough examination of the facts proved to be an ongoing problem during the first months after the quake.

In the initial days following the quake, William McGee was the only scientist who could speak with any real scientific authority about the catastrophe because he had been the first to arrive on the scene. In the first hours after his arrival on Friday, he met with city officials, took a brief tour of the city, and read the latest edition of the *News and Courier* for background information. This gave him a broad overview of the magnitude of the earthquake and its intensity. To determine the effects of the event on the surrounding countryside, he boarded a special engine furnished by the South Carolina Railway Company. Since Mendenhall and Hillers would not arrive until the next day, McGee traveled alone on his first visit to Summerville.[31]

In Hillers' absence, and later, when Hillers' full supply of equipment failed to arrive, McGee illustrated many of his observations with extensive pencil sketches, which would have provided a better record had they been photographs. In his final report to the USGS, McGee referenced the numerous photographs made by Hillers and a number of others taken by

C. C. Jones; Dr. E. P. Howland from Washington, D.C.; Aiken's J. A. Palmer; and Charleston's own George LaGrange Cook.

As McGee's South Carolina Railway engine left Charleston on its way to Summerville on September 3, he must have been surprised that the land in the narrow part of the peninsula, known as Charleston Neck, showed so little damage when compared to the central part of the city. Inside the city itself, two enormous sixty-foot-by-four-hundred-foot warehouses belonging to the South Carolina Railway, each containing several thousand tons of phosphate fertilizer, had been moved ten feet south of their original location and had been thrown vertically at least three inches.[32] But just a mile to the north, the rails and their roadbed appeared to be unaffected. Those facts immediately suggested to McGee that the type of ground on which a structure sat had a pronounced effect on the type and extent of the damage it received.

Within the first three miles outside the city limits, he noted, "The buildings are of wood and mainly uninjured, so far as external appearances indicate and very few of the chimnies [*sic*] are displaced. Moreover, the tall brick smoke stacks of the different chemical establishments [factories] here are uninjured."[33] He found no significant distortion of the roadbed for almost four miles outside Charleston.[34]

McGee had the engine proceed slowly so that he could closely examine the natural landscape, the railway roadbed, and the buildings he passed. As he continued to move away from Charleston, McGee started to see occasional fissures in the ground and in the ditches alongside the tracks from which water and sand had been forced to the surface, creating craterlets. "The fissures extend in various directions and strong streams of water have sometimes issued from them bringing up generally yellow and some blueish from greater depths. So numerous are the fissures and craterlets that in some cases the entire surface of extensive fields appears to have been flooded."[35] A bit further on, he noticed that rails were noticeably bent, and the track had shifted on a north-south axis, which opened the joints between the rails.[36]

In the area near the four-mile post, McGee noted increased fissuring of the ground and more craterlets. At the five-mile post, the track again showed great stress. There, the fish-plates—steel strips with bolt-holes in both ends used to bind the thirty-foot sections of rail together—had been torn apart when their steel bolts sheared off. These sections of rail showed seven-inch gaps.[37]

As McGee continued on towards Summerville, the track damage increased. Near the six-mile post, the fish-plates had again snapped, leaving gaps in the rails. In addition, the roadbed had sunk six inches.[38] By the seven-mile post, McGee noted that there were fewer ground fissures and that wooden houses had generally lost their chimneys. Around the nine-mile post, the ground was literally pockmarked with craterlets, and water was still flowing from them in the late afternoon of September 3. The area was also marked by numerous fissures, one of which was two feet wide, three feet deep, and twelve feet long.[39] One of the hundreds of craterlets near Ten-Mile Hill

A large craterlet on Ten-Mile Hill

station measured twenty-one feet across. McGee noticed that there was a strong taste and smell of hydrogen sulfide—the odor of rotten eggs—from the freshly spouting water. The same odor was noted in Charleston, where the fumes tarnished silver throughout the city in the first days of September.

Sinkhole under a house on Ten-Mile Hill

At the cabin of a black family that lived in the cratered area about three hundred yards from the Ten-Mile Hill railroad station, McGee's notes reflect that he found a crater about fifteen feet square and sixteen feet deep that "swallowed up two of the three brick piers supporting the house... and a peach tree 6 or 8 feet high with the exception of the topmost twigs. [When it was later photographed], the crater was sounded to the depth of 4 or 5 feet without finding bottom; and it was reported by the proprietor of the house and the adjacent store, Mr. Lee, that during the morning following the earthquake attempts were made to find the piers with a 15 foot pole, but that bottom was not reached."[40] The location of Ten-Mile Hill, northwest of the city, can now be found only on old maps. This area of extremely heavy seismic damage and severe cratering is now the site of Charleston International Airport and Charleston Air Force Base.

McGee asked the engineer to stop the engine numerous times as they proceeded onward. A few hundred feet beyond the nine-mile post, he found a serious flexing of the track, but the worst distortions were seen between the ten-mile and eleven-mile posts. Here the rails had been distorted in extreme ways. Not only was the track displaced laterally, it was also twisted severely in alternating directions, creating deadly S curves. McGee and others found that hundreds of yards of track had been shoved forcibly southward, and the roadbed was often alternately depressed or elevated.[41] Nearby was the site of one of the four serious railroad accidents that occurred on the night of the earthquake. In his field notes on this wreck, McGee wrote that the evidence indicated that the train was actually thrown off the rails by the second shock of the quake itself, not by broken or bent track. The bent and broken rails at the site might have resulted when the locomotive was hurled off the track.[42]

As McGee journeyed slowly up the South Carolina Railway toward Summerville, between Ten-Mile Hill and Ladson's Station, eighteen miles outside of Charleston, he found a dramatically fissured area. One fissure ran for several hundred yards through the forest, sharply changed course as it intersected a steeply sloped railroad escarpment, and ran an additional two hundred yards parallel to the track. Contrary to McGee's expectations, the narrow openings in the ground's surface seemed to point in all directions.[43]

Near the eleven-mile post, the track was wrenched apart, leaving gaps of seven inches between the rails.[44] At the fourteen-mile post, the retaining-wall bulkheads at each end of a thirty-foot railway trestle had been shaken so violently that "the earth in the embankment had been molded or pressed away from the bulkhead six inches at the northern end and five inches at the southern end of the trestle."[45]

At the little station called Ladson's, near the eighteen-mile post, the land slowly rose a few feet more above sea level onto firmer ground. There, the few houses near the tracks showed signs of intense vertical shocks. The chimneys had all been toppled, and the wooden structures were "severely shaken and strained in a northerly direction."[46] At Lincolnville, just south of Summerville, McGee found that many chimneys and some houses were thrown down.[47] The property damage increased significantly the closer the expedition got to Summerville. In the area between the eighteen-mile marker and Summerville itself, which lay just short of the twenty-two-mile marker, McGee found the rails in every condition except straight and level. One set of the twenty-nine-foot-long rail sections at the eighteen-and-one-half-mile point had been bent into an "S."

Professor McGee arrived in Summerville about 5:00 p.m. on Friday, September 3. Having already seen the massive devastation in Charleston, what he was to experience in Summerville would astound him. When the morning after the earthquake had dawned, the ruin and desolation of the village seemed complete. All of the houses had been destroyed or damaged. Virtually all the chimneys were down. "Walls were rent in twain, ceilings had fallen, and in numerous cases the houses that rested on wooden blocks or masonry were leveled to the ground. Other houses were split from top to bottom and left yawning chasms in the buildings." The falling ruins of a house had killed two black residents, Thomas Ellis and John Allen. They were Summerville's first earthquake fatalities.

By the time a reporter from the *News and Courier* reached the village on September 2, all the stores were closed, and the few people who walked the streets "wandered about in an aimless way not knowing what next to expect. All the inhabitants had abandoned their houses after the shock on Tuesday night, and but few of them had the temerity to return."

From the increasing levels of geological disturbance and destruction of manmade structures he saw as he approached Summerville, Professor McGee quickly realized that he had reached the settlement closest to the earthquake's epicenter. For all practical purposes, Summerville was Ground Zero.

11

Ground Zero

Send three hundred tents direct to this place. The shocks still continue, and many families are homeless.
—F. B. Fishburne

Summerville, South Carolina
Friday, September 3, 1886

The scene that faced Professor McGee as his locomotive arrived in Summerville on September 3 was altogether different from the calm, restful little village that greeted businessmen as they returned home on a normal day. As soon as he descended to the station platform, McGee was immediately struck by Summerville's natural beauty—and the savagery of the earthquake damage. From the station, he could look down Main Street across the long, broad village green, thirty yards wide, with trees and decorative shrubs along its length. The foliage looked normal, but all of the houses were abandoned, their occupants having left Summerville or fled into the streets. The stores near the station all showed the typical signs of the earthquake: broken posts, piles of bricks fallen from chimneys, building frames twisted and torn.

Word of the scientist's visit had been telegraphed to Summerville. Upon his arrival, McGee quickly stepped into a waiting carriage to begin his investigations. Everywhere he looked, he found devastation, but his trained eyes quickly noticed significant differences between the earthquake's effects on Summerville and Charleston. "The direction of destructive motion [in Summerville] was vertical rather than horizontal and…chimneys seldom appeared to have been thrown but to have been simply crushed and then to have toppled over."[1]

McGee estimated that three-fourths of all chimneys in the village were thrown down, and about three-fourths of these were thrown either to the north or the south.[2] The portions of the chimneys located below the roofs suffered the most serious damage. "A very large number of them [chimneys] were crushed at their bases; the bricks as well as the mortar being disintegrated and shattered, allowing the whole column to sink down, carrying fire-places, mantels, and hearthstones with it through the floors. All this indicates a direction of motion more nearly vertical than horizontal."[3] This was a clear indication that Summerville was closer to the epicenter than Charleston because it had received the main force of its shocks from almost directly below. In Charleston, the chimneys had been broken off chiefly by lateral, shaking motions rather than vertical thrusts.

If McGee held any notions that the earthquake was done with South Carolina, those beliefs were immediately dispelled. No sooner had he stepped off the train, which ran through the center of the village, than he was greeted with the explosive sounds and shaking that had terrified Summervillians for three days since the first shock. "Detonations were heard at intervals averaging perhaps half an hour," McGee wrote. "From that time until 9.30, occasional and very slight spasmodic tremors of an instant's duration accompanied the detonations. I endeavored to determine the direction from which the sound appeared to come and had others make the same effort; but the result was that no two individuals agreed as to the direction whence the sound proceeded.... The detonations were very much like, but somewhat more muffled than, peals of thunder at a distance of a half mile or more, or perhaps more like the discharge of a blast in a mine or quarry at a little distance. It is my impression that the sound was about as grave as the ear can conceive."[4] As he proceeded with his survey, McGee made a careful record in his field notes of the numerous shocks felt in Summerville following the main one at 9:51 p.m. on Tuesday night. The people in the village had a right to be terrified. There had been seventeen additional shocks between the first one on Tuesday evening and 8:24 a.m. on Wednesday morning.[5]

In Charleston, the shocks were usually preceded or accompanied by roaring or rumbling, but in Summerville, the soundscape was different. Loud, muffled explosions, which were sometimes followed by shocks, became the unique auditory "fingerprint" of the Summerville earthquake experience. Of Summerville, the scientists noted, "Although a roaring sound is the almost invariable accompaniment of an earthquake shock exceeding a very moderate degree of vigor, the sounds at Summerville appear to have been highly exceptional and perhaps unprecedented. They were heard throughout the entire epicentral tract, though it is difficult to determine whether they were as loud and frequent in other parts of that tract as at Summerville."[6]

Unlike Charleston, which had thousands of brick buildings, virtually all the structures in Summerville were made of wood. Most of the buildings were usually at least partially surrounded by a piazza supported on relatively slender columns, as the piazza supports did not have to bear much weight. The houses themselves were supported by posts driven into the ground or by brick piers, which raised the house from four to six feet above the ground to provide good air circulation. The chimneys were usually independently supported by arches or piers built up from the ground.

McGee and his colleagues believed that this style of elevated construction was fodder for a lateral shock because the slender piers or posts would be ripped out from under the larger, heavier building they supported. The resulting disruption would cause the unsupported building to crash to the ground. Homes would be split from top to bottom and left with yawning chasms. Nine houses, those belonging to Gen. John C. Minott, Captain Vose, Mrs. B. F. Tieghe, L. DeTreville, E. J. Limehouse, Percy Guerard, Benjamin Perry, the Nettles family, and Edward Fishburne, had been completely demolished or severely wrecked this way.[7] Mr. Brown's "commodious hotel," was later added to the list.[8]

Mr. Brown's "commodious hotel"

Although most Summerville houses had moved on their supporting piers, the majority did not fall. Instead, their piers were severely damaged by nearly vertical movement from below, rather than from lateral forces, which would produce shaking. "The brick piers were found to be more or less disintegrated, as if pounded; while the wooden ones were driven deeper into the soil, as if they had been hammered by a pile-driver. Many of them were inclined, though still supporting the houses."[9]

The Minott house was a worst-case example of the damage. It looked as if it had been hit by a tornado. The porch had collapsed entirely, and the rest of the one-and-a-half story cottage was ripped off its piers and smashed into the ground, leaving it tottering thirty degrees from vertical and ready to crumble. The chimney and fireplace were pulverized by the shockwaves, peeled off the side of the house, and crashed to the ground, throwing bricks twenty-five feet from the foundation. An addition had been separated entirely from the cottage itself, and half the shingles had been stripped from the roof.

Professor McGee was apparently undaunted by the thought of being injured or killed while inside a damaged home. On the evening of September 3, he was the overnight guest of a

Gen. John C. Minott's House

Mrs. Johnston in Summerville.[10] Despite the fact that her wooden house was built on brick piers about six feet high, which, he found, were "considerably crushed," McGee slept inside. During the night, he gained more first-hand knowledge of a Summerville aftershock.

He was awakened by a huge roar, and the house shook. The vertical vibrations were so intense that his mattress bounced, pummeling him in the back three times every second. The bed leaped up and down so quickly that its feet made a continuous rapping sound on the floor. Despite having been trapped on the furniture equivalent of a bucking bronco, which would have frightened a lesser man to distraction, McGee later described the event with his usual dispassionate, analytical precision.

> The blows produced by the falling bedstead were, judging from the sound, somewhat unequal. Other articles of furniture in the room were tossed up and down in like manner, some plastering fell, the entire building appeared to creak and groan, windows and blinds rattled, and above all rose an indescribable dull roaring sound as if from a very rapid succession of detonations. This phase of the shock lasted but a

short time—15 or 20 seconds by estimate, when the sound rapidly died away and the vertical motion gave place to a peculiar twisting one combined with the vertical in such manner that while I was yet thrown upward by a rapid succession of impulses, my head was carried to the west and my feet to the east through an angle of perhaps 15°. This also gave way to a gentle rolling movement east and west unaccompanied by sound. The whole disturbance from the time of awakening (which was in all probability directly simultaneous with the commencement of the shock) did not last more than 20 or 30 seconds. The hour was noted and found to be 11.02 corrected to 75th meridian. It was then found that a considerable quantity of water had splashed out of the tumbler, and that the slender glass vial was yet upright though it had evidently danced over a considerable portion of the dressing case and was more than a foot from the point at which I placed it before retiring. Other light articles had also moved around not only on the dressing case but on the wash stand and floor; and a chair (previously oriented) was found to have suffered torsional displacement with the sun (i.e., in the direction which I myself was turned) through about 45°.

As the roar of the shock died away it was almost immediately followed by a murmuring sound rapidly increasing in loudness which was soon traced to the people and animals of the town; dogs barked and whined; chickens cackled; ducks quacked; horses neighed; and above all rose the sound of voices of the colored population, screaming, shouting, praying, sometimes in articulate but generally in inarticulate tones, the whole blending in an indescribable wail.[11]

After that, detonations, sometimes accompanied by slight tremors, occurred frequently during the night, awakening McGee from time to time. Then, a little after 4 a.m., a sudden, loud detonation, like a clap of thunder about five hundred yards away, got his attention. The tremor that followed was less intense than the first one, and although the building creaked and things rattled, nothing shifted or was lifted. Again, the scientific observer in McGee came to the fore.

> The entire duration of the shock from the first indications to the last faint east and west undulation could not have exceeded 20 seconds, and was probably less. The time was noted as soon as possible and found to be 4.17 corrected. This shock was followed by the crowing of cocks, the cackling of ducks and general animation among the domestic fowls and animals.... In Mrs. Johnston's house...the injuries were confined to the crushing and toppling over of the chimneys, the crushing of the pillars, beneath the corner posts and other heavier portions of the building, the cracking and shaking down of plastering particularly that from ceilings, and the breaking of the upper hinges of several doors. A number of fissures occurred in the yard.... From [one] water flowed with such violence as to flood an area of at least an acre several inches deep.[12]

When he arose the next morning, McGee discovered more houses that had been thrown off their supporting piles, smashed onto the ground, torn apart, fallen over, or crushed by the weight of their collapsed upper stories. Captain Carsten Vose's inn and residence, which was flourishing as a guest home by 1860, became the poster child for the Summerville earthquake damage. Its rear piers had been crushed, but two of the three front piers remained intact. This threw the entire two-and-a-half-story wooden building backwards, splitting the

Capt. Carsten Vose's inn and residence

house open at the seams and shearing off a two-story attachment, whose piers also collapsed. Had all of the piers been shattered, the house might have simply dropped to the ground with much less structural damage. Shortly after Charleston photographer George L. Cook made his photo of five women posing with the wreck (probably in early September 1886), the house was declared unsafe and irreparable. The Vose building was quickly stripped of its valuables, abandoned, and left to deteriorate.[13] It sat behind present-day Ambler Hall, the parish house for St. Paul's Episcopal Church.

Earthquake bolts in St. Paul's Episcopal Church

Invigorated by his close scientific encounters with the great quake, McGee set off to examine more damaged buildings in the village. He diagrammed the destruction of Mr. W. S. Hastie's house and made copious notes. Then, he proceeded to one of his major targets, St. Paul's Episcopal Church, one of the first substantial buildings erected in Summerville. The spiritual heir to the parish church of St. George, Dorchester, St. Paul's first sanctuary was completed in 1832. In less than thirty years, the congregation outgrew its original home, and the present wooden-frame church, built in the Greek Revival style, was completed in 1857, with a twenty-foot extension added in 1878. McGee noted, "The Episcopal church, in the south-western part of the town, a wooden structure 30 by 50 feet, resting on 36 piers of brick, each 2-½ feet square and 4 feet high, fronting N. 70° E., has been displaced northward 2-½ inches at the west end, 1-¾ inches in the middle, and 1 inch on the east end." None of the nine pillars under the south wall was displaced, but one or two under the north wall and several directly under the church had moved. Several had crushed tops, and a few had fissures running obliquely from south to north.[14]

The tombstone of Dr. Benjamin Burgh Smith

The building itself suffered significant structural damage, as witnessed by the earthquake bolts that were used to square and stabilize it afterwards. Three iron rods were run under the floorboards of the sanctuary, but the only signs of their existence today are gib plates and nuts on the outer walls, three feet above ground level. Inside the sanctuary, another three rods run from one wall to the opposing wall, each piercing two arched supports, which are held up by a colonnade of Doric columns. St. Paul's Church is one of the finest South Carolina examples of how earthquake bolts, usually associated with masonry structures, were also used to resquare, repair, and reinforce wooden buildings damaged by the earthquake.

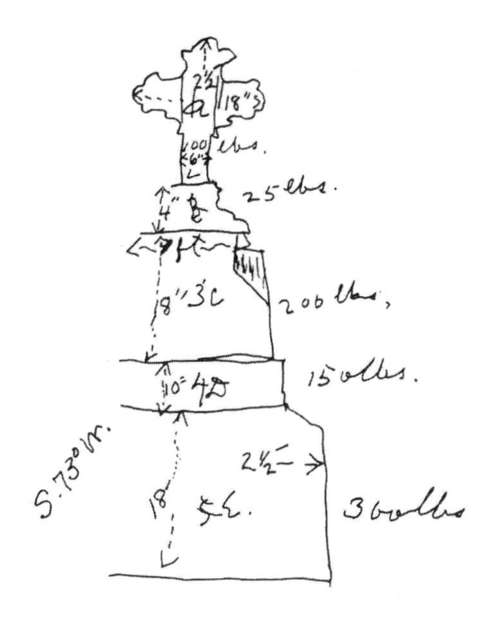

William McGee's field drawing of the tombstone

In St. Paul's churchyard, McGee found evidence of rotational twisting produced by the shock waves. During the quake, a one-hundred-pound stone cross ("too heavy," McGee felt) atop the burial monument marking the grave of Dr. Benjamin Burgh Smith broke off and fell directly to the west. Because McGee's photographer was not available, the scientist made a detailed, measured drawing of the monument, noting that one section of its base had been rotated two-and-one-half inches clockwise.[15]

To document the damage in Summerville in his final report to the USGS, McGee ultimately used photographs made by Dr. E. P. Howland, a Washington, D.C., dentist who was also a skilled photographer. Howland was one of a half-dozen photographers who roamed the epicentral tract of the earthquake in September and October, making photographs to sell to publications and tourists.

The parish church of St. George, Dorchester

A few miles away, McGee found evidence of the earthquake in the abandoned village of Dorchester. The shocks had split the tower of the parish church of St. George in two vertically, hurling large blocks of brick and mortar in four directions, with the largest blocks falling to the northeast and southwest. Although the tower was not completely destroyed, the quake left little of the once-handsome church intact,

Earthquake crack in a tabby wall at Fort Dorchester

and it left behind indelible clues about the direction the shockwaves had taken.[16]

The earthquake also produced two remarkable cracks in the nearly impregnable tabby walls of ancient Fort Dorchester, which were two-and-a-half feet thick and seven feet high. About a week after McGee left Charleston, geologist Earle Sloan wrote in his field notes, "Old Fort walls of shell concrete 8 ft high with thickness battered from 3 ft at base to 2 ft at top cracked through E[ast] wall at SE corner also badly cracked in two places at N.W. corner." The cracks showed that the walls had been shifted about four inches, suggesting an extreme shearing motion at nearly right angles to most of the rest of the shock damage in Summerville.[17] This would later be a significant clue in solving the mystery of where the shocks came from and determining the shape of a geological fault line that was later shown to run through the Summerville area.

McGee spent a good part of September 4 touring the phosphate mining area along the Ashley River with Colonel Gregg, the manager of a phosphate mining plant. From Summerville, the pair traveled eight miles south to the phosphate mining works, crossing the river to its west bank in the process. McGee saw forty to fifty fissures along the way, some of which were still producing water. Unlike the water ejected from the craterlets near Ten-Mile Hill, which tasted and smelled of sulfur, Summerville's craterlet water was clean, tasteless, and odorless. At the phosphate plant, the well-made brick foundation was intact, but two chimneys approximately twenty-five feet apart had been thrown down. McGee documented the chimney destruction patterns at the plant and at Gregg's house and studied the damage to the railroad tracks on the site.

A damaged house at Lincolnville

That afternoon McGee and the colonel traveled back through Summerville en route to Lincolnville, a hamlet two or three miles to the southeast, and the only other settlement of any size in the area of the epicenter. Lincolnville contained a few hundred residents, several well-built wooden cottages, and numerous cabins occupied by the poor. The violence of the shocks there appeared to be more intense than at Summerville. A larger proportion of the buildings were wrecked, and in some cases, the destruction was more complete. Signs of damage were similar to those in Summerville—evidence of "vertical movement of great power," along with some horizontal swaying—and the piles on which the houses rested were hammered into the ground.[18]

In the outlying areas, McGee also inspected Prof. Charles U. Shepard's tea farm plantation and returned to Ten-Mile Hill Station, where he examined the site of the fatal northbound South Carolina Railway train wreck and explored dozens of craterlets. Satisfied that his tour of the areas in the epicenter was complete, at least for the moment, Professor McGee again boarded the train and returned to Charleston to finish gathering data and to compile his copious notes and sketches.

Back in Charleston, McGee seemed to be everywhere at once. He wasted no time during his visit and was continually on the move, observing, measuring, sketching, and making notes. He studied several dozen residences and examined water flowing out of ground fissures on Savage, Queen, and Tradd streets; thrown parapets and walls at 157 Tradd Street; a fractured wooden house on Council Street; the thrown-out gable walls of a house on the corner of Queen and Mazÿck streets; damage to the Catholic bishop's residence on Broad Street; and the dislocation of chimneys on Beaufain Street. He measured twisted monuments at the Powell Marble Yard near St.

Michael's Church and at White's Marble Yard; examined the damage to two large granite gateposts at George W. William's Meeting Street mansion; studied the cracked flagstone sidewalks along the Battery and at 84 King Street; and made a detailed sketch of the unique X-shaped stress fracture patterns that typify how walls with multiple window openings crack when a building is subjected to a strong twisting motion.[19] He also visited Kahal Kadosh Beth Elohim Synagogue on Hasell Street, the First (Scots) Presbyterian Church on Meeting Street, and the Unitarian Church on Archdale Street.

McGee appeared in many of the photographs that Jack Hillers made to document the earthquake's aftereffects. A fastidious man, his work attire consisted of a three-piece black suit with buttoned vest, black shoes, a white shirt, a black bow tie, and a black bowler hat. Given that the high temperatures in Charleston during his September visit averaged eighty-two degrees, he must have been sweating most of the time. In almost all the photographs in which he appears, he is holding his field research notebook and a small leather case, which contained his compass and writing instruments.

William J. McGee in the churchyard of the First (Scots) Presbyterian Church

Although his contributions to the scientific study of the earthquake were invaluable, McGee's greatest gift to Charleston was something intangible—hope. His willingness to talk to the press proved to be a major asset in a region wracked by insecurity, for he was able to give the people some sense of authoritative assurance that things were getting better, not worse. In addition to appearing in the local papers, McGee's opinions reached people in affected areas in other states. Captain Evan P. Howell, editor-in-chief of the *Atlanta Constitution*, interviewed him, and Howell or one of his reporters accompanied McGee on his trip around Summerville on September 4.

On September 5, the *Constitution* reported that McGee had returned to Summerville. Howell provided his readers with a detailed report of McGee's research at Ten-Mile Hill, noting that the scientist collected extensive sand and water samples and made detailed notes about the craterlets in the region. He quoted David Ebaugh, a Summerville resident and owner of Ebaugh's Phosphate Works, on the northeast bank of the Ashley River, who stated that McGee "talked freely with the gentlemen there."[20] Ebaugh was quoted so much that he seems almost to have become McGee's unofficial spokesman. The scientist reportedly told him that "the ornamental work and gingerbread work in Charleston was, in many instances, out of proportion to the size of the building and these parts would be apt to fall and carry away portions of the general structure." This is exactly what had happened, of course, but McGee had already seen that for himself when he arrived on September 3.

McGee also was reported to have said that, when he left Washington, he thought he would have very little difficulty in determining the approximate cause of the earthquake, but he confessed to have been much puzzled by his observations.[21] McGee was confused because the fissures were not uniform in their direction. "He did, however, give the opinion that the shocks were the result of local landslides," Ebaugh told the newspaper, but "by the term 'local' he does not mean that the shakes can be traced to any particular place at Summerville, Charleston, or elsewhere, and remarked that there was no connection between the shakes and volcanic action. His impression is that the area covered by the landslide theory has been from forty to one hundred miles under the bed of the sea."[22]

Ebaugh went on to say that McGee "was confident that the worst was passed." This statement, the *Constitution* noted wryly, "was made prior to the severe shock at 11 o'clock last [Friday] night." Nevertheless, the paper went on, "Professor McGee's remarks here had excellent effect. Up to yesterday about 1,000 persons, or one third of the whole population, had left Summerville, and it is expected that Professor McGee's statement will stop any further exodus."[23] The *Constitution* was correct.

On September 5, Mendenhall and Hillers accompanied McGee back to Summerville via the South Carolina Railway, examining the Ten-Mile Hill area and Lincolnville as McGee had done previously. "Some strange freaks of the earthquake were found and photographed," the *Constitution* wrote. It noted, "Professors Mendenhall and McGee take friendly issue as to the cause of the earthquake. The latter maintains that they are the result of land slides, while the former believes that they are the result of a readjustment of the earth's crust to the basin of the earth, or a conforming of the exterior to the interior of the earth, and points to the geysers as evidence that the earth is settling and the surface lowering."[24]

Mendenhall, the physicist, then went on to make prognostications about aftershocks. "Professor Mendenhall anticipates another shock tonight, between 10:30 and 2 a.m., when the tide, which rises here to a height of six feet, is at full flood. He thinks the weight of the water along the shore line of Charleston, which is over five miles in length, is the immediate determining power or cause of the earthquake. The fact of the tide being in, and the pressure off the coastline, he says, might also provoke a shock."[25] Mendenhall offered no explanation as to why the weight of the tides—which came and went twice a day—would not also trigger shocks at each following high tide or along other parts of the South Carolina coast. Later research eliminated any possible connection between tides, weather, and earthquakes.

In less than a week, McGee, Mendenhall, and Hillers compiled a staggering amount of data. They worked at a frantic pace, seeking out local experts as guides, visiting numerous sites, taking extensive measurements and field notes, and making time to grant interviews to reporters. In the lee of their whirlwind visit, the three men left behind their team of talented local experts, ready to complete the work that the federal scientists had begun: to accurately locate the epicenter of the earthquake, its hypocenter deep in the earth below, and the geological mechanism that triggered it.

Just prior to his departure at noon on September 8, William McGee wrote a reassuring and compassionate letter to the *News and Courier*'s editor, Francis W. Dawson.

> Before taking final leave of your city, I desire to add another word of reassurance to the good people of Charleston in addition to what I have already said. I feel quite satisfied that there is no reason to fear future shocks of greater severity than those which have occurred during the past sixty hours [September 6-8]. I am, therefore, decidedly of the opinion that the citizens can safely return to wooden houses in which the chimneys or plastering are not so severely shattered as to be in constant danger of falling.
>
> Many of the less seriously shaken brick houses, too, are perfectly safe. I have myself slept in the third story of a brick house [the Charleston Hotel] every night since my arrival, with the exception of one night spent in Summerville, where also I slept in a house, but it cannot be too often repeated that, in case of a shock, however severe, *those indoors should remain there until the disturbance is over.* I also desire to reiterate my firm conviction that there is not the slightest danger of tidal waves, volcanic eruptions or other catastrophic disturbances. There is every probability that slight shocks will continue for some days; indeed such shocks ought to be regarded as favorable indications rather than otherwise, as they indicate that the accumulating stresses, to which the earthquake is so far due, are relieved from time to time.

> As a resident of a distant State I beg, dear sir, to express my heartfelt sympathy for stricken Charleston, and once more repeating my firm conviction that the worst is over, I am, yours obediently, W. J. McGee, Geologist, U.S. Geological Survey.

In the wake of the visit by the science team, and amid the detonations—as many as a half-dozen a day, and nearly as many aftershocks—Summerville was determined to rebuild. The evening that Professors McGee and Mendenhall left the village for the last time, Colonel Averill sent a telegram to the *News and Courier*. "We have passed a quiet day here. A better feeling prevails among our people, and if we escape any shocks to-night many who are camping out will return to their houses. Many of our citizens will commence repairing their houses on Monday. The Council met this evening and the intendant [mayor] announced the receipt of valuable aid for the sufferers from Orangeburg, Bamberg, Williston, and other interior towns. A relief committee was appointed, with the intendant as chairman, who will give prompt relief to the needy. While our town is very badly damaged, we don't propose to abandon it, and hope soon to show that entire confidence has returned, and that the village in the pines is firmly on its feet again."

Although thousands had fled the Charleston-Summerville area, the informative discussions of the earthquake by the visiting federal scientists, coupled with aggressive rumor control by the Charleston and Atlanta newspapers, quickly reversed the flow of the outbound human tide, and people started returning home to literally pick up the pieces of their lives. And the search for the scientific Holy Grail continued.

12

IN SEARCH OF THE HOLY GRAIL

The work I have found to be hydra-headed.
—Earle Sloan, Assistant USGS Geologist

Charleston, South Carolina
September 8, 1886

The extent of the devastation in Charleston and Summerville and the enormous size of the earthquake's epicentral tract amazed its chief investigators. Realizing that the task facing him was much larger than anyone had foreseen, William McGee worked at a fevered pace trying to find the geological Holy Grail of the earthquake: its epicenter. Hampered by the lack of adequate time, his work was confined solely to Charleston, the Summerville-Ashley River-Lincolnville area, and the area adjacent to the tracks of the South Carolina Railway. In all, McGee put in only five full working days of research before returning to his duties in Washington.

His colleague, Thomas Mendenhall, had his own goals: studying the wave motion produced by the earthquake, determining where the earthquake had come from, and calculating how fast it traveled outward through the earth. Ultimately, Mendenhall covered roughly the same territory as McGee, but he only worked a total of three full days before he left. Neither scientist walked the land southwest along the Charleston and Savannah Railroad tracks nor up the coast on the Northeastern Railroad. The only glimpse they got of the region northeast of Charleston was the craterlet-pocked countryside they saw through the window of their passenger train as they arrived and departed on the Northeastern. The on-site work of the two official members of the earthquake commission gave them only a preliminary overview of the types and severity of the damage. They did not have time to range far enough into the field to determine the size or shape of the epicentral tract, the area of maximum damage.

Despite the fact that the Charleston earthquake was the greatest national seismic event since the New Madrid series of 1811-1812, McGee and Mendenhall were able to spend less than a week in the epicentral area. Although the 1886 earthquake was a major phenomenon, it struck when most of the Geological Survey's talent, time, and resources were being invested in the West, with its vast untapped mineral resources that were so important to the developing American economy.

From the outset, McGee realized the unique scientific opportunity offered by the earthquake. The magnitude of the research needed to answer the key questions about the event was enormous, and McGee was exceedingly frustrated by the amount of fieldwork left undone when it was time for him to return to Washington. With his usual analytical precision, he reviewed his list of qualified local scientists to find someone to lead the in-depth investigation needed to successfully complete the mission to find the epicenter. Salvation was named Earle Sloan. A hard-working, twenty-eight-year-old, unmarried mining engineer and geologist, the South Carolina native came with an impressive list of credentials. For McGee, Sloan was the answer to his prayers. For Sloan, McGee's assignment was a once-in-a-lifetime opportunity to explore one of the greatest mysteries of the natural world. Sloan worked closely with the federal scientists

Earle Sloan

in Charleston, and a few days after McGee arrived back in Washington, Sloan was appointed an assistant USGS geologist. His mandate was to survey the entire epicentral tract, determine its borders, and find the epicenter, the scientific Holy Grail. The epicenter, the spot directly above the place in the earth where the earthquake originated, would lead the investigators to the hypocenter, the belowground focus of the earthquake.

The assignment was an enormous honor for Sloan, and he appreciated the magnitude of the trust invested in him. In return, he unleashed his boundless energy on the project throughout September and October 1886. McGee's choice was well made. Sloan proved to be an "even more serious and thorough investigator of the Charleston earthquake than was his supervisor."[1]

The son of Col. John Bayliss Earle Sloan and Mary (Seaborn) Sloan, Earle Sloan was born at Cherry Hill Plantation near Pendleton, South Carolina, in 1858. His father was a planter and cotton factor (broker), and served as colonel of the Fourth Regiment, South Carolina Volunteers, during the Civil War.[2] His ancestors represented the old breed of Southern, plantation-based, slave-owning landed gentry.

Born only three years before the start of the Civil War, Sloan was the product of the new, post-war generation: an educated analytical man of science, not a student of Homer, Plato, rhetoric, and the classics or a planter devoted to agriculture. His meticulous observations would yield a wealth of information critical to answering the most important scientific questions posed by the earthquake: What caused it? Where was the epicenter? The hypocenter? Or—and this soon became a tantalizing question—were there more than one of each?

In his youth, Sloan was fond of reading and outdoor activities but was "possessed of an ingrained aversion for school."[3] His father strongly influenced his studies, and Sloan soon devoted himself to gaining a solid education, the inspiration being his interest in nature. He attended country schools and later enrolled in Prof. Augustus Sachtleben's highly regarded private academy after his parents moved to Charleston.

Too young to serve in the Civil War, he entered the Carolina Military Institute in Charlotte, North Carolina, in 1878. He ultimately graduated from the University of Virginia in 1881, having been inducted into Phi Beta Kappa, the oldest and most respected undergraduate honors organization in the United States. His interests included petrography, engineering, and mining, and he studied chemistry and geology at a post-graduate level. After leaving school, Sloan became a consultant. For one project, he made detailed surveys of the routes of the Charleston and Savannah Railroad and the South Carolina Railway and conducted a survey of Summerville. This work gave him an intimate knowledge of the area affected by the earthquake and made him eminently qualified to become the chief on-site researcher.

A man with good financial instincts, Sloan invested his money wisely. He became involved in the phosphate mining industry and was a director or officer of three major Charleston businesses: the Edisto Phosphate Company, the Carolina Bagging Manufacturing Co. (which was severely damaged in the earthquake), and the People's National Bank.[4] At the time of the earthquake, Sloan was working chiefly as a chemist in the phosphate industry. He was noted as being a man of "pronounced and independent convictions," and in politics, "believes the question of white supremacy to be paramount in importance, appreciation of all other political and issues being largely suspended by the exigencies of the absorbing question of race preservation." His long-standing interests in hunting, fishing, boating, and horseback riding made him well suited to rigorous work in the outdoors.[5]

After his colleague, McGee, left Charleston on September 8, Sloan set off the find the Grail. He concentrated on two avenues of investigation. He would examine damaged man-made objects that could provide clues about the direction, force, and duration of the earthquake waves that struck them. He would also travel through a wide area of the countryside to find out how far the disruption reached and where the most intense areas of damage were. The first investigation technique would, he hoped, tell him the direction from which the shocks emanated. The second, he believed, would pinpoint the actual epicenter of the earthquake.

During an investigation of which direction the earthquake sounds came from in Charleston, Sloan was alert to the fallibility of human perception. He found that the numerous accounts were conflicting, especially when he interviewed people who were indoors at the time of the shock. Each person invariably stated that the sound came from the nearest open door or window, regardless of which direction it happened to be facing. Discarding those accounts, he stated that "[The] opinion of cool-headed men out of doors at instant of shock, very generally agrees that noise was heard approaching from northerly direction a few seconds prior to experiencing the shock, which during its first period shook matter in a manner which in culminating in utmost violence suddenly yielded to [a] phase characterized [as] more undulatory."[6]

During the early days of Sloan's investigations, the evidence seemed to indicate that the force of the earthquake had come from the northwest, the area that included Middleton Place, Dorchester, and Summerville. The Charleston Gas-Light Company's massive storage tank near the Cooper River seemed to give a clear indication of the enormity and direction of the force. The tank's cylindrical masonry wall had a circumference of 402 feet and extended 25 feet below the ground. The earthquake forced the wall eight inches southeast, and when the wall rebounded, it pushed the earth two inches northwest, leaving earth molds that could be accurately measured.[7] Similarly, six of the eight enormous warehouses belonging to the Northeastern Railroad were shifted six to eighteen inches south and three to six inches east.[8] The displacement of these huge structures also suggested that the shockwaves emanated from a hypocenter northwest of Charleston. However, if Sloan had any thoughts that determining the source and direction of the earthquake waves would be quick or easy, he was soon proven wrong. Describing his research to his chief, William McGee, Sloan ruefully admitted, "The work I have found to be hydra-headed."[9]

During September, Sloan examined nearly one thousand funeral monuments in Charleston. After plotting the direction of displacement of about one hundred of them, he concluded that the pattern was nearly random, with "no simple indication of direction of seismic force," the result of "two rapidly successive forces involving most varied concurrences and interferences." Those forces randomly tipped some monuments, while nearly identical monuments close by were undisturbed. Both McGee and Sloan found sections of memorial shafts that had rotated, indicating that they had been twisted by forces coming from different directions. McGee found one in Powell's Marble Yard near St. Michael's church that had been turned twenty degrees to the southwest and another turned twelve degrees to the northwest.[10] So complex was the interlacing of shocks and their effects that Sloan described the Charleston tombstones as having "danced about in a manner most erratic."[11] This interlacing of shocks was the first firm clue that the earthquake had multiple epicenters.

The erratic tombstone displacements were mirrored by the patterns that resulted when thousands of fallen chimneys in Charleston and Summerville were compared. A good example was the two identical, well-built chimneys observed by McGee at a two-story wooden house at Gregg's Phosphate Works on the Ashley River near Summerville. The long axis of the house ran northwest to southeast, and one chimney was thrown almost due east, while the other landed almost due north. This indicated

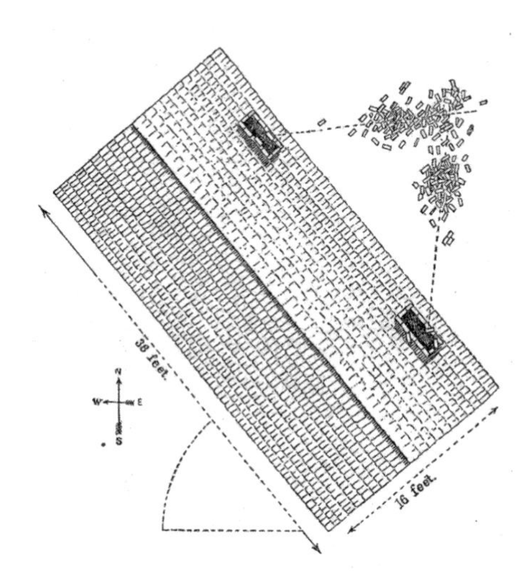

The fallen chimneys at Gregg's Phosphate Works

that the shocks arrived at the house from places about seventy-five degrees apart.

When and how the city's most precise clocks stopped during the first shock added another layer of evidence to the multiple-epicenter theory. Virtually every clock that was stopped by the earthquake was silenced by the first shock, at 9:51 p.m. Sloan wrote, "There are three fine time pieces in this city, the accuracy of which is vouched for in daily telegraphic regulation from Washington. Two of these pieces [one owned by the Northeastern Railroad] closely sustain each other in recording [the] instant of arrest at 9^{hr} 51^{min} $15^{\prime\prime pm}$, [the] accepted instant of transit [time of arrival of the first earthquake shock]."[12] These two clocks had their pendulums swinging forty degrees and sixty-six degrees northeast, respectively, "requiring for their arrest," Sloan wrote, "an approximately normal force say NW-SE."[13]

The third clock had a pendulum that swung from thirty degrees northwest to thirty degrees southeast, which would have been interrupted by a force striking from west of Charleston. It belonged to the South Carolina Railway, and stopped at 9 hours, 51 minutes, 48 seconds p.m., thirty-three seconds after the first two clocks. Sloan concluded from this data that the fatal first shock at 9:51 p.m., which was generally agreed upon to have lasted just under one minute, consisted of two separate components. The first struck at fifteen seconds after 9:51 and came from the northwest; the second struck at forty-eight seconds after 9:51, and came more from the west.

Sloan was amazed by his findings. The evidence seemed to indicate something none of the scientists had expected: The

Charleston earthquake had multiple epicenters! Now he had a general compass to guide his fieldwork. Since he knew the directions from which the shocks seemed to have come, he could focus on those areas and assess the levels of geological and structural damage along the way. The areas of maximum damage, he logically believed, would be the epicenters, which sat directly above the hypocenters where the earthquakes were generated.

Sloan faced a major challenge when he began his explorations of the countryside. With the exception of Charleston and Summerville, the epicentral tract was very thinly populated, and man-made objects that could show the effects of the earthquake were few and far between. The chief exceptions were the railroads, which fanned out in three directions from Charleston. Their roadbeds, rails, trestles, bridges, water towers, and stationhouses made up most of the built environment in the epicentral area. For that reason, they were the targets of intense scrutiny. To gather evidence, however, Sloan and his predecessors had to move quickly because all three railroads had brought in large work crews as early as September 1 to restore service, and their repairs were rapidly obliterating many of the earthquake's telltale signs.

On September 13, Sloan left Charleston on the South Carolina Railway to make detailed observations and measurements of the craters and fissures in the earth and the horizontal and vertical displacements of railroad tracks and trestles. As a geologist, he was especially interested in any changes in the level of the earth.[14] Once past the five-mile marker, where the South Carolina Railway continued on northwest towards Summerville and the Northeastern Railroad turned to the north towards Florence, Sloan was literally out in the woods.

The work of a field geologist in the 1880s was not for ivory-tower scholars. Sloan, with his love of the outdoors, was ideally suited to the task. The area most severely affected by the earthquake consisted mostly of uninhabited pine forests, cypress swamps, and uncleared land filled with shrubs, thorny bushes, weeds, wildflowers, wild raspberry and blueberry bushes, and wild animals.

In the late fall, it was a hunter's paradise, filled with white-tailed deer, wild turkey, rabbits, bobwhite quail, and mourning doves. However, in the steaming heat of late summer, when Sloan was combing through the fields and forests for clues to the earthquake, his days were filled with blood-sucking leeches, chiggers, and swarms of tiny black gnats (known locally as "no-see-ums"), mosquitoes, and large blue-bottle flies that have a bite like the kick of a mule. Despite the ninety-degree heat and ninety percent humidity, he would have worn a hat to shade his eyes from the bright sun, a long-sleeved shirt and heavy cotton pants to protect himself from insects, and knee-high leather boots to repel bites from timber rattlesnakes, which were abundant in the area. He would also have taken a handgun similar to William McGee's powerful .45 caliber, long-barreled revolver in case he was charged by a wild boar in the woods or encountered an alligator in the swamps.

A craterlet at Ten-Mile Hill

As had McGee, Sloan found that the number and size of the craters and fissures and the severity of the railroad track displacement peaked in the area north of Ten-Mile Hill. At the South Carolina Railway's seventeen-mile marker, between Summerville and Charleston, he obtained one of the steel rails that the railroad repair crews had removed and shipped it to the Smithsonian Institution. It was described as "one of the most remarkable evidences of the subterranean force exerted on the night of the earthquake. The rail in question is bent into a reverse curve, very much resembling the letter S, only drawn out somewhat. It is also drawn out or extended an inch and one eighth beyond its original length. The curves in the rail were evidently produced by lateral thrusts, and the fact that the curves are reversed is taken as an evidence of the irregular or zigzag currents of this mysterious underground force."

Sloan continued north on the railroad past the Summerville, Lincolnville, and Woodstock stations, where severe cratering, fissuring, and structural damage had already been identified by McGee and Mendenhall. At each stop, he annotated a large-scale map with the craters, fissures, and other damage he found.

Along a clay ridge near Woodstock Station, he found strong evidence of a major geological disruption. A seven-hundred-foot-long fissure that was eight to fourteen inches wide connected groups of craterlets that had spewed out water and flooded the surrounding land to a depth of seven inches. Because this was the most severe faulting and cratering he had seen in the region, he concluded that Woodstock was probably an epicenter of the earthquake. To be certain that he had seen the worst damage in that region, Sloan continued as far as Jedburg, twenty-two miles northwest of Charleston. There the degree of cratering and fissuring had dropped off considerably, leading him to conclude that he had reached the far northwest edge of the epicentral tract. That left the areas west, east, and northeast of Summerville to cover before he could make any decisions about the earthquake's epicenters. On September 14, Sloan ventured out onto the yet-unexplored tracks of the Charleston and Savannah Railroad, which ran southwest out of the city. Using a two-man-powered railroad handcar, he investigated the area from John's Island Ferry to Adam's Run, both west of Charleston. As before, he was watching for the area of greatest damage, and he found it along a line that lay between fifteen and twenty-seven miles from the city. At the eleven-mile marker, he discovered that two trestle approaches to a drawbridge across the upper Ashley River had been shoved in opposite directions towards the center of the stream, jamming the drawbridge. West of the river, the joints of the rails were torn open by the tension produced by the bridge movement. The bridge itself had been pushed thirty-seven inches south of its original position.[15]

At the sixteen-mile point west of Charleston, the earthquake had created a twenty-seven-inch curve in the tracks, which flexed to the south. Rantowles Station lay just two hundred feet short of the eighteen-mile marker. A few hundred feet west of it, numerous craterlets had undermined the track so heavily that the roadbed rippled from depressions of up to two feet. In addition, the track was bent into a forty-foot double-reverse curve. Because of the severity and extent of the damage, Sloan felt that he was close to an epicenter. After trekking deep into the countryside on either side of the roadbed, he concluded that he was right. He defined its location as about one or two miles northwest of Rantowles Station. As before, to make sure that his conclusion was valid, he proceeded further down the line to the west. By the time the railroad reached the Adams Run station, twenty-nine miles west of Charleston, the damage had died out almost entirely, and he returned to Charleston. On September 15, he worked in Charleston, making further observations and measurements, and then made a day trip to Augusta, Georgia, which, like Summerville, had been severely shaken by the earthquake.

When Sloan started out on the third and final leg of his journeys on September 17, it was aboard the Northeastern Railroad, where he found evidence of the earthquake all along the line. At the station in Charleston, the cylindrical masonry wall supporting the roundhouse turntable had been pulled to the southeast, just like the Charleston Gas-Light Company's immense reservoir. By the time Sloan was ten miles northeast of the city, track flexures were common, and he found an embankment that had been forced four-and-a-half feet to the east. At the twelve-mile marker, he encountered a belt of narrow earth fissures that was two hundred feet wide and seven hundred feet long and single fissures as wide as twenty-one inches.[16] Nearby, a trestle and its embankment had been shoved a little more than eight feet to the west. Another mile up the track, the main house of Otranto Plantation, then being used by a hunting club as their clubhouse, was damaged. Its west gable and east wall were completely destroyed; a piazza had collapsed; and a wing had been sheared from the building.

Another mile or so beyond Otranto lay one of the saddest wrecks of the earthquake outside Charleston. The magnificent old Anglican parish church of St. James, Goose Creek, which had been built in 1713 to replace a 1707 structure, had been totally ravaged. Made from the highest quality brick and mortar, which is still a testament to the magnificent craftsmanship of the early eighteenth century, it was nevertheless heavily

The parish Church of St. James, Goose Creek

damaged. Both the west and east gables were completely destroyed; the walls were badly cracked; and the northeast corner was severely injured. Inside, most of the handsome altar survived, but above it on both sides, gaping holes from the demolished gables left the church wide open to the elements. The building had been struck by forces from almost directly below it, as evidenced by the fact that when the gables shattered, the bricks fell in two neat piles directly below the damaged areas, and not to either side. This suggested that the church was sitting extremely close to the epicenter of the shock that struck it. His first examination of the Woodstock area proceeded east from the tracks of the South Carolina Railway. His second proceeded west from the tracks of the Northeastern Railway. Ultimately, a close examination of the region from both sides, including St. James' Church, which lay between the two railroads, convinced Sloan that a point half a mile northeast of Woodstock Station on the South Carolina Railway was the northernmost of the two earthquake epicenters.

By the time Sloan completed his two-month investigation of Charleston and the regions surrounding the three rail lines that ran into it, he was convinced that there were at least two components to the main catastrophic shockwave, each of which was produced by a separate hypocenter. He wrote, "The experience of individuals as well as the testimony of objects tends to inference that at the City of Charleston the great shock of night of Aug. 31st was a compound [multi-part] shock; a second component being so quickly successive upon the first as to have been almost simultaneous therewith." The short (thirty-three-second) time differential between the first and second components of the 9:51 p.m. shock suggested that the first component came from the Woodstock hypocenter, which then almost instantaneously triggered a second earthquake from the Rantowles hypocenter, located about thirteen miles to the southwest of the first.

The twin epicenters that Sloan initially proposed, when connected by a straight line, formed an axis that ran northeast to southwest through or within a mile or two of all the main

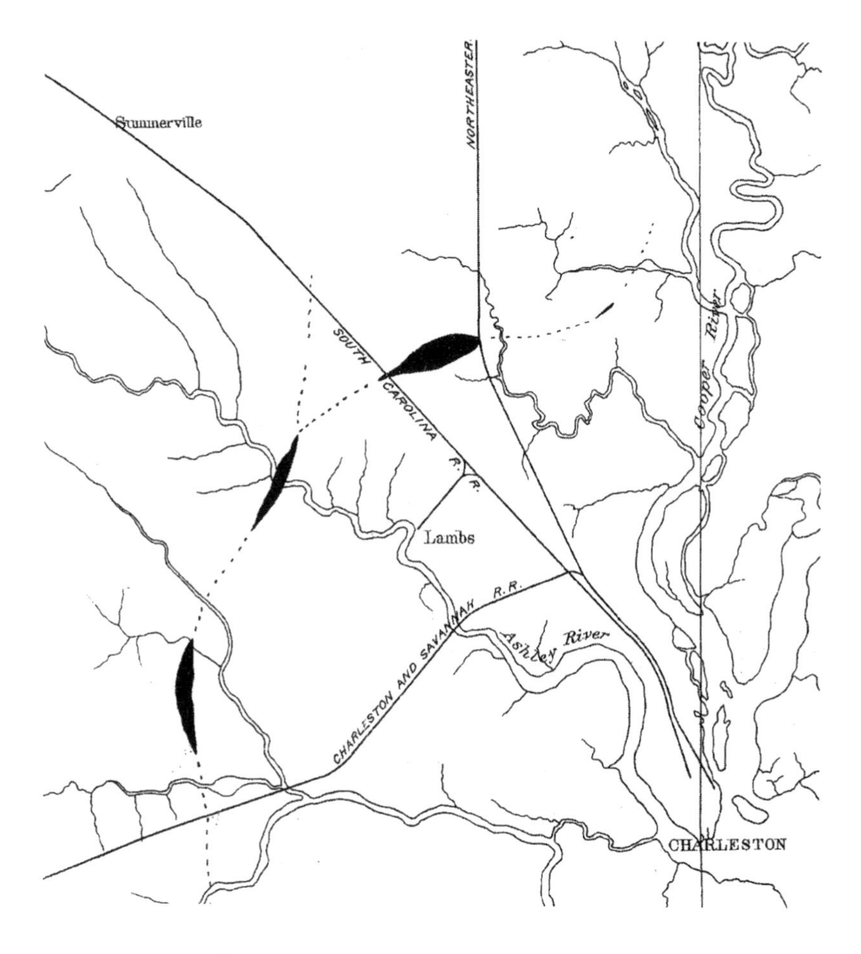

Sloan's field map of the damage areas

areas of heavy damage outside of Charleston: St. James' Church, Woodstock Station, Lincolnville, Summerville, Dorchester, Middleton Place, Magnolia Plantation, and Rantowles Station. He drew a preliminary isoseismal map to indicate the levels of damage at each place that he and McGee visited. Its concentric lines created a large oval area about twenty-five miles long and fifteen miles wide. One hundred and twenty years later, that oval is now known as the Middleton Place-Summerville Seismic Zone, and the northeast-southwest line is known as the Woodstock Fault Line.

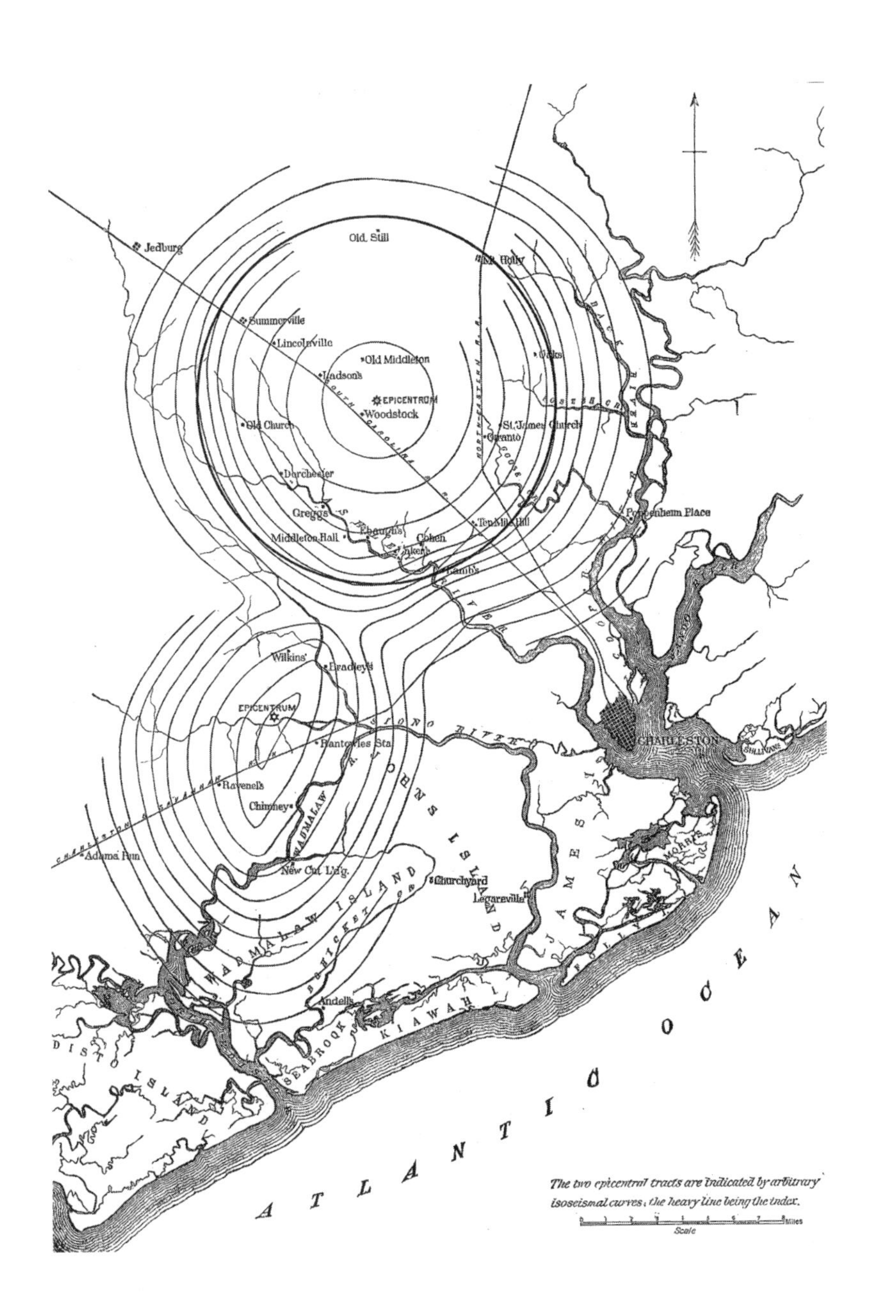

Sloan's 1886 isoseismal map

The two-hypocenter theory made sense to Sloan because if all the energy had originated from one source, the earthquake waves would have all moved in the same direction. In that case, tall, slender objects, such as tombstones and chimneys, would all have fallen either toward the shock, or, when the earth rebounded, in the opposite direction. A two-center earthquake, with shocks coming from two directions, would have tended to produce the results Sloan and his predecessor, McGee, had observed: tombstones, chimneys, and similar structures falling in all directions. Sloan's two months of patient and exhaustive field research and his insightful analysis of his findings would withstand the scrutiny of hundreds of other scientists. In only slightly modified form, his basic conclusions about the twin hypocenters and their locations are still accepted by scientific researchers.

Sloan sent his copious research notes, report, and annotated photographs to William McGee, in October 1886. McGee reviewed the data and turned it over to Capt. Clarence E. Dutton, who wrote the official USGS report on the earthquake in 1887. In the preface of his report, Dutton described Sloan's contributions in glowing terms. "I must mention with special gratitude the work of Mr. Earle Sloan, of Charleston, who undertook the investigation of the epicentral tract of the earthquake. With great labor and patience he studied the ground for two months, and subsequently reviewed portions of it, gathering together a large amount of information of the kind that was wanted, and submitting it in a form capable of being used intelligently. Without his researches some of the most valuable results of the study of this earthquake would never have been realized.... Certainly the most valuable characteristic of Mr. Sloan's work is the candid, impartial spirit with which every observation was made and the just weight which is attached to every fact. It is impossible to bestow higher praise upon an observer."[17]

In all, more than half of the technical information and many of the conclusions in the "Dutton Report," as it came to be known, came directly from Sloan's work, which constituted the single greatest contribution to the earthquake commission's study of the earthquake. In the end, it took a native South Carolinian to fully understand the natural catastrophe that destroyed Charleston. But it would take one of Charleston's own to put the city back together.

13

The Measure of the Man

Many thanks for your telegram reporting my dear family safe. Do not be disheartened; let us meet our great disaster bravely. We own the land the city is built on and we have our heritage there.
—William A. Courtenay to Francis W. Dawson

Aboard the R.M.S. *Etruria*
In the North Atlantic Ocean

Charleston's most famous citizen was the last to learn about the catastrophe that plunged his city into despair. Mayor William Ashmead Courtenay had been abroad since mid-May and was enjoying the last day of his well-earned European vacation aboard the elegant R.M.S. *Etruria,* now approaching the U.S. coast. The Marconi transatlantic radio telegraph had yet to be invented, making the gleaming, state-of-the-art steamship an idyllic floating island, totally out of communication with the rest of the world.

In private life, Courtenay was the Charleston agent for the Clyde & Company shipping line. As a saloon-class [first-class] passenger, he had enjoyed a spacious stateroom and extraordinary service during the voyage. On the morning of

September 5, five days after the earthquake, his ship would dock in New York, where he planned to spend a convivial evening with his business and personal friends. On Monday, if his schedule held, he would transfer to a Clyde Line coastal steamer and set sail for the two-day trip to Charleston.

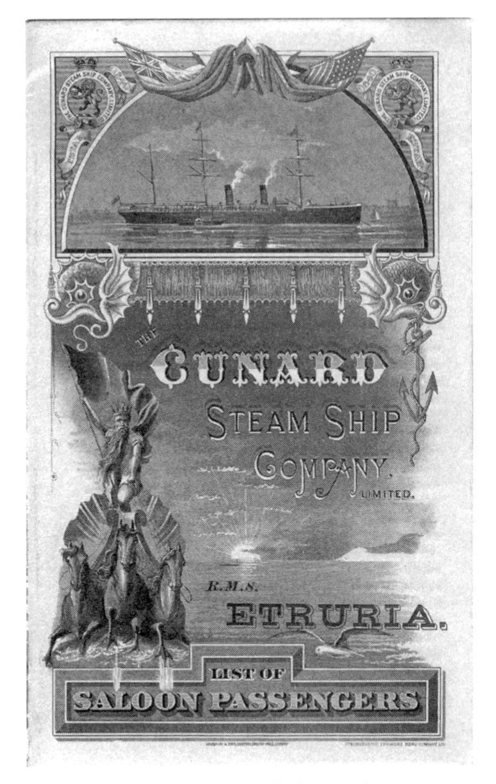

The Etruria's *Saloon-class passenger list*

Courtenay was not the first of his family to make the voyage from Ireland to America. In 1791, his grandfather, Edward Courtenay, Jr., and his great-uncle, John, had sailed to the United States on a small cargo ship. Both were natives of Newry, near Ulster, in northern Ireland. The ship had been loaded with Irish linens, beer, and whiskey, and the two brothers were immigrating to take up the merchant's trade. Edward settled in Charleston; became a successful merchant; married Lydia Smith, a native of Newburyport, Massachusetts; and fathered seven children, including William Courtenay's father, also named Edward.[1]

William Ashmead Courtenay was born in Charleston in 1831, the son of Edward Courtenay, a schoolteacher, and Elizabeth Storer Wade, a native of New York.[2] For a man who would become an avid scholar, antiquarian, and devotee of the arts, Courtenay received only a limited education in his early years, for his parents could not afford better. Taught at home by his father and female relatives until he was twelve, he was taken under the scholarly wing of Dr. John C. Faber, principal of one of Charleston's finest private academies. At Dr. Faber's school, he received a classical education to prepare him for college. His classes included Greek, Latin, rhetoric, mathematics, history, and geography, all provided to Courtenay without charge. Courtenay never forgot the largess of his mentor, and although it took him twenty years to repay the cost of the tuition, he had fully discharged the debt by the time he was thirty-five years old and had returned from the Civil War.[3]

His family's economic circumstances cut short his formal studies with Dr. Faber and forced him to leave school at the age of fifteen to go to work. In 1850, when he was nineteen, Courtenay and his younger brother, Alexander, a carpenter, were living in Charleston with his parents. He took a position as a clerk at the bookselling and publishing business run by his older brother, Samuel Gilman Courtenay, on Broad Street.[4] There, over a ten-year period, his future literary tastes formed and flourished.

In 1854, Courtenay, then twenty-three, married a teen-aged heiress. Blue-eyed, raven-haired, Charleston-born Julia Anna Francis, a lively, petite seventeen-year-old, was the only child of George Francis, a wealthy Englishman, who had emigrated to South Carolina and bought Feddersgrassen, a plantation on the Cooper River, where he raised sheep. George died young of yellow fever. His wife, Julia Durrett Francis, died in 1847, also of fever, leaving Julia Anna a ten-year-old orphan with a handsome inheritance of $85,000 ($1,708,500 in 2006 dollars).[5] Upon his marriage, Courtenay received control of his wife's fortune, making him a very wealthy man. In 1860, on the eve of the Civil War, Courtenay owned real estate valued at $6,000 ($120,600) in Charleston's Fourth Ward. His personal property was valued at $3,500 ($70,350), most of which was invested in his three adult and two juvenile slaves, who were domestic servants.[6]

Like other wealthy entrepreneurs, Courtenay was a dynamic man who needed to be involved in the professional life of his community. On October 1, 1860, he became the business manager for the Charleston *Mercury* newspaper, and he was soon a full-blown member of Charleston's *literati*. He was also a member of the Washington Light Artillery, a volunteer militia unit comprised of men from Charleston's better families.

When the Civil War broke out, he promptly volunteered for active duty and enlisted as a private on May 13, 1861, in Company B of the 8th Regiment, South Carolina Infantry, which ultimately became part of Gen. Robert E. Lee's Army of Northern Virginia.[7] Elected a second lieutenant, and later promoted to captain, he rendered distinguished service to the Confederacy as a quartermaster and infantry officer at the battles of First Bull Run, Fredericksburg, Gettysburg, and Chickamauga.

In December 1861, while home on leave from his unit, then serving in Virginia, he witnessed the Great Fire of 1861, which burned across Charleston from river to river. Demonstrating the talent for writing that would become a hallmark of his later life, Courtney became an instant journalist and prepared a detailed narrative of the calamity. After recruiting two assistants, he surveyed the burnt district block by block, and in twenty hours handed the printers of the Charleston *Mercury* a comprehensive report, noting every burned building and the names of the owners and occupants. It was described as "the most prompt, extensive, and complete newspaper work known at that date in Charleston."[8]

The Civil War brought challenges that further shaped Courtenay's life. The first was a challenge to his honor. In February 1863, a fellow officer and fellow South Carolinian, Capt. William L. DePass, was accused by his commanding officer of seriously neglecting the health of the horses that transported his artillery. DePass defended himself by stating that his supply officer, Lieutenant Courtenay, had not given him sufficient food for the horses for six weeks. With the war raging all around him, Courtenay did not hear of DePass' testimony until many weeks later. When Courtenay confronted his accuser, DePass denied that he had made any such statement. Courtenay then queried the officers who had examined the charges, and they confirmed his suspicions. DePass had lied and blamed him for the problem. Courtenay's good name had been tarnished, and when DePass refused to apologize, Courtenay challenged him to a duel. As the party who had been challenged, DePass had the right to choose the weapons. He sent word to Courtenay that "of course, the dueling pistol will be used."[9]

After a protracted correspondence, seconds were chosen to negotiate the details of time and place, but fate temporarily intervened. Courtenay and DePass were each captured and held by the enemy. Courtenay was taken prisoner near Opequan Creek, Virginia, on September 13, 1864, and was held at Johnson's Island prison camp near Sandusky, Ohio, for ten months. The conditions there were grim. Confederate captives were denied access to newspapers, books, and religious texts—a special hardship for a highly literate man like Courtenay.[10] However, the prison commander did permit him to receive one parcel of clothing from an aunt in Baltimore.

When the war ended, Courtenay took the detested oath of allegiance to the Union on June 6, 1865. He was released from prison on June 16. With no horse and no possessions, Courtenay walked home—as did tens of thousands of his fellow comrades in arms—only to find that DePass, described as a "celebrated duelist" by the *Charleston Courier*, was still ready to meet him on the field of honor.[11] After trading indignant letters, the two men decided that they had better things to do and agreed to "let the matter rest."[12]

On the long walk back to civilian life, the thirty-four-year-old soldier had plenty of time to think about his future. His prospects were grim. A true Southern patriot, he had invested his wife's inheritance in Confederate bonds, which were now worthless.[13] He and his family were destitute, and he was unemployed. When he reached Spartanburg, South Carolina, he encountered another ex-Confederate soldier, who was in charge of turning over confiscated mules to the Union army. After reminiscing about the battles they had fought, and bemoaning their lost cause, Courtenay persuaded his newfound friend to give him two of the animals in lieu of the two years' back pay he was owed for his military service.[14] This act, primarily intended to rest his weary feet, would return him to prosperity.

On his way home to Charleston, he found that South Carolina's railroad lines had been destroyed during the war. Using his mules, the resourceful newspaperman became an entrepreneur and started what quickly became a lucrative wagon transportation service, hauling cotton and other supplies eighty-four miles between Newberry and Orangeburg, South Carolina. Resources in the interior of the state were so scarce that he had to provide his own fodder for the mules, and the costs were so high that the price for hauling a fifty-pound sack of salt over this route by wagon was $2 ($40.20 today). Although he started out with active competition, Courtenay's reliability and honesty soon won him virtually all the business, and his long wagon trains arriving and departing from Newberry and Orangeburg attracted great attention.[15]

When the railroad was rebuilt in the spring of 1866, his mule trains were no longer required, so Courtenay moved back to Charleston and invested his profits in the transportation business. Soon he was shipping goods by steamship to Baltimore, Philadelphia, New York, and foreign capitals. He also became involved in the management of several steamship lines and served on the board of directors of the Charleston and Savannah Railroad.

His strong military ties continued when he began his community service work in 1866 by joining the Washington Light Infantry Charitable Association, which aided the widows and orphans of his former militia unit. During Reconstruction, private militias were banned, but many white paramilitary organizations sprang up to take their place. In 1873, Courtenay became president of the Washington Light Infantry Rifle Club, established "for the protection of social order." He also led successful efforts to establish permanent memorials to the unit's dead in Magnolia Cemetery and to erect an obelisk in memory of the infantrymen in City Hall Park, which was renamed Washington Square.

Only five years after limping back to South Carolina broke and unemployed, Courtenay was wealthy enough to buy a handsome, three-and-a-half story Greek Revival-style brick single house. It was located at 75 Lynch Street (later renamed and renumbered as 95 Ashley Avenue) in the village of Harleston, an early nineteenth-century residential suburb northwest of Charleston.[16] His real estate was valued at $10,000 ($201,000), and his personal property was worth an additional $6,000 ($120,600).

The wealthy businessman and Freemason quickly became a leader in the business community. He was elected president of the Charleston Chamber of Commerce in 1875 and served six years, to the great satisfaction of its members.[17] That same year he made a strong impression on conservative men of the North when he and the Washington Light Infantry he led accepted an invitation from Boston to participate in the centennial celebration of the Battle of Bunker Hill on June 17, 1875. Most of the infantry members had served in the Confederate army, and the color sergeant who carried the crimson battle flag used by patriots at the Revolutionary War battle of Eutaw Springs, South Carolina, had suffered multiple wounds for the Lost Cause. The presence of the historic Charleston unit "created a profound impression throughout the North." Courtenay and his men accepted a similar invitation for the centennial celebrations in Philadelphia on July 4, 1876. Courtenay's willingness to participate in events that honored the country as a whole, even when it meant cooperating with men who had formerly been his enemies, established a base of Northern good will that would prove invaluable when the earthquake struck Charleston a decade later.[18]

In 1879, to further serve the city he loved, he decided to enter politics. In his first bid for mayor of Charleston, Courtenay, a Democrat, garnered 4,463 votes (53.6 percent of the total), defeating the former Republican mayor, William W. Sayle. Sayle, who had been elected during the hated Reconstruction period, ran for re-election as an independent and received only 1,660 votes. A third candidate, W. J. Gayer, received 2,191 votes.

At that time, Courtenay was forty-eight years old and his family was still living at 75 Lynch St. The household included his wife, Julia, thirty-eight; his widowed mother, Elizabeth; and seven children: three daughters (Edith, Julia, and Elizabeth) and four sons (Campbell, Carlisle, Ashmead, and St. John), ages three to thirteen. At least three children had died in infancy before the Civil War. A white boarder, Belle Wilson, age forty, and perhaps a relative, lived with them. The prosperous family employed four servants: Elizabeth Mills, a sixty-year-old mulatto cook; Joseph Mills, a twenty-one-year-old mulatto servant-waiter; Richard Mills, an eighteen-year-old mulatto servant-waiter; and Duncan Thompson, an eighteen-year-old black house boy.[19] On Sundays, the family attended Grace Episcopal Church on Wentworth Street.[20]

William A. Courtenay

A conservative man of unquestionable personal and financial integrity, Mayor Courtenay set the tone for his administration in an address to the city council when he took office in December 1879. "The functions which we are to exercise are committed to us as a sacred trust," he said. "The government which we control as public officers is not our own; it belongs to those who placed us here. The laws we enact do not express our will; they are the voice of the people." He concluded with a statement of his financial policy, "The money which we handle belongs to them, and not to us. We can only take it from them for the legitimate expenses of government. *More than this is robbery. Official generosity is official crime.*"[21] In 1883, he ran for reelection, this time unopposed, and served until December 1887.

During his two terms, his accomplishments included reducing the city's annual budget, paying down its debt, and carrying out many significant public improvements. In the ten years before Courtenay took office, the city's budget had averaged $794,000 and did not include any funds for public improvements. Courtenay's fiscal conservatism and highly developed business skills reduced the budget to $661,000. When he took office in 1879 the city's debt stood at $4,544,000. Convinced that he could improve Charleston's fiscal health, he persuaded the city council to give up the right to increase the city's public debt by issuing bonds. This reduced the debt to $4,000,000 by 1887. Much of the savings came from his cuts in the municipal budget.

Courtenay believed that the city government should be open and honest with its citizens. To keep his constituents actively informed of Charleston's financial status and the work done by civic employees, he published an annual *Yearbook* containing detailed reports of all city income and expenses, as well as descriptions of the work of each department. Reflecting his deep love of history, the *Yearbook* also included long, authoritative articles and maps portraying the city's rich heritage. In addition, Courtenay distributed complimentary copies of the *Yearbook* annually to highly placed federal, state, and municipal leaders throughout the eastern U.S.; to the most prestigious libraries, archives, historical, lineage, and patriotic societies, and to a number of foreign consuls. Indeed, there was almost nothing the mayor would not do to project the image of Charleston as a progressive, bustling, and progressive business-minded city with a rich and noble history.

In addition to introducing stringent cost-cutting measures, Courtenay made major improvements to the city's appearance, sanitation, and services to the poor and elderly. He had the main sand, cobblestone, and wood-plank roads replaced with permanent granite block roadways. The major streets also received slate flagstone sidewalks with granite copings, which are still in use today. To improve sanitation, ironstone drains were laid under the sidewalks to replace open gutters.[22] Courtenay also proposed that animals be slaughtered in an abattoir outside of town instead of in the City Market in the heart of the commercial and residential areas. To improve the city's parks and business districts, he had electric lights installed in Marion Square and the City Market and provided "iron sofas" (iron-and-wooden-slat park benches) facing the Ashley River at Battery Park (White Point Gardens) and at Washington Park.[23]

One of his innovations had already paid heavy dividends. Against entrenched opposition from the members of the prestigious, century-old, often ineffectual, volunteer fire-fighting companies, he created an efficient, full-time paid municipal fire department and installed a modern telegraphic fire alarm system.[24]

The mayor faced his first major unforeseen civic challenge in August 1885, when the immense cyclone ravaged Charleston. Courtenay directed the Herculean efforts to clean and rebuild the city. Under his leadership, without significant financial aid from outside the city, virtually all of the hurricane damage was repaired in less than a year, and Charleston was again running smoothly.

In the summer of 1886, with the storm's fury a year behind him, the wealthy, fifty-five-year-old patriarch was enjoying his well-earned vacation. As his ship slipped gracefully through the starry night and dark waters of the north Atlantic, the mayor enjoyed his last hours of freedom from civic responsibility. The band played waltzes in the main ballroom for the sophisticated revelry as Courtenay chatted with friends and colleagues in the grand saloon—the main social cabin—enjoying his last night of uninterrupted pleasure before the voyage ended. Late that night, his ship was spotted off Fire Island, halfway between Long Island's Montauk Point and the entrance to New York Harbor. Near the Sandy Hook Lighthouse in Lower New York Bay, a tugboat dispatched by the Cunard Line came alongside the *Etruria* and sent aboard a large mailbag.[25]

Mayor Courtenay's idyll ended when the special bag was brought to him that night. Alarming telegrams and frantic letters from his friends and colleagues in South Carolina spilled out, along with newspapers from New York City.[26] The first telegram he opened left him speechless. Captain Dawson,

A New York pilot boat greets the Etruria's *arrival*

editor of Charleston's *News and Courier*, had laid out the bare bones of the catastrophe: "Your family all safe and uninjured by earthquake your residence badly damaged."[27]

Shock and disbelief set in as the mayor read through the papers. When his fellow passengers heard of the catastrophe and came to comfort him, tears came to his eyes, and he was unable to speak. Horrified, he scanned the newspaper reports, grimly absorbing the details of the destruction and imagining the terror of his constituents. Then he fumed in anger over a *New York Herald* article, which alleged that the city government's response to the tragedy had been inefficient and that the police department was demoralized.[28] Courtenay knew better. He immediately retired to his cabin for the rest of the night, planning his response.

The *Etruria* docked on Sunday morning, September 5, at 7:30 a.m. Courtenay was among the first to disembark. The crew quickly located his baggage in the hold, and customs officials, briefed in advance of the need, gave the mayor priority processing. One of his traveling companions, James Sprunt, of Wilmington, North Carolina, recalled the moment. "At the quarantine station in New York Harbor we were handed several telegrams, and, looking up in dismay from the reading of one addressed to me, I saw that Mr. Courtenay had suddenly vanished without a word. Panic-stricken by the terrifying news, he had hurried ashore to catch the first train to Charleston."[29]

Despite having just received catastrophic news, Courtenay immediately began to take control of the situation. He hit the ground running. J. B. Lanneau, formerly of Charleston but now a New Yorker, met the mayor at the gangway, and the two were driven to the New-York Hotel. There, Courtenay dashed off a telegram to Captain Dawson. "Many thanks for your telegram reporting my dear family safe. Do not be disheartened; let us meet our great disaster bravely. We own the land the city is built on and we have our heritage there."

Next, Courtenay called a meeting of merchants with Charleston connections to facilitate relief efforts for his stricken citizens. Then he inaugurated what would become a fourteen-month mission, which used the power of the national press to bring financial aid to his beloved and beleaguered city. Disgusted by the negative coverage in the *New York Herald*, he granted his first newspaper interview to the *New York Tribune*.

Courtenay was a quick study. He had already begun to establish his priorities. As Charleston's chief spokesman, he had to make the world aware of its situation. His interview sent a message that embraced the enormity of the city's loss, indicated what assistance was needed, and ended with a positive resolution to make a full and rapid recovery.

> I don't know what to say. The news is incredible. I am overwhelmed with anxiety; how to meet the exigencies of the hour, I am not now prepared to say. The disaster, from what I can learn, must be one from which months will be required to recover. I do not worry about the business or commerce of the city. The pressing question is as to provision for the homeless. Ours has been a sorely tried community, but of

> one thing I am certain and that is that, whatever can be done by the people of Charleston for themselves, will surely be done.
>
> I see that some correspondents say that the city government is weak and inefficient, but what, I must ask, can be done at a time of such disaster. Time will be required to right matters. The city government has its hands full. There are inmates of hospitals to care for besides other matters connected with the interests of a large city. Add to all else the fact that about forty prisoners have escaped, and there can be no wonder that a theft is going on even at this critical time. In a few days, however, order will be restored.
>
> I cannot see what is to be done, besides getting all the assistance possible in our time of need. What is needed most now is means of shelter until the wrecked buildings can be restored. The people need canvas for tents, and many I am informed, are starving because food cannot be got into the city. The worst is that much suffering will ensue among the people of moderate means who are too proud to make their wants known. The people of Charleston must have prompt aid to tide them over the next few days. After that they will recover rapidly, for they are plucky and will do all they can to gain their feet again.

The telegraphed news of Courtenay's imminent arrival in Charleston produced jubilation in the city. The *News and Courier* exclaimed, "At such a time as this, Mayor Courtenay is the one man needed to organize and execute, to think and act, to plan and perform. There will be new life everywhere when he is at his post again. He is worth a thousand men to us in this crisis of our fortunes."

On Sunday, September 5, at 4:30 p.m., Mayor Courtenay boarded a train for Washington, D.C., and went from there to Charleston. The leisurely voyage aboard the *Etruria* was the last period of peace and quiet that he would experience as mayor. Now, his responsibilities were clear: retrieve the reins of government, provide compassionate aid to the victims, and rebuild his beloved city—again.

One week after the earthquake, Charlestonians were still reeling from the devastation that had left them disoriented and grief-stricken from the loss of family, friends, neighbors, homes, businesses, jobs, and money. But the citizens had never lost hope. Despite frustration, privation, and sorrow, Charlestonians kept their spirits up and projected a confidence born of experience with numerous previous disasters. Their attitude could be summed up simply: Though the burden is heavy, this, too, will pass.

The best news since the disaster struck was Mayor Courtenay's eagerly awaited return early on Tuesday morning, September 7. The *News and Courier* wrote, "No higher tribute has ever been paid to any public officer than that paid to Mayor Courtenay by the sense and expression of relief with which all classes recognised and acknowledged the fact that a strong hand was felt to be at the helm of city affairs and that order would be surely and promptly brought out of confusion by his clear-headed, cool judgment and wise management."

When he arrived at the Northeastern Railroad Station early on Tuesday morning, Courtenay was greeted by his family, a large delegation from the city council, members of the relief committee, and a horde of grateful citizens. The *News and Courier* proclaimed, "[Courtenay] is the man of men for such an emergency as this." On the day of Courtenay's return, Julian M. Bacot wrote in his diary, "I saw old W. A. Courtenay driving through Broad Street this morning & was delighted at his return, as there will now be a head sufficiently cool and collected to check the demoralization with which the whole community has been affected, and I feel confident that with him at the helm poor old Charleston will pull through this apparently overwhelming calamity and arise once again from her ruins like the Phoenix."[30]

Upon his arrival, Courtenay immediately thanked Acting Mayor Huger, the members of the relief committee, and other civic leaders for their heroic work in the face of the appalling difficulties they had confronted. After briefly spending time with his family to make sure they were well and unharmed, Courtenay quickly set to work, making his own tour of the city. He learned first-hand the extent of the damage and evaluated the initial disaster response. He was totally appalled by the former and greatly gratified by the latter.

On September 8, he arrived at his office in city hall before dawn and promptly began to put in place the relief and recovery plan he had drawn up on the long railroad trip from New York to Charleston. As usual, his plans were ambitious. He would expand the original relief committee to provide more comprehensive coverage for the community; launch a nationwide appeal for financial assistance; pursue state and federal funds; seek a large, low-interest government rebuilding loan; clear out the refugee camps; and inspire the residents to rebuild their own homes and businesses. He launched his plans with an address to the citizens, published in the *News and Courier* on the day of his arrival. A master at public relations, Courtenay knew the nation would be waiting for his response to the disaster, and his words were calculated to elicit support while reassuring commercial interests that Charleston was open for business.

> To my fellow-citizens of Charleston: I have this day returned to my loved city, amid its widespread desolation, its homes shaken to their foundations, many of them utterly wrecked and few without serious injuries, and I find many of you, my fellow citizens, with your dear and tender families living and sleeping still under frail shelters, and some under the sky, with the recent terrible calamity and its awful suspense still lingering in your minds. Amid the ruins of

this far-reaching and terrible calamity, I am profoundly thankful that so much of life and property has been spared, and I rejoice that the same fortitude and heroic patience that, in the trust of God and His providence, has always characterized his people, is now their stay in this time of dire trouble, and I am thankful to add that, in this past week of disaster, the good order and helpful cooperation of all classes of our citizens has conduced to the maintenance of the public weal.

It is inspiring to behold amid these grave difficulties the resumption of the business life of our city and the *quenchless faith of our people in the future*. The open exchanges, banks and leading houses in all departments of business—the uninterrupted commerce of railroads, steamships and sailing fleet, and the ready wharves, the busy workshop and the usual routine of all the avocations and employments of our city, proclaim that we are already going forward into a new future. That future is based on *work*, not idleness, and I call upon everyone to seek *work* in any and every way possible.

Although the situation is critical, it is not insurmountable. It demands from every citizen in our midst calm judgment—the broadest charity—a resolute determination in word and act, an unfaltering trust in God, to tide over the unparalleled calamity that has so unexpectedly come upon us. This disaster, that reaches every home and every part of our city, can only be met and overcome by the moral courage and the united effort of the whole people. The immediate and serious duty before us is the protection and succor of the houseless, the sick, and the indigent, the unfortunates and the helpless, that crowd around us. As the Executive of the city it is my

> duty to create an organization, looking to dealing with the different problems before us, and I have to act at once. I have therefore initiated this work, by asking service of some of my fellow-citizens, in a worthy attempt to organize such plans as will mitigate the suffering and distress, so universally around us, and which threatens such serious consequences to many of our people. These plans will be expanded if found in any way wanting in completeness as the work progresses.
>
> In this effort it is a source of great gratitude to us to know that we are not grappling with this unspeakable disaster alone. The sympathy of the whole Union of States has touched us deeply, and the spontaneous giving of practical and speedy aid in this our struggle shows that the large and true heart of the people of this great Country beats with us now as it will hereafter.
>
> In this hope, and cheered with this promising future as part of a great people whose helping hands are outstretched to us, let us turn manfully to our heritage, and, as many times in the past on this very spot, work out, under the blessing of God, a new future for our shattered but dearly loved city.

The citizens of Charleston, young and old, black and white, were ecstatic. Their beloved leader and inspirational icon was again at the helm of the city's ship of state. They knew the measure of the man. Their confidence in him was boundless, and he would not disappoint them.

14

COURTENAY TAKES COMMAND

The countless acts of bravery and devotion of the men and women who, in the darkness, returned to and rescued from the still quaking and crumbling ruins, the children, the sick, the aged and the helpless, will never be told. —William A. Courtenay

September 8, 1886

As Mayor Courtenay sat at his desk on the day he returned to Charleston, the tasks that lay before him must have seemed overwhelming. Countless people had been severely injured. Numerous lives had been lost, and people were still dying. Emergency food rations were being issued to more than five thousand people a day. Forty thousand people—two thirds of the city's population—did not feel safe sleeping in their houses and were living in homemade tents in their yards or in the more than twenty refugee camps that had sprung up all over the city. Lack of adequate protection from the elements, insufficient clean water, and woefully inadequate sanitation facilities in the foul-smelling tent cities were already killing people, and the city was on the verge of being ravaged by disease epidemics. Thousands of buildings were damaged or destroyed. Virtually every business was damaged, closed, or both. Almost every citizen was unemployed. And Charleston's treasury, which had no emergency disaster funds, was running dry.

A lesser man would have thrown up his hands, turned the problems over to the relief committee, and gone home to his family. Courtenay could not. His hometown was an integral part of him. He had extolled its virtues in numerous widely distributed publications. He had built up its trade and commerce. At the cost of his own peace of mind, he had saved it after the hurricane in 1885, and he could not bear to see all of his work wasted now. By the time he had been back on the job for three days, Courtenay had thoroughly assessed the magnitude of the disaster he was dealing with, and the tools he had to work with, and had formulated an aggressive social and financial recovery plan.

The first decision he made was to put his own imprimatur on the relief process. The day he arrived in Charleston, Courtenay asked the city council to replace the relief committee set up by Huger with an Executive Relief Committee (E.R.C.). In accordance with the acting mayor's wishes, all of those who were serving on Huger's committee resigned. However, Courtenay showed his support for his predecessor's successful efforts by immediately reappointing all of the relief committee members that were still able to serve.

Because of the heroic actions of Acting Mayor Huger, the first relief committee, and numerous civic leaders, no one was starving, housing was being set up, and many public services had already been restored by the time Courtenay returned on September 7. A comprehensive, effective food rationing system was in place. A soup kitchen was in full operation, and food was efficiently delivered to all who needed it. For those who could afford to pay, newspaper advertisements demonstrated that Charleston's usual wide variety of food and beverages was readily available. On September 2, J. S. Terry & Co., a dealer in food and ice at 9 Market Street, one of the most heavily damaged areas in the city, advertised rice birds (bobolinks), a local delicacy, "Cheaper than ever. Fresh arrivals every day."[1] Donated tents, which were now flooding into the city, were being erected as fast as they arrived, and the city had started building hundreds of wooden booths to shelter more of the homeless. Patients from City Hospital and Roper Hospital were housed in Agricultural Society Hall, which was providing full medical services to the community.

Despite occasional disturbances and incidents of petty theft, the police department was able to maintain law and order. Additional temporary officers had been recruited, and volunteers from the Washington Light Infantry and the German Hussars provided mounted patrols to help keep order in the streets and watch for fires. The prison break from the earthquake-damaged city jail had proven to be more of an embarrassment than a public safety threat.

Every firehouse in the lower part of the peninsula had been evacuated because of serious damage. Fortunately, all of the steam engines and ladder trucks had been spared. The firehouses north of Calhoun Street were relatively untouched and remained in use. Until September 10, when the telegraphic fire alarm system was restored, all firefighters remained at their posts at all times.[2] Each man was required to stay by his engine except when he ate. The engines and crews were repositioned next to the larger refugee camps throughout the city, and fire horses that had stampeded were rounded up and stabled near their engines.

Public utilities were also returning to service. The Electric Light Works on Hayne Street was not injured during the quake, although downed poles and wires led to outages during the week that followed. Because few people had electricity in their homes, the only places that were really affected by the lack of electricity were Central Market and Marion Square. Even there, the expensive electric lighting was primarily an ornament of civic pride, rather than a public necessity.

The City of Charleston Waterworks Company's main pumping station on George Street sat on high ground and had sustained major damage. The waterlines, however, were broken in countless places, and it took a week to make basic repairs. The mains were pressurized from 9:00 to 10:00 p.m. on September 7 "for the purpose of filling the tanks in private houses," and by September 10, the water system was officially declared safe for use and was back in operation.

The city's gas supply was more of a problem. The Charleston Gas-Light Company's street mains were so badly broken, and the joints so strained and dislocated, that the distribution system had to be shut down to avoid explosions. The company had to repair more than one hundred breaks in the mains and examine and reseal twenty miles of gas pipes. Even as repairs were made, aftershocks reopened many joints. Although the system was partially operational by Christmas, gas lines were still being repaired six months after the earthquake in March 1887.[3]

Within three days after the earthquake, full communication within the city and with the outside world had been reestablished. The newspaper noted that the officials and clerks of the post office "stuck to their posts like heroes" during the earthquake, enabling the postal service to continue without interruption, although receipt of mail from outside the city was cut off for three days while the train tracks were repaired. The *News and Courier* of September 3 was also elated to report that both the Western Union and Southern Telegraph companies had repaired their lines to Columbia, Augusta, Savannah, and New York and expected to be in contact with all points by the end of the day. The Southern Bell telephone system was also operating again within the city.

Public transportation had been restored. For two days after the earthquake, railroad work gangs labored day and night, with a success that was greeted with "great jubilance." By September 2, trains were rolling into and out of Charleston again, although they often had to slow down on sections of track that had been patched but not yet fully repaired. The inbound excursion train from Columbia, whose shaken and frightened passengers had spent the night in Summerville, arrived at 6:30 p.m. on September 2, and the first Charleston and Savannah Railroad train from Augusta arrived at 11:00 a.m. the next day. On Friday,

September 3, the newspaper stated, "All of the [rail]roads have made great progress in repairing damages and expect to run their regular schedules today." The trains brought three days' worth of mail from the North on Friday night.

The city's two horse-drawn trolley systems were operational by the second day after the quake. City workers had cleared the rubble from the streetcar lines, and the trolleys were running on their normal schedules. One problem area for the trolleys was the intersection of Broad and East Bay streets, where extensive wooden bracing had been erected to keep the post office (Old Exchange Building) from collapsing. Travel was also difficult on lower Meeting Street, where the rails were blocked by the massive wooden timbers used to stabilize the portico of St. Michael's Church.

The iconic church, whose bells inspired the whole city, had undergone a close inspection on September 5, and the investigators found both good and bad news. The building had been heavily damaged, and the portico had separated from the steeple and the body of the church. Although it had settled eight inches into the earth, the steeple and its foundation were sound.[4] A trip inside the steeple revealed that the clockworks, powered by a pendulum, appeared to be in good condition and that the bells seemed intact. Charles N. Beesley, the church sexton, encouraged Charles Lockwood, the watchmaker charged with regulating the clockworks, to climb the seventy-seven steps to the clock works in the steeple and restart the clock, whose pendulum had been stopped at 9:51 p.m. by the shock of the earthquake.[5] Lockwood then climbed up to change the hour and minute hands, and restarted the clock at 2:00 p.m. on September 8.

Stabilizing the portico of St. Michael's Church

Despite these successes, Mayor Courtenay still faced two major challenges. First, he had to get people out of the squalid, overcrowded refugee camps and back into their homes before massive epidemics swept through the population. Then, he needed to raise huge sums of money to continue to feed the homeless and to repair or replace thousands of damaged or destroyed houses and businesses.

Public health topped the list of Courtenay's immediate concerns. According to the official record of deaths, more than fifty people had died by the time he returned, and the death toll was mounting daily. By midnight on August 31, at least twenty-seven people (seven whites, twenty blacks) had been killed by falling buildings alone. That exceeded the total number of people killed by all other U.S. earthquakes in recorded history.

The first earthquake death to be entered into the official record on August 31 was that of Charles Albrecht, a fifty-two-year-old native of Germany who had lived in Charleston for thirteen years. The cause of his death was listed as "Accident – earthquake." C. H. Rivers, Charleston's deputy coroner, signed his death record, and he was buried in Bethany (Lutheran) Cemetery in the northeast part of Charleston Neck. For two months following the first shock, the city carefully kept two separate tallies for deaths: one for those resulting from the earthquake and one for those resulting from other causes. Forty out of the eighty-three officially recorded earthquake deaths were caused by blunt-force trauma from falling bricks and masonry. Of those, twenty-seven took place in the first two hours after the main shock, and the rest were victims who died in the first two weeks of September from the same injuries.

In September, fifteen whites and thirty-four blacks died. However the causes of those fatalities rapidly changed from "injuries received - earthquake" to "exposure to the elements." Of the forty-nine September deaths, only fourteen had "injuries" listed as the primary cause. The rest of the deceased died from exposure to the elements brought on by being forced to live in tents or shelter sheds or on the streets. Once the blunt-force injuries tapered off, the most frequent causes of death (the 1886 terms in italics are no longer in use) included shock; premature or still births; malaria and other fevers; cholera; *scrofula*, a form of tuberculosis characterized by swelling of the lymph glands; *marasmus*, extreme malnutrition and emaciation, especially in children; and *trismus nascentium*, a form of tetanus seen only in infants, probably due to infection of the umbilical stump. The clouds of brick and mortar dust stirred up by the earthquake and the heavy palls of smoke from the eight fires undoubtedly contributed to the increase in deaths from pulmonary problems, including asthma, congestion, and pneumonia.

The stress of the earthquake also pushed some mentally fragile people over the edge. The newspaper described the case of Ann George, a fourteen-year-old girl who was transported to the South Carolina Lunatic Asylum in Columbia for treatment in the first week of September. "The earthquake shock of Tuesday completely shattered her mind, and she is now raving with a strong homicidal tendency." The tragedies were not confined to South Carolina. The *Atlanta Constitution* reported that shoe salesman C. H. Murphy had been terribly frightened by the initial earthquake. On September 4, "he sat on the floor of his room, coolly loaded a thirty-eight calibre pistol, and deliberately blew out his brains. His mother was the only person in the house at the time, and this shock, together with the earthquake scare, will end her life in all probability."[6] The same newspaper reported that believing the end of the world had come, three members of a family bound themselves together with rope and jumped off a bridge into Horse Creek near Rocky Ford, sixty miles northwest of Savannah, Georgia. All were drowned.[7]

As the temperatures dropped, hypothermia from exposure to the increasingly cold, wet nights took its toll. By the end of September, ninety percent of the earthquake-related deaths were caused by exposure or the foul sanitary conditions in the camps. As the *News and Courier* noted on September 7, "The earthquake was the indirect cause of many deaths attributed to disease." By October, there were only seven earthquake-related deaths. All of the deceased were black, and they died from liver inflammation, enteritis, or malarial fever—all caused or aggravated by exposure to the elements. In all, thirty blacks and nine whites died exposure-related deaths during the two months after the main shock. The city recorded its last official earthquake death on October 8, 1886. The victim was Eddie Jenkins, a black man whose age was not noted. He died of exposure and malaria.

Of the people whose deaths were officially recorded, twenty-seven percent were white and seventy-three percent were black. At the time, the population of the city was forty-six percent white and fifty-four percent black. Although blacks outnumbered whites by only a small majority, 2.45 blacks died for every white. The reason for the disparity was simple: whites had much better shelter available to them than blacks.

The eighty-three earthquake-related fatalities recorded in the Return of Deaths were the official death toll for the 1886 earthquake. However, that record never captured the names of many who died as a result of the disaster. The reason for this disparity was also simple, though not immediately obvious. The only people whose deaths were officially recorded were those who died within the city limits of Charleston *and* whose deaths were certified by a licensed physician, the coroner, or the deputy coroner.

Although the earthquake affected a 2.5-million-square-mile area, the official death records only listed those persons known to have died within the 1886 city limits of Charleston, South Carolina, a five-and-a-half square-mile area.[8] Earthquake-caused deaths in the other ninety-nine percent of the area affected by the disaster were not included. The known dead from Summerville and McClellanville were not counted, and the railroad men who died in train wrecks near Summerville and Aiken and the earthquake-induced suicides in Georgia were never listed as "official deaths." In addition, many deaths, those of blacks in particular, and rural people in general, were substantially under-reported because people died without the official oversight of a physician and were buried without a burial permit signed by an authorized Charleston city official.

As a result, the city's official death register accounts for only about two-thirds of the earthquake deaths. Also, none of the severe injuries caused by the earthquake was ever officially recorded. A new count of deaths and casualties taken from the *News and Courier*; the *Atlanta Constitution*; other regional newspapers; private correspondence; and the city's surviving church, public, and private cemetery, death, and burial records revealed that the earthquake resulted in at least 124 fatalities and 132 additional major injuries, many of which were undoubtedly fatal but were never recorded as deaths. Taking into account the massive area affected by the earthquake, the total casualties (deaths and injuries) could easily have topped five hundred people. One addition to the September Return of Deaths was especially poignant for the mayor. At 8:30 a.m., September 19, 1886, his eighty-one-year-old mother, Elizabeth Storer Courtenay, died quietly at her son's home of "old age." She was buried in Magnolia Cemetery. Because her death came about as a result of natural causes, she was not included in the city's count of "earthquake deaths."

To stop the deaths from exposure and disease, Courtenay knew he had to persuade his constituents to return permanently to their homes. He was fighting an uphill battle. Prior to Courtenay's return, William B. Guerard, Charleston's superintendent of streets and a civil engineer, published a stern warning in the newspaper on September 5.

> A constant danger menaces us ever to a greater extent, in my judgment, than an anticipated renewal of fatal shocks. This danger consists in the numberless fractured walls in the city. As they now stand tottering, they are a menace to our lives, but if we should have a rain of any considerable extent, even if not followed by a high wind, either of which phenomenon is frequently amongst us at this season of the year, the consequence would be that the water-laden brick and mortar would lose their present

> unstable tenacity and walls would crumble and be precipitated to the streets, with probably awful fatality to all in the vicinity. It would be impossible to get down all the insecure buildings before the next rain storm, may it not then be advisable to have ropes stretched around all such dangerous localities?

On September 7, the day Courtenay returned to Charleston, the Reverend Edward T. Horn, pastor of St. John's Lutheran Church in Charleston, underscored the feelings of many Charlestonians in a letter to a friend in Pennsylvania. "I arrived here yesterday at 6:30 A.M.," he wrote. "I found our whole family asleep in the dining room—the first sleep they had had, and that a nap. They were extremely grateful for the tent I succeeded in purchasing on Saturday night, and last night insisted on sleeping in it on the croquet ground. Ten were in it on mattresses, and there was room to spare. I hope my Father will send some tents to me; I have many applications for them. They are afraid to sleep in the houses. Brick houses are [selling] at a discount. Some persons felt shocks yesterday, but I think they would have felt them anywhere: I believe there were none."[9]

The mayor realized he had to reassure Charlestonians that the city's buildings were safe if he was going to successfully entice them into returning to their homes and businesses. He immediately initiated an aggressive program of building inspections to encourage people to get out of the camps and off the streets. Acting Mayor Huger had already laid the groundwork for this effort when he asked the city's aldermen to make an official assessment of the damage in the first days after the earthquake. In addition, by September 5, William Aiken Kelly, the city assessor, with the assistance of Street Superintendent Guerard, had made a thorough preliminary inspection to assess the cost of rebuilding the city. Kelly and Guerard indicated that a much more detailed survey would be needed to estimate the expected repair costs.

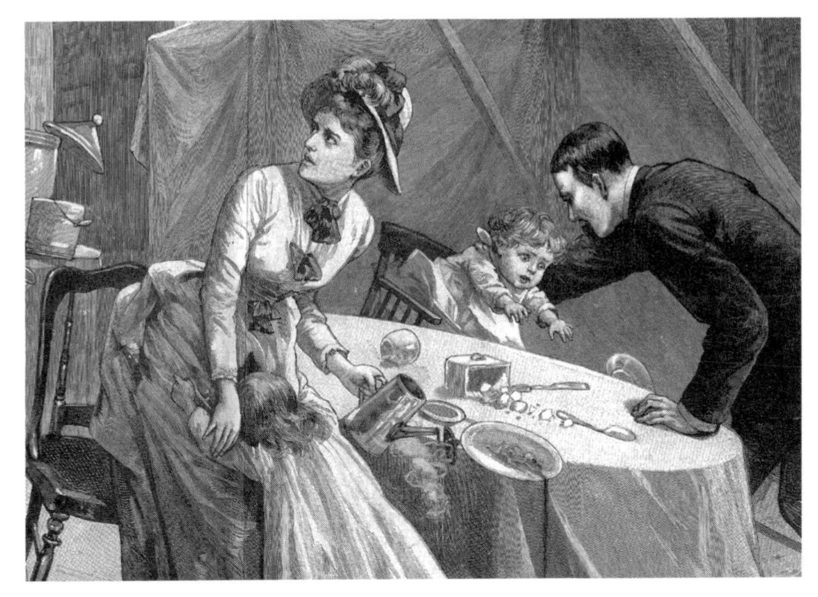

"The constant aftershocks kept everyone in constant fear"

As soon as the enormity of the damage was clear, Acting Mayor Huger had turned to the federal government for help. He asked General R. C. Drum, acting Secretary of War, for permission to use the services of Lt. Frederick V. Abbott of the Army Corps of Engineers, who was stationed in Charleston, and his colleague, Capt. William H. Bixby, stationed at Wilmington, North Carolina, to examine the city's buildings and evaluate their safety.[10] United States Congressman Samuel Dibble, a South Carolinian who chaired the federal government's Committee on Public Buildings and Grounds, requested a senior architect to assist with the formal damage assessment. The request was immediately approved, and Special Agent William E. Speir, Inspector of Public Buildings in the Supervising Architect's Office, Department of the Treasury, arrived in Charleston on September 8 to inspect federal buildings for damage.

On September 8, at the office of the city auditor, Mayor Courtenay formally named the members of the new building inspection team, which became known as the Federal Engineering

Commission. The team included the three federal officers—Speir, Bixby, and Abbott—assisted by Capt. George D. Bryan, the city's corporate counsel; City Engineer Louis J. Barbot; and two civil engineers—Street Superintendent William B. Guerard and John Henry Devereaux. The team made an immediate disclaimer, "[We] are here as a commission to advise the Mayor and city officials as to the danger and safety of buildings, and are in no sense responsible for any action the city authorities may see fit to take."

Having washed their hands of legal responsibility, the commission members immediately set their inspection priorities. First, they would view public buildings and those serving the public, such as post, telegraph, and express offices and municipal buildings. Next, they would inspect factories and similar structures that provided employment for large numbers of people. Third, they would look at buildings vacated because of real or supposed danger. Finally, they would get around to the remaining dwellings and businesses. A separate committee of builders and building contractors was appointed to supervise and direct the removal of dangerous walls and structures.

By September 11, three days after its formation, the commission had finished inspecting Charleston's major federal, state, and municipal buildings and had moved on to examining privately owned properties. Its daily inspection results were published each morning in the newspaper. Eventually, the commission confirmed that the majority of wood frame houses in the city, though damaged, could safely be reoccupied.

Courtenay, who had followed every move the federal inspectors made, could now prove that the majority of homes were habitable. He had to bolster the confidence of his constituents and begin the campaign to get them back indoors. An astute politician, he knew that optimism and proof that things were getting better were his best weapons. On Friday, September 10, he published an address that praised the courageous survivors and the civic heroes and showed them they had nothing to fear.

To my Fellow-citizens: I have been three days at home—have been over most of the city, and can testify to the large damage throughout its whole extent. Frame buildings which appear unhurt, except as to chimnies [*sic*], are wrecked inside as to plastering and wood work.

I have, by official report and otherwise, informed myself of the unparalleled events of the past week, and I deem it my duty to address you at this time, in the belief that what I can truthfully say to you will give you increased confidence from now on. In the midst of the most appalling circumstances, when the very foundations of our city were shaking, it may be truthfully said of our community as a whole that their moral courage and heroism was equal to their great disaster. Conspicuous in their calm and brave facing of imminent peril were our noble women, the frailest of them suddenly becoming strong, and by their faith and endurance and good work giving courage and help everywhere.

Among our colored population, whose very natures are emotional, there was the quick resort to prayer and song of their multitudes, which was but the natural cry of the creature to the Creator in time of sudden tribulation to them unknown and mysterious. With time for reflection, all undue excitement passed away, and the colored people of Charleston can point with satisfaction to their own exemplary conduct in these trying circumstances.

Everything that could be done by public or private agency to meet the sudden, startling and threatening events, this community did. The extensive City Hospitals were wrecked and rendered uninhabitable, but the faithful employees first rescued the numerous sick and wounded and helpless, and then removed the entire furniture and outfit to comfortable and convenient quarters, and within thirty-six hours this extensive charitable work of our city was moving on in a most satisfactory way. Our large number of orphans were promptly quartered in tents as a precautionary measure, but their spacious homes, the Orphan-houses, were substantially uninjured and will now be reoccupied. The old and infirm and unfortunate in our Almshouse were carefully looked after and removed from their wrecked quarters to a safe temporary home.

As evidence of the faith and courage of all our people, I refer to a few incidents only, not to all, in these trying scenes. I mention the devotion to duty of the firemen, who amidst crumbling buildings and walls and entangling wires, moved their apparatus to the several scenes of conflagration, and carried their hose by hand over ground strewn with ruins and trembling under earthquake shocks; and by this desperate struggle, saved the city from further disastrous destruction by fire. Amidst scenes like this the police force, driven from their wrecked Main Station-house, with many of their families in peril and trouble, still stood to their post of duty; and their chief reports no single case of unauthorized absence.

The telegraph operators stood by their instruments amid the ruins of their office. The municipal medical corps and the private physicians of the city were immediate and constant in their attendance on the wounded and sick on their perilous rounds in the terror of that night. The many nurses, both regular and volunteer, administered to the general suffering and distress around them, while all the clergy, in the comfort and consolation of their faith, sustained and strengthened the many awe-stricken gatherings in open places of the city. The countless acts of bravery and devotion of the men and women who, in the darkness, returned to and rescued from the still quaking and crumbling ruins, the children, the sick, the aged and the helpless, will never be told.

In the midst of this sudden and almost overwhelming disaster, I rejoice to find that the City Government was equal to every emergency, and not only continued uninterrupted and unimpaired in its official functions, but also came actively to the relief of the suffering and distress of the community. Acting Mayor Huger, on whom this grave responsibility was suddenly cast, was ceaseless and untiring in the discharge of his official duties. What has been done under this action will be found in the full report of the committees which I have requested shall be published in full herewith, that our people may know the labor of the citizens who have undertaken and so faithfully discharged this large and beneficent public service.

The special executive committee, appointed by Acting Mayor Huger, is hereby reappointed, together with the sub-committees on shelter and subsistence, of which Major C. F. Hard and Dr. A. B. Rose respectively are chairmen. It may be necessary, in view of the magnitude of the work, to reorganize and rearrange the relief organization, but it will not be forgotten by the public or the Mayor of the city that the committees, which were suddenly called into existence in the midst of a great emergency, proved themselves, with limited means and comparatively small opportunities, fully equal to the demands that were made upon them.

I have said enough to show the high endurance of our people in these trials. I now wish to offer you some advice for the future which I believe to be for your best interests. At the end of sixty days we must surely expect cold weather, and I am alarmed at the consequences to follow the use of tents and frail shelters in our streets and public squares, which must lead surely to sickness and calamity. We have it on the highest authority of science and experience that there is no likelihood of further violent earthquake shocks. Their gradual decrease of itself assures us of their total cessation.

And now, my fellow-citizens, I deem it my imperative duty to advise the escape of living under thin shelter and on the open ground at this season of the year, when our rainy weather is usual. Nearly every frame building in Charleston has been declared at least safe for habitation; the chimneys may be rebuilt while occupied, the plastering may be restored at some future time. Many of the brick buildings have also been also pronounced safe in part and can be occupied in part. All such frame and brick buildings should be immediately occupied. I return this day to occupy the uninjured part of my brick house, and in all such cases where return to their homes is possible I earnestly invoke the immediate and united action of all my fellow-citizens. Where immediate return is impossible by reason of the dangerous condition of the home, I recommend a temporary refuge in the interior of the State or elsewhere, or the hiring of some safe dwelling in the city, a number of which are today un-occupied.

> A word to those who can of themselves put their own homes in repair: Let them do so without an hour's delay; if the labor is not ample enough here, it is offered us from surrounding cities, and can be had. The next sixty days in Charleston should be busy days in preparation for the winter. Whatever the discomforts may be under a roof, the penalties to follow a continued use of the present arrangements will make them un[en]durable.
>
> What our people want is relief, immediate, permanent relief from the terrible nervous strain to which they have been suddenly subjected, and which will certainly continue in the tent life which many are leading in the streets and public squares. A renewal of the *home life* of the city alone will restore confidence, rest, and quiet. The same characteristics which have carried our people through the trials of the past ten days, if exerted under the advice I have earnestly given them, I am sure will bring in safety and relief of mind, *permanent benefit to them.* In conclusion, I earnestly invoke all the clergy, physicians and heads of families throughout the city to co-operate in this effort to restore the healthful repose of all our people.

Then to set an example for his fellow citizens, the mayor moved his wife, his children, his ailing mother, and their servants back into their brick house. Courtenay's inspirational speech and personal commitment galvanized Charlestonians. The combination of the mayor's urgings, the newspaper's support, and the positive reports of the engineers worked a miracle. Almost immediately, the refugees who populated the tent cities started flocking back to their homes.

There were still holdouts who could not face the prospect of living within four potentially shaky walls. With the imminent threat of epidemics firmly in mind, city officials ordered the total evacuation of the most dangerous camps. Without question, Washington Park topped the list. For the more than 1,500 refugees crowded into half an acre, it was a hellhole, yet some of them had patched together one-story wooden shelters, intending to stay permanently out of reach of falling walls. Police helped to enforce the evacuation order, and by dusk on September 11, virtually all of the refugee campers in Washington Square had either returned to their homes or relocated to wooden sheds erected by the city. The evacuation was reported by the newspaper, which applauded the fact that the city had "obliterated a bed of pestilence and by forcing the squatters to return to their homes contributed to a large extent to the restoring of confidence."

The camps cleared out rapidly over the next week. The newspaper reported on September 14 that heavy winds and the absence of aftershocks for three days enticed many more people to return to their homes. All tents had been cleared from Marion Square and Linguard Street, leaving only the shed-dwellers. Camp Duffy was almost deserted, and the camp on the Battery had lost some of its inhabitants. The E.R.C. started hauling in unused lumber intended for the construction of wooden shelters and began to dismantle partially completed sheds.

As the tents came down and the sheds were abandoned, the overworked E.R.C. staff's employees had to quickly begin the grueling and odious work of cleaning up the filthy sites to stave off disease. Reporters from the *Atlanta Constitution* inspected the former camps and reported on September 16 that "all the negligent habits of the tenants had been foretailed by proper remedies of prevention" and that the abandoned refugee areas had been made "scrupulously free of anything that might engender disease in any shape or form."

The final countdown on the end of the tent cities began on September 18, when the E.R.C. ordered that all donated tents and temporary shelters erected by the committee were to be vacated by the end of September, implying that refugees who remained would face eviction. The committee held true to its word. On September 30, all the wooden booths were sold at auction. The successful bidders had to remove the empty sheds by the first of October, one month to the day since the terrified people of Charleston evacuated their homes. The police ensured that the private tents were also evacuated and dismantled. The tent cities disappeared almost as quickly as they had sprung up. Courtenay's goal of saving the city from disease epidemics had succeeded. The proof was found in the clear, black ink of the city's Return of Deaths. By the end of October, the people of Charleston were once again dying at the same rates and from the same causes as they had in recent years before the 1885 cyclone and the 1886 earthquake. The former tent city residents were, indeed, forced to shoehorn themselves into houses with more than the usual number of occupants, but that reality was far healthier and humane than had they tried to survive the cold, rainy winter in the stinking refugee tents and shed cities.

The evacuation of the tent cities did not mean that the need for food rations had ended, however. As of September 14, the E.R.C. still had twenty-four canvassers seeking out people in need of food, and five to nine wagons went out daily to distribute rations. Each wagon could haul fifty weekly rations of food and made four trips out each day. By September 19, the newspaper noted, more than 80,000 one-day rations had been distributed, "a fact which better than anything else indicates that the city is now well taken care of by the subsistence committee."

The E.R.C. was under constant scrutiny by conservatives such as the Reverend Anthony Toomer Porter, whose voice was raised in complaint once again. He accused the E.R.C. of dispensing unnecessary welfare and coddling the unworthy by giving food to unqualified people. In the committee's defense, the *News and Courier* ran an article on September 18 that was designed to defuse the criticism. It shows the lengths that the city administration went to in aiding starving earthquake survivors.

Very few people have any fair idea of the wants of a community of 60,000 people, a majority of whom have actually no means of subsistence whatever. The work of the subsistence committee reaches a large class of people, white and black, to whom the relief offered is of infinitely more advantage than its money value. At first a great many undeserving people were furnished with rations but this was unavoidable. The new system adopted by the committee has done much to prevent this, and out of the one thousand two hundred and seventy rations distributed by the eight relief wagons yesterday, it is probable that at least twelve hundred went to persons who really stood in need of them. No system of charity could accomplish better results. The canvassers are doing good work.

In several instances complaints have reached the committee that rations have been delivered to undeserving persons, but these cases are always investigated, and in this way the committee are gradually weeding out their list. The number of applicants for relief at the commissariat is gradually decreasing. No one, however, is turned away unheard. When applications are made the address is taken and a canvasser is detailed to go to the place and investigate. If the party is found to be deserving of help, whether his or her poverty existed before or after the earthquake, the relief is afforded. The amount of the money spent by the committee is comparatively small compared with the relief afforded, and this was proved to the satisfaction of the executive committee yesterday. About two-thirds of the persons relieved by this committee are colored.

> The soup kitchen, which is an attachment of the commissariat, is still continued, and supplies nourishing food to about 500 persons daily. Capt. Hyde, who is in charge of the commissariat, said yesterday, 'I wish to say that a more industrious, hard-working set of men than the clerks in this establishment cannot be found in any part of the city. These men work hard from 7 A.M. till 6 P.M., and their labors, as you can see, are of the hardest kind. There is not a man among them who does not honestly earn the pay, that he gets, and there is not a man among them who does not stand in need of his pay.

While the E.R.C. purchased the majority of food for distribution, it also received occasional donations from benefactors. In mid-September, a barrel of rice came in from Georgetown; some hams arrived from Baltimore; and a keg of wine was donated by Mr. W. G. Lawrence of Fayetteville, North Carolina. The wine was turned over to the City Hospital for the use of the patients. By the end of November, the demand for food had subsided enough that the E.R.C.'s food distribution system could be phased out, and those persons who still needed food and medicines were accommodated by the city's existing charity caregivers, such as the Alms House and the Old Folks' Home.

Having successfully forestalled the potential public health crisis, Mayor Courtenay turned to his next challenge: finding the money. As a successful businessman, he was acutely aware that restoring Charleston's commerce and trade was the key to the city's long-term survival. If there were any hint that the city was not open for business, maritime trade, a mainstay of Charleston's economy, would soon disappear, taking income and jobs with it. The day after the quake, rumors had circulated nationwide that Charleston's harbor, channels, and docks had been damaged and the city itself had been completely destroyed. Fortunately, most of these stories had been debunked by the time Courtenay returned. Although a number of the warehouses near the docks had been damaged, few were totally destroyed, and the docks and shipping channels were intact. Two days after the earthquake, two British steamships and two schooners arrived in Charleston, and on the following days, steamships and schooners continued to dock, drop off, or pick up cargo as if the disaster had never happened.

Knowing that numerous commercial interests would be watching the maritime situation closely, the *News and Courier* supported the public relations drive to reassure businesses that everything was normal in Charleston. On September 6, it boasted to the world, "The wharves and docks are in fine condition to receive and accommodate all the vessels that may arrive. Not a vessel in port sustained the slightest injury. The pilot boats have been cruising as usual, and Charleston's large fleet of powerful tugs have been actively at work."

The formal surveys of the physical condition of the city were beginning to provide the mayor with initial estimates of the costs he would be facing. The Federal Engineering Commission had inspected buildings that the federal government was responsible for: the U. S. Custom House, the post office (The Old Exchange Building), and a structure on Meeting Street used as headquarters for the lighthouse and engineering departments. The commission found the Custom House damaged but safe and repairable. Commission member Speir deemed the post office to be seriously damaged and advised its "instant removal" if a suitable replacement building could be found to rent. The newspaper stated that the building would be propped up temporarily and rendered safe. The U.S. government assured that it would pay for demolishing or rebuilding these federal buildings.

That left Courtenay with a sizable chunk of non-federal real estate to rehabilitate or pull down. The preliminary inspection made by City Assessor William Kelly prior to the mayor's return painted a grim financial portrait of the damage. Kelly stated that the assessed value of the city's real estate as of August 31, 1886, was $20 million ($402 million today), excluding

The U.S. Post Office (Old Exchange Building)

government, state, county, city, and other public property and property belonging to churches, schools, and other charitable institutions that were exempt from taxation. He estimated the value of the non-taxable property at $6 million ($120.6 million). Kelly counted approximately 8,500 structures in the built-up area of the city. Of these, about 7,000 were houses, including 1,000 small wooden dwellings recently built by poor blacks, and the rest were businesses. He had visited more than three-fourths of the city and found that every house was damaged. The brick houses were seriously damaged and, in his opinion, many of them would have to be pulled down. Many buildings that at first appeared entirely safe were discovered to be dangerous, and "persons who thought that they had escaped without injury find themselves confronted by damages which are entirely beyond their means to repair." Kelly's initial estimate was that the damage would exceed one-fourth of the total assessed value of the real estate, or in round figures, $5,000,000 ($100,500,000 today).

The city's coffers had been severely drained to support the relief efforts. However, within a few days after the news of Charleston's catastrophe was published in the nation's newspapers, spontaneous donations of money started to arrive in the city. They came through many channels and were delivered to many different custodians. This made it impossible to distribute funds according to need, and no one person or group could account for all the money that arrived. For a financially fastidious man such as Courtenay, this unregulated accounting system was unacceptable. With that in mind, as Charleston's chief executive, he immediately took charge of all incoming funds designated for the city's use. In the hands of a dishonest, unprincipled man, this could have led to uncontrolled graft and corruption. Under Courtenay's supervision, the system was a godsend.

Taking the financial reins, Courtenay's E.R.C. "[assumed] the duty of receiving money contributions for the relief of the necessitous sufferers by the recent earthquake shocks" and named a subcommittee, headed by William L. Campbell, the city treasurer, to accept and disburse the funds. The Charleston Exchange, the Merchants' Exchange, and the Chamber of Commerce immediately pledged that they would funnel all funds received by them directly to the E.R.C. Everyone prayed that the nation would be generous. The devastated area would need every penny it could get. All the positive thinking of the citizens, coupled with the massive public relations work of the mayor and the *News and Courier* combined could not mask the ugly truth: Charleston had been bludgeoned and shattered by the earthquake.

With the flow of spontaneous donations under control, the mayor began to consider his next step. A man of astonishing knowledge, Courtenay could have involved himself in every detail of the relief operations, but he had the sense not to. He realized that he had an excellent team to handle the chores of providing food and shelter, so he left that work for the E.R.C. staff. He needed to lead and inspire the people and to do the one thing that he could do better than anyone else—raise the immense amount of money it would take to pay for the relief operations and physically rebuild the city. His background as a hard-working, successful, self-made businessman with national and international connections gave him the confidence that he needed for the work that lay ahead.

In a letter to Dr. Andrew Simonds, president of the First National Bank of Charleston, then visiting in New York, Courtenay frankly laid out the magnitude of the catastrophe and the urgency of rebuilding. "Our losses are certainly $5,000,000 to $6,000,000. Of course, the majority falls on those who must stand it [that is, those who suffered the losses must themselves bear the burden of paying for their own recovery].

But you know how universal has been the effort through loan associations to secure homes. There are 2,000 such damaged [homes, each requiring] $200 to $600 ($4,020 to $12,060 in present dollars)—mostly helpless. Their houses must be restored within sixty days or cold weather will overtake them. All these people are our business force. Now we can use $500,000 to $700,000 ($10,050,000 to $14,070,000). The question is, 'Can we get it?' You can use this cold, dry statement of facts in your own way. I am looking over the situation dispassionately, and I am not at all scared."[11]

15

FINDING THE MONEY

The practical way in which the outside world evinces their sympathy for us has touched our very heart strings.... hereafter there will be no North nor South but a "Love-Knit Union."
—L. E. "Ned" Cantwell, to his cousin, Marie

Charleston, South Carolina
September 5, 1886

After being shaken and terrorized by the shocks for days, the people of the Lowcountry could only cringe to think what was coming next. To their great relief, some good news finally appeared in the newspaper. Americans had spontaneously opened up their hearts and wallets to the victims of what was already being called "The Great Charleston Earthquake." As soon as the telegraphs were operating again, the newspaper reported that money orders started pouring into the city from every part of the country. So numerous were the donations received in the first two days that the *News and Courier* devoted a full column on page one of the September 5 issue to list them. "It is a consolation in our dire distress to feel that we have the tender sympathy of our fellow men in all parts of the world, and that everywhere the noble-hearted are extending us a helping hand. First, of course, comes the generous offers of our sister cities, Savannah, Augusta, Columbia and Wilmington, and from them the word of sympathy has spread." By Christmas, every state and territory in the United States, almost every city in the nation, and seven foreign countries had donated hundreds of thousands of dollars to the Charleston relief fund.

At first, people wanting to help contacted Charleston officials to inquire about the city's specific needs. On September 1, Mayor Rufus E. Lester of Savannah wrote to Charleston's mayor, "The citizens of Savannah in mass meeting assembled, request me to express their deep sympathy for the citizens of Charleston afflicted by the disastrous earthquake of last night, and to tender them whatever aid may be needed."

From North Carolina, Dr. W. G. Thomas notified Charleston's mayor that one hundred volunteers from his home city of Wilmington were ready to help in whatever capacity needed. A delegation from Wilmington, headed by its mayor, and another, consisting of John Daniel, Albert Gore, and James Crowly, reached Charleston only a few days after the first shock. The latter three were given a carriage tour of the city by Dr. W. P. O'Neill, and while making the rounds, "spent every cent of money they had with them relieving such suffering as was brought to their attention. It is a fact that these gentlemen had to borrow enough money to pay their way home." Ultimately, the people of the Tarheel State dug deep into their meager resources and were able to contribute $13,329.86.

One of the most poignant responses was voiced by the *New Orleans Picayune*, which wrote of Charleston, "when we were visited with the pestilence [yellow fever epidemics] there was no more generous hand [that] came to our aid." Louisiana, still economically weak after the war, nevertheless found $8,543.42 to share with the earthquake victims.

Virginians were also among the first to give. From Richmond, the former Confederate capital, the response was strong and immediate. Dr. John G. S. Kelton, president of the Richmond

Medical and Surgical Society, wired Captain Dawson in Charleston that his society "tenders sympathies and professional services if any doctors are needed. Telegraph us how many. We will bear [the] expenses." The Old Dominion State quickly collected $13,365.45 to share with its neighbor to the south.

As heart-warming as these spontaneous demonstrations of goodwill were, Mayor Courtenay knew that, unless Charleston acted quickly, the nation's attention would soon move on to other concerns. He and Dawson, the city's mighty megaphone, worked as a team. Together, these two influential men launched an unprecedented public relations campaign. Their goal: restore the nation's confidence that Charleston was ready, willing, and able to transact all its normal business immediately. The city would be rebuilt as soon as the money was raised.

In 1886, Francis Warrington Dawson, editor and publisher of the *News and Courier*, was by far one of the most influential publishers in the cotton states. A bold, principled man of unquenchable energy, he had demonstrated his courage and vigorous spirit by the time he was twenty-one years old.[1]

Capt. Francis W. Dawson

Dawson was born in London in 1840 and christened Austin John Reeks. His parents became impoverished as a result of his father's financial speculations. His early education was made possible by a widowed aunt, but instead of attending college, he acquired a broad cultural and artistic education by touring Europe. By his late teens, the fair-complexioned, handsome young man with curly red hair had already showed talent as a playwright. In the latter part of 1861, after serving for a time as a clerk in a real estate office in London's Haymarket district, he decided to follow his fortune and seek adventure by emigrating to America and entering the service of the Confederacy. Just before doing so, he changed his name to Francis Warrington Dawson, a reflection of his gratitude to an uncle, William A. Dawson, whose widow had helped finance his education.

With the assistance of A. Dudley Mann, a Confederate agent in London, Dawson persuaded the captain of the C.S.S. *Nashville,* a Confederate steamer taking on supplies at Southampton, to let Dawson join the crew as a common sailor. He made himself popular with the crew. His ship ran the Union blockade and entered the harbor of Beaufort, North Carolina. Shortly thereafter, he was commissioned a master's mate in the Confederate Navy. Sent to serve on a ship at New Orleans, he learned of its capture by the Union Army and went to Richmond, Virginia. Wanting to stay in the middle of the fray, he volunteered for duty in a Confederate artillery unit and was wounded in the leg in 1862. His bravery won him an officer's commission, but he was captured, freed in a prisoner exchange, resumed active duty, participated in numerous battles, was promoted to captain, and finished out his career as an officer in the Army of Northern Virginia.

At the end of the war, he joined the staff of the Richmond *Examiner* and later the Richmond *Dispatch.* In 1866 he moved to Charleston, where he worked for the Charleston *Mercury,* and then the *News.* He and a friend, Bartholomew R. Riordan, bought a one-third interest in the *News* in 1867. That same year, he became a citizen of the United States and married Virginia Fouregard of Charleston, who died of tuberculosis in 1873. Dawson and Riordan bought the Charleston *Courier* in 1873 and formed the *News and Courier.* A political conservative but a progressive businessman, he turned the *News and Courier* into one of the leading newspapers of the South, championing industrialization, diversification, and scientific farming.

In 1874, the widowed Dawson, forty-one years old, was married to Sarah Morgan, thirty-one, by Patrick N. Lynch, bishop of the Catholic Diocese of Charleston. Sarah was a semi-invalid, but the Dawsons had a happy marriage, and they had three children, two of whom survived to adulthood.

A vocal advocate for the business-oriented Bourbon class—as was his friend, Courtenay—Dawson was more liberal than many of his contemporaries. In 1876, he endured severe criticism by declining a challenge to duel Gen. Matthew C. Butler, as dueling was contrary to his principles. In 1880 he instigated a concerted effort to ban dueling in South Carolina and secured passage of a South Carolina law barring it. For his fight, Pope Leo XIII made him a knight of the Order of St. Gregory the Great, a papal honor that would stand in bizarre contrast to his later demise.

His devotion to duty on the night of August 31 - September 1, 1886, when he rallied a few stout-hearted employees to print the September 1 edition of the newspaper under life-threatening conditions in a building ominously near collapse, earned him nationwide recognition for valor. Dawson became Charleston's chief publicity agent on September 3, when he wrote a nationally syndicated article titled "Charleston Ready for Business," which clearly stated that, despite the earthquake, all of the city's main business facilities were operational.

Charleston, a proud and independent city, had graciously turned away numerous offers of aid after the Great Cyclone of 1885. But now the devastation was far too great, and the city far beyond its capacity to heal itself unaided. Even before reaching Charleston, Mayor Courtenay had taken the lead in securing aid to rebuild the city. In his September 5 interview in the *New York Tribune*, which was reprinted throughout the nation, he had made his agenda clear: "The people of Charleston must have prompt aid to tide them over the next few days. After that they will recover rapidly, for they are plucky and will do all they can to gain their feet again." The Charleston earthquake immediately became national news. Not only did newspapers nationwide send reporters to Charleston to cover it, but national magazines, such as *Harper's Weekly* and *Frank Leslie's Illustrated Magazine*, put out special earthquake editions in September.

In sending an earthquake to visit Charleston, fate had dealt the city a doubly cruel blow. The first was the massive damage itself. The second was the fact that even those prudent property owners who had bought insurance policies covering the standard hazards (wind, fire, and water damage) found out that earthquake damage was not included. Charleston would have been much better off if its buildings had burned rather than been damaged by earthquake. Had that been the case, insurance would have paid to repair the damage. After the earthquake, the only people who collected on their general hazard insurance policies were those whose buildings were destroyed by fire.

When it came to obtaining emergency funds for relief and rebuilding, thoughts naturally turned to the state government in Columbia. Yet from the day of the earthquake until the E.R.C. formally closed its books on July 25, 1887, and despite numerous appeals from mayors, delegations, and influential businessmen, no agency of the South Carolina state government ever sent aid of any kind—food, tents, or money—to the earthquake victims. The reason was simple: politics.

The wholesale abandonment of the earthquake victims by their own state government started at the top. On July 10, 1886, six weeks before the earthquake, Colonel John C. Sheppard of Edgefield, a second-term lieutenant governor, took his seat as interim governor of South Carolina. The governor's chair had been vacated when President Grover Cleveland appointed Governor Hugh S. Thompson assistant secretary of the treasury.

An Edgefield correspondent for the *Atlanta Constitution*, which was always ready to leap on a juicy story and pop the balloons of over-inflated egos, wrote that "Colonel Sheppard has long had a burning desire to be governor of South Carolina and now that longing is to be gratified. While it is an accident, yet he is not the less proud of his temporary reign. Colonel Sheppard's chances of political preferment were excellent until he became mixed up with the unfortunate Culbreath murder case. His connection with this case destroyed his political future. But, nevertheless, he is going to be governor of this state for nearly six months. His friends here [in Edgefield] are very proud of his advancement."[2]

Sheppard's political future had been presumed doomed by his involvement as an attorney in the notorious Edgefield lynching of a black man, O. T. Culbreath, who had been accused of killing a white man. Culbreath was attacked in the offices of his defense attorneys, dragged into the woods, beaten severely, and left to die by a thirty-man lynch mob. He survived long enough to crawl back to his lawyers' office and name his attackers, all well-known men of Edgefield. Colonel Sheppard, while serving as lieutenant governor, was one of the lawyers who defended the members of the accused lynch mob.

The same month that Sheppard ascended to the governor's chair, the South Carolina Democratic party was nominating delegates to its state convention, to be held in Columbia in August. Few of the delegates seemed interested in nominating Sheppard to run for a full term of his own. Col. John Peter Richardson, of Clarendon County, then the state's third-term treasurer and an announced candidate, was said to be "a man of the people, and everybody likes him." However, the previous November, the *Atlanta Constitution* had pointed out that another highly respected man—one who was not officially in the race—could easily win if persuaded to run: William A. Courtenay. "He has made a great success of his administration as mayor, and has the ability to be as good a governor."[3] Again, on July 6, 1886, the *Constitution* came out strongly for Courtenay. "There is one thing certain—Captain Courtenay would make the best governor South Carolina has had since the war."[4] The importance of Courtenay's popularity was not lost on Sheppard. If he wanted to be elected governor for a full term of his own, Col. John Richardson posed a threat, but Courtenay, if drafted, would be pure poison for Sheppard's tottering career.

In the aftermath of the earthquake, Governor Sheppard seemed sympathetic. He telegraphed Charleston on September 1, "I am appalled at the advices of destruction of life and property in Charleston. Inform me of the extent of the calamity. What can I do to alleviate the distress?" His query reflected the reality that there was no state government emergency response plan to deal with natural disasters. Two days later, Sheppard wrote to Captain Dawson that he was extremely distressed by the reports of mass suffering he had read, and that as soon as he heard from the mayor, he would issue a proclamation and "ask the whole people to come promptly to the rescue." Sheppard was careful not to imply that the state government itself might be a source of direct aid.

Sheppard arrived in Charleston at half-past twelve on September 5 to make a personal evaluation of the damage. He met with Acting Mayor Huger and other city officials, and was driven through the city by carriage, returning to Columbia the same day on the afternoon train. Appalled as he was by the damage, he had three constraints. The state constitution did not provide for disaster relief funds of any kind. It prohibited the use of state funds for the benefit of any private party, and limited expenditures to be used solely for state-owned property, such as the Citadel. It specifically forbade the use of state money on property not owned by the state. That use would have required an amendment to the state constitution, a fight he was unwilling to undertake. Although some legislators supported calling a special session of the legislature to change the constitution, most of those from the Upcountry—Sheppard's homeland and primary constituent base—did not. The support only weakened as time wore on, as it became clear that the humanitarian aid—temporary food and shelter—was being financed by outside donations, and that the bulk of any new funds would be spent repairing private property. Sheppard seemed "intimidated to a degree by the state's powerful legislature," and "powerless and maybe unwilling to mount any full-scale campaign of state aid."[5] Despite support from leading daily newspapers, including the *Columbia Register* and the Orangeburg *Times and Democrat*, Sheppard never made any attempt to call an extra session of the legislature to amend the constitution and get relief funds for the earthquake victims.

Never a leader in the relief efforts, Sheppard nevertheless offered moral support through a proclamation on September 6. He noted that "the people of South Carolina have heard of the calamity that has befallen Charleston and Summerville; the representations of the distress there have not been exaggerated." After describing the magnitude of the damage, he appealed to individual South Carolinians "to contribute as promptly and generously as their means will permit to the relief of our afflicted fellow-citizens," and assured all that their money would be wisely expended. Ultimately, the South Carolina state government struck a benevolent pose, urged help, but never furnished a dollar, a tent, or a box of food to the earthquake relief effort.

In 1886, no agency of the federal government had any official responsibility to provide aid or relief to victims of "acts of God." As a Massachusetts newspaper put it, "Civilization has made no provision yet for spreading the burden of such a calamity as a serious earthquake. For the most part, it has to rest where it falls."[6] Some South Carolina legislators suggested that an aid appeal be made to Congress. The prospect of congressional support was quickly doomed, however, as soon as it became obvious that South Carolina was seeking federal relief for a state disaster that the state's own government would not assist.

The U. S. Army tents that had arrived in Charleston the first week of September proved to be the only physical aid to come from Washington. Sensing that no federal money would be forthcoming, the Charleston bankers and businessmen who made up the Merchants' Exchange and the Cotton Exchange met in joint session and adopted a resolution to petition the president and Congress for a federal loan of $10 million ($201 million today) to be used by owners of Charleston residences and business buildings to finance rebuilding at low rates over a long period of time. They wrote to the president, "A city has been wrecked and its people are without means to rebuild it. We conceive it our duty to lay the situation before you, as the head of the nation, to the end that if possible steps may be taken to afford a prompt national aid by loan to restore this city."[7] On September 7, the president of the Charleston Merchants' Exchange, told his counterpart at the New York Chamber of Commerce, "We trust that the next Congress will grant the same."

The federal loan proposal failed, for it faced some strong and unexpected opponents. One of them was William L. Trenholm, a Summerville resident and son of George A. Trenholm, the second Confederate Secretary of the Treasury. After the war, he entered his father's banking business, and after being appointed to the Civil Service Commission by President Cleveland, served from 1886 to 1899 as U.S. Comptroller of the Currency. Trenholm had many allies in the world of private banking and viewed the possibility of low-interest federal loans as competition for commercial lenders.[8] With such a highly placed native son opposing federal help, supporters of the proposed federal loan found few friends, and the necessary legislation was never enacted.[9] Even Hugh S. Thompson, a former South Carolina governor, then assistant secretary of the treasury, could do little more than seek personal donations from the department's employees, which would be passed on to the Charleston relief committee.

Ultimately, the federal aid to Charleston consisted of the one hundred tents; the use of a revenue cutter to transport relief supplies from Wilmington, North Carolina; and the use of U.S.S. *Wistaria* as temporary shelter for some of the refugees. The government also sent the team of scientists chosen by the Federal Earthquake Commission to study the event and provided architects and engineers to inspect both federal and private buildings for safety. No federal relief or rebuilding loans or grants were ever approved. Nevertheless, the private citizens of the nation's capital collected $11,767.89 for the Charleston cause.

Charleston's financial salvation, it was soon clear, would have to come from private, rather than public, pockets. In an independent initiative, a group of major Charleston businessmen met on September 9 with representatives of the New York Chamber of Commerce to organize a trust fund, secured by property mortgages, to raise funds for the rebuilding of Charleston. Loans at only three percent annual interest would be available to repair business, church, charitable, fraternal, and residential structures.

Of the private venture, the *New York Star* stated, "It would be a simple matter for a syndicate of Northern capitalists to raise the money for a loan of $5,000,000 to the city of Charleston for the purpose of rebuilding the city and it would be a loan based on the amplest security, to which unbounded gratitude would give additional strength.... If such men as G. W. Childs, Andrew Carnegie, and a dozen others easily named, who are known to the world as large-hearted philanthropists, would set the ball rolling, a few months would see the thing accomplished." The problem with the privately financed loan, however, was that the city would not be the guarantor of repayment. That would come from mortgages on the buildings for which money was loaned—and a large number of the borrowers would be poor people who had little or no ability to repay. Faced with this lack of repayment assurance, the private loan initiative never materialized. Mayor Courtenay was now fully convinced that he had to take his appeal directly to the American people.

Most of the members of the county council had been adamant about not seeking outside aid after the 1885 cyclone. In order to assure that he now had their support to solicit aid nationwide, Courtenay brought the subject up for discussion and ratification at the council's meeting on September 14. This time, there was little opposition. After a brief discussion, Alderman Augustus W. Eckel, a King Street druggist, offered a motion that was unanimously adopted by all twenty council members present. It noted that the generous donations already received "will be wholly insufficient to meet our unexpected exigencies," and authorized the mayor "to prepare and issue an address to the public setting forth our condition and invoking additional aid for this stricken city."[10]

On September 17, Mayor Courtenay took his proposal to the world. In a simple and businesslike proclamation, published in all the city's newspapers the next day and distributed nationwide by the Associated Press telegraph wires, the mayor repeated the city council's authorization for the appeal, adding only that "in making known to the general public this declaration of the municipal government as to the condition of our city at this time, it seems to me unnecessary that I should add any words of my own. The unfortunate facts are before the country by the statements of disinterested visitors, from different parts of the land, after personal observations, and are known here and deeply felt."

The simple, unembellished request triggered an outpouring of warmth, sympathy, and financial support the likes of which had never been seen before in the South. People of every state, region, religion, occupation, and age group dug deep into their pockets and sent money. Thanks to the enormous publicity provided by newspapers nationwide, charity fundraising events were launched immediately in every state and continued for months. The immense outpouring of sympathy and money was probably aided by the geographic expanse where the earthquake was felt. Two-thirds of the population of the United States actually felt the quake to some degree. In the months to follow, the *News and Courier* glowed with thanks, praise, and appreciation for the kindnesses bestowed upon the suffering city.

While the aid and comfort of South Carolina's Southern neighbors was as heart-warming as Charleston might have expected, the outpouring of warm sentiments and cash donations from another source was more than amazing. From the most unlikely of places—the heart of the Union, whose troops fought in combat against the Confederates for five gruesome, bloody years—there was a massive outpouring of sympathy. Almost immediately, Northern money flowed like water.

One explanation for the unexplained blessing was put forth in the *New York Mail and Express* on September 3. It noted the South's mourning of the 1881 death of President James A. Garfield of Ohio, a former Union general, and in 1885, that of Ulysses S. Grant, another former Union general who, as president, had presided over most of the Reconstruction period so hated by Southerners. The *Mail and Express* wrote, "so the North, rising above any recollection of those portions of Charleston's history which were most odious to us, should make the terrible calamity the occasion of an outburst of tender friendship and generous aid that will be worthy to be chronicled and remembered to all time, as the most patriotic and noblest demonstration of the greatest height and breadth to which human nature, in the mass and on a large scale, has ever attained." That dream was fulfilled. The South's former Civil War adversaries lavished upon Charleston every form of goodwill, sympathy, and financial aid imaginable.

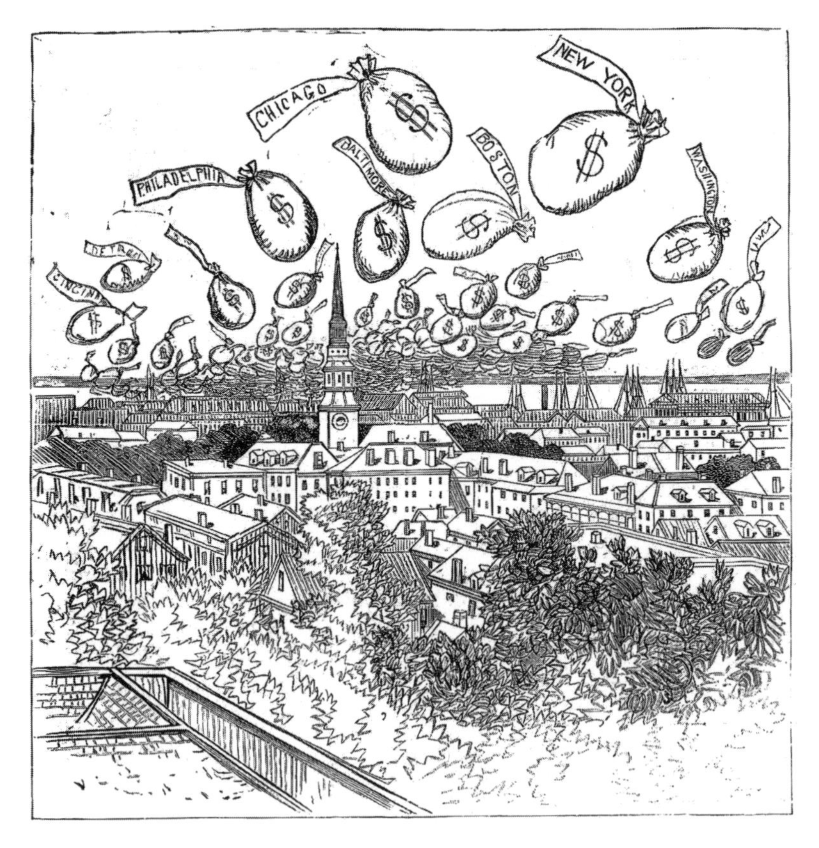

"As Charleston has had an earthquake, let us all 'chip in' and give her a charity cyclone."

In early September, the Rev. E. C. L. Browne, pastor of Charleston's Unitarian Church, received a letter from Edward L. Galvin, superintendent of the Young Men's Christian Association in Chicago, asking the extent of Charleston's need and offering to inform the American Red Cross. Browne answered that "money was the great comforter now needed in Charleston."

As soon as she heard of the catastrophe, Clara Barton, then sixty-four years old and internationally famous as the founder and head of the American Red Cross, came to Charleston to see if she could be of service. A native of Oxford, Massachusetts, she was born on Christmas Day, 1821, and spent her entire adult life collecting and distributing medical supplies and nursing the sick and wounded. Barton was no stranger to the South. Operating first from a base in Washington, D.C., she provided relief services to the soldiers at numerous Civil War battle sites in Virginia. In 1863, she moved her base of operations to Hilton Head Island to be closer to her brother, U.S. Army captain David Barton, who was stationed at the federal stronghold and assault base there. There she cared for the Union soldiers wounded during the furious and futile attack on Battery Wagner (located at the mouth of Charleston Harbor on Morris Island) and distributed supplies to them after the failed siege of Charleston. In 1881, she founded the American Red Cross and continued her wide-scale nursing efforts.

Clara Barton

Barton arrived in Charleston on September 27 in the company of Dr. J. L. Hubbell, the national field agent of the Red Cross. They stayed at the Charleston Hotel, where she gave an interview to a reporter from the *News and Courier*. After describing the organization's international humanitarian activities, she carefully pointed out that "the Red Cross was not a secret society, but the powerful, peaceful sign of the widest, humane charities the world has ever seen." She had been in the West when the earthquake occurred, she said, noting that "if I had been at home in Washington when this thing occurred, we should have been here within twenty-four hours." She told the reporter that the Chicago Red Cross Relief Society had already forwarded $2,000. After visiting the improvised City Hospital in the Agricultural Hall, "she expressed much gratification at the self-reliance and courageous conduct of the people of Charleston," wrote its medical director, Dr. Horlbeck, and "said that she had nothing to suggest, as everything seemed to be admirably organized. As an expression of her sympathy, she donated $100 to the inmates, which was distributed by Miss Alma Jenkins, one silver dollar to each patient." Barton also donated similar amounts to the Orphan House, Confederate Home, House of Rest, and Old Folks' Home, for a total gift to Charleston's caring institutions of $500.[11]

Upon her return to Washington, Clara Barton wrote to Mayor Carter H. Harrison, Sr., of Chicago, confirming the extent of Charleston's losses and encouraging him and his city to be generous in their relief efforts. He did not disappoint her. The Elgin Watch Company, the nation's largest manufacturer of clocks and timepieces, sent a donation of $630; Armour & Co., the nation's largest meatpacker, sent $1,000; and the Produce Exchange wired $125. Within months, Illinois had contributed $13,992.47 to Charleston's relief.

Oddly enough, one of the first boosts to Charleston's economy was tourism. In the first week of September, as Charlestonians were tending their wounded and clearing fallen bricks and timbers from their homes and businesses, a new curiosity arrived on the scene: earthquake tourists. Although an estimated 2,500 white and 3,500 black residents had left the city in fear of their lives, others flocked in to see the carnage.[12] Within a few days after the first shock, with the earth still shaking on a daily basis, excursion trains full of fascinated gawkers rolled into the city in the morning, were driven through the rubble-strewn streets by eager carriage drivers, and left by return train the same afternoon.

The *Atlanta Constitution* wrote that on Sunday, September 5, 1886, "The streets were dusty and hot. . . . Had it not been for the arrival of two excursion trains bearing 1,200 people from Savannah and intermediate points and religious services in the open air, the day would have been an exceedingly quiet one."[13]

The visitors were curious to see the effects of the earthquake. "All during the morning and afternoon the streets were alive with people inspecting the ruins both in carriages and on foot. The demolished buildings, gaping walls, novel street scenes and other evidence of the earthquake to be seen on all sides were viewed with interest and wonder not unminimized by awe," the newspaper reported. "Nearly one thousand visitors were brought over on the Charleston and Savannah Railway alone, arriving here about 12 o'clock, two trains of twenty cars. About 800 of them returned at 4:50 P.M., while the remainder stayed over until 12 o'clock last night and went back by special train." From September 10 to 14, the railroads, and the lines they connected to, ran special excursion trains in from Gainesville and Jacksonville, Florida; Bainbridge, Jessup, and Savannah, Georgia; and Chattahoochee, Tennessee. The lines generously donated half of these ticket sales to the E.R.C. relief fund. Earthquake tourism remained a strong business throughout the fall of 1886.

As morbid as some Charlestonians viewed it, the disaster-seekers served several useful purposes. The railroads had taken heavy financial losses after the earthquake, transporting thousands of evacuees out of the Lowcountry for free or at greatly reduced fares. In addition, the emaciated condition of the city's trade had cut off most of their usual freight revenues. The earthquake excursion trains, with their fare-paying passengers, helped fill the financial void. On September 27, the newspaper noted that a single excursion train run by the Charleston and Savannah Railroad brought in five hundred people. "The crowd was so large that the seven regular coaches comprising the train were insufficient to accommodate it, and many of them were obliged to get into the baggage car." Another, run by the Atlantic Coast Line on September 31, was billed as "the monster excursion train." The railroad donated all proceeds to the earthquake sufferers. The trains started out from Richmond, and the newspaper reported that "the entire motive power of the Coast Line has been prepared for service on this occasion, extra cars and locomotives being parked at different points along the line, where they can be used at a moment's notice."

In the city itself, the tourist influx helped fill hotel beds now unoccupied by the usual hordes of business travelers, who, with commerce at a near-standstill, were now scarce. Stable owners could not keep up with the demand for horse and buggy rentals, and carriage services were booked solid. The waves of people also boosted revenues for the city's two trolley services, as well

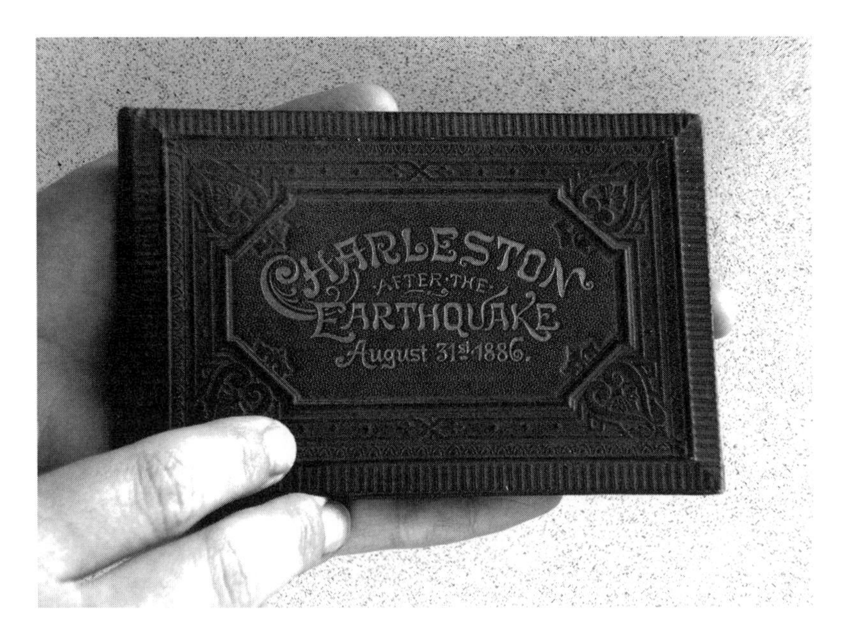

An earthquake tourist booklet

as its restaurants, who again had hungry, financially solvent mouths to feed. On any given day, nearly four hundred people dined at the Charleston and Pavilion hotels.

Innovation followed close on the heels of the disaster. As the number of tourists swelled into the tens of thousands, alert entrepreneurs came up with products to satisfy the eternal need for tourist souvenirs. In this case, at least three varieties soon appeared: earthquake photos, earthquake brochures, and even earthquake beer.

On September 14, 1886, the city council adopted a resolution authorizing Mayor Courtenay "to have large photographs taken of the public buildings in their present condition as shattered by the earthquake of the 31st August ultimo, with the hope of hereafter having these views preserved illustrative of this momentous epoch in the history of Charleston."[14] Charleston's most prolific photographer at the time was George LaGrange Cook, whose well-known studio on King Street turned out thousands of portrait prints each year. Almost as soon as the earth stopped heaving, whether by commission or on speculation, Cook was out with his half-plate view camera. He ultimately recorded over two hundred views of damaged and collapsed buildings, the destruction of railroad track beds, and exotic earthquake phenomena, such as craterlets, as far as twenty-five miles from Charleston. His laboratory made thousands of copies of the pictures, which were widely advertised and sold throughout the city and by mail order.

Cook had plenty of competition. It was soon clear that printed booklets of earthquake views would also be best-sellers, and a gaggle of other photographers joined the tourism-fueled frenzy. They included W. H. Fairchild of Charleston; Dr. E. P. Howland, a Washington, D.C., dentist who was also a skilled photographer; J. A. Palmer, of Aiken; J. H. Wisser; and William Ernest Wilson, a portrait and landscape photographer from Savannah. From the pictures they all produced, busy printers including Charleston's Walker, Evans, and Cogswell; A. Williamson, of New York (sold through their exclusive Charleston agent, C. C. Righter); and the Kensington

The Palmetto Brewery's Earthquake Beer

Art Studio, a busy portrait photography studio in Brooklyn, New York, all created picture albums for sale. Even the railroads got into the act, producing a sixty-four-page, heavily illustrated booklet titled *Charleston As It Is After The Earthquake Shock of August 31, 1886*, only one copy of which is known to have survived. One especially popular tourist trinket was small corked bottles of multi-colored "earthquake sand," harvested by enterprising local people who lived near the numerous sand blows and sold the geological curiosities for up to one dollar a bottle. But by far the most refreshing of all the tourist goods was Earthquake Beer, "the only Beer made entirely out of Cistern (Rain) Water, Filtered and Condensed, and the Choicest Hops and Malt." It was brewed exclusively by Cramer & Kersten's Palmetto Brewery on Hayne Street, which had narrowly escaped the total destruction suffered by its next-door neighbor.[15]

On August 31, in Hartford, Connecticut, the nation's insurance capital, the *Courant* reported that in one house, a rumbling was heard, followed by a swaying of the house, which moved chairs, and stopped a clock. At another home, a vase was upset, chandeliers swung, and one of the residents became nauseated. Elsewhere in the city, plaster fell, picture frames were shoved ajar, billiard balls rolled across their tables, and many frightened people rushed from tall buildings in their night clothes. At the U.S. Signal Service office, atop a seven-story building, many people were made ill by the rocking.[16] Citizens pledged $1,000 within forty-eight hours of hearing of Charleston's disaster. The First Regiment, Connecticut Volunteers, which had fought in the War of the Rebellion, as it was known in the North, immediately collected and forwarded $100. In all, the Nutmeg State contributed $11,946.06.

Throughout Massachusetts, slight shocks were felt in some places, chiefly in tall buildings. Boston and Charleston had had close commercial ties since well before the Revolution, and Boston had played host to William Courtenay and his fellow delegates during the reenactment of the Bunker Hill in 1885. The city was extraordinarily generous, sending $25,000 in aid, earmarking $2,500 for the Confederate Home, $2,500 for the Medical College, and asking that some be used to rebuild the public schools. At a meeting of the E.R.C. on November 24, Boston's wishes were approved, and $10,000 was designated to rebuild the schools.[17] By the time the E.R.C. closed its books, citizens of the Bay State had donated an amazing $99,878.61, of which Boston contributed $76,902.35.

During the night of August 31, people sleeping in second- and third-story rooms in Albany, New York, distinctly felt four shocks, which woke many, produced a rolling sensation, shook beds, rattled windows, and scared children. Although the earthquake had not been felt at all in the small town of Oswego, the members of the 29th Separate Company of the New York State National Guard nevertheless organized an evening of entertainment at the State Armory there on September 13. Tickets were sold for only fifteen cents, and a total of $201.75 was raised from all sources there.

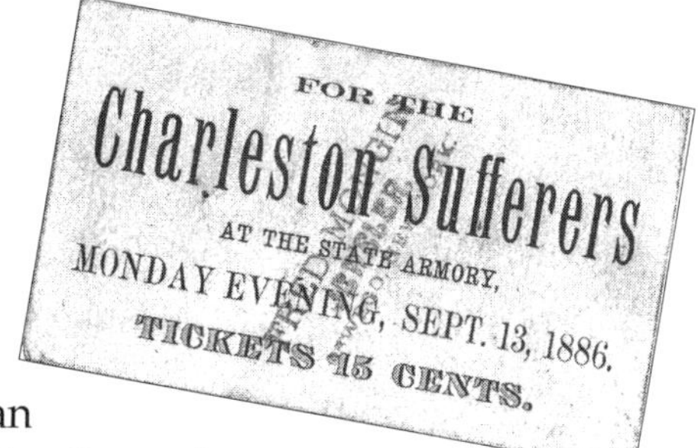
FOR THE
Charleston Sufferers
AT THE STATE ARMORY,
MONDAY EVENING, SEPT. 13, 1886.
TICKETS 15 CENTS.

In New York City, Prof. D. E. Bardwell was walking on the Brooklyn Bridge with two ladies when "all felt the walk[way] yield and rise two or three times as though the cable had been tightened and loosened; the river, previously smooth, became covered with white-caps."[18] The city's wealthy capitalists and their banking and commercial firms all opened their wallets after hearing of the catastrophe. The New York Stock Exchange's handsome gift of $5,000 (equal to $100,500 in present funds) was a shining light for its colleagues. Three major financial firms donated $2,500 apiece. Fourteen $1000 donations were made by millionaires, including William K., Frederick W., and Cornelius Vanderbilt and John Jacob and William Astor. Soap maker Enoch Morgan's Sons & Co. sent $500. Miss Annie and Master Peter Wagner were able to donate twenty-five cents apiece, and J. A. Perkins of New York wrote to the E.R.C., "Please credit the enclosed [five cents] to Fund of Sufferers. It was given to the undersigned for that purpose by a little boy of 5 years."[19] In all, New York State contributed $181,827.04, of which the Chamber of Commerce alone raised $83,000.[20] Through its largesse, New York became the E.R.C.'s largest donor, contributing just over twenty-eight percent of all funds donated.

Given its moderate size, Baltimore raised an astonishing amount of money for Charleston. There had been no mistaking the earthquake there, and most people immediately recognized it for what it was. The *Baltimore Sun* reported that citizens "rushed affrighted into the streets," and one lady "ran from her house *en dishabille*."[21] At the office of the *Baltimore American*, the printers were frightened when the building heaved, moving the heavy type cases six inches. At Mrs. M. M. Piggot's house, her sewing machine rolled back and forth on the floor and the tassels on her furniture swayed.[22] Mr. G. W. C. Krebs was reading on the second floor of his house when the first shock hit, rattling the door, throwing the chandelier swinging in an arc, and leaving him feeling nauseous.[23] From Baltimore, Feliz Agnus, publisher of the *Baltimore American and Commercial Advertiser*, wrote that they were soliciting donations for the relief of Charleston's sufferers, noting that the newspaper itself had started it with a $500 donation of its own, which quickly swelled to $1,500. "The hearts of the people of Baltimore, a sister city, go out to you and will not fail you in this the hour of your great misfortune," he assured Charlestonians. Rieman & Co., a Baltimore provision and commission merchant firm, telegraphed on September 2, "expressing our sincere sympathy for your community," and sending $100 for the victims.

Jacob Frye, Baltimore's police marshal, deeply admired the devotion to duty of Charleston's police force. He forwarded to Mayor Courtenay $700 donated by his men, writing, "The fact that your force never missed roll-call, although their own families shared the common peril, and that they were never absent from duty, makes the action of our own force a greater pleasure." The Baltimore police also sold tickets to a comedy performed at Ford's Opera House on September 8, which netted $5,000. Their work was so efficient, and the people of Baltimore so generous, that over three thousand people wanting tickets had to be turned away at the theater. When Mayor Courtenay endorsed the check, he wrote on it, "The momentous question that has come down to us through the centuries—'Who is my neighbor?'—has been answered." In the first two weeks after the quake, Baltimore officials estimated that as much as $25,000 might be raised. In all, Maryland gave $39,216.42, of which Baltimore contributed $37,966.04.[24]

In Canada, the quake was felt in several places in Ontario, and in Toronto, it was recorded by instruments. The meteorological service of Canada had installed a set of magnetographs—instruments that measured the earth's magnetic field intensity. These also functioned as seismoscopes, measuring, with great precision, the arrival time of earthquake shocks on August 31.[25] Canadians, few of whom felt the shocks, donated $116 to the cause.

The event attracted such worldwide attention that the *Illustrated London News* and Rome's *L'Illustrazione Italiana* ran major stories on it. From London on September 3, Her Majesty Queen Victoria sent a note of condolence to President Cleveland, writing "I desire to express my professed sympathy with the sufferers by the late earth quake & await with anxiety fuller intelligence which I hope may show the effects to have been less disastrous than expected." In addition, the Lord Mayor of

London opened a subscription to aid the earthquake victims, and collected $15,466.15. In all, foreign contributions from Canada, England, France, Germany, Holland, Shanghai, and Spain totaled $19,317.89, just slightly more than total contributions from the citizens of South Carolina, which came to $19,235.69.[26]

By far the most insightful article in any foreign publication was written by José Martí. A thirty-three-year-old Cuban-born, anti-colonialist intellectual, writer, and political mentor for Latin America, Martí published an extensive, two-part description of the earthquake in Argentina's *La Nación* newspaper on October 14 and 15, 1886. He created vivid portraits of the events, which he had read of in newspaper reports from New York. It is possible they may have been forwarded to him by the Charleston consul for the Republic of Argentina, Motte Alston Pringle, or the Spanish consul, Nicanor Lopez y Chacon.

His article described in gripping detail horrors of the night as experienced by the Charlestonians. "The ground was wavy, the walls cracking; the houses shook from side to side; the people—almost naked—were kissing the earth: 'Oh Lord, Oh, my beautiful Lord!' smothered voices would pray, crying, 'A whole portico came down!' Courage fled from the heart and all thought was blurred; it's stopping now; it's shaking less; it's stopping! The dust from the fallen houses rose above the trees and the roofs." Marti devoted most of his space to the idea expressed at the start of his article: that earthquakes are indispensable land movements, necessary to bring the earth back into equilibrium, and that "man is a warrior who rises once again during disasters to continue struggling for life." In the closing lines, he portrays the event as an example of new life arising from the rubble through the tale of the birth of twins in one of the emergency tents put up for the victims.[27]

From every walk and station of life, people helped the earthquake victims. One touching contribution to the relief fund was $1, sent by Miss Lily Logan, a poor working girl of New York. A nine-year-old girl from Beaufort, North Carolina, asked her grandfather to send a five-dollar bill to Charleston as her offering. She had found it when she was five years old, and when she heard of the distress of the children in Charleston, insisted that the money be sent there. In the same spirit, the children's class of the Bainbridge, Georgia, Baptist Sunday school sent $1.05. Another little girl, whose savings consisted of $5, was prepared to donate $4 to the sufferers, but had a change of heart. "No, I will send it all," she wrote, saying that "of course it wouldn't be very acceptable to God" if she kept part of the money to use solely for her own enjoyment.

The Reverend E. T. Curnick, pastor of the Uxbridge, Massachusetts, Methodist Episcopal Church, sent in a contribution of $2 on September 20. He said it was from "two poor colored women, who work as domestics.... They gave out of their poverty, & I am sure the donations will be owned of [esteemed by] the Father of us all."[28] From Danbury, Connecticut, Miss Eva Bates wrote that she and her cousin, Annie Servoss, "felt so sorry for the Charleston sufferers we wanted to do something to help them, so we have been making paper flowers and selling them, and in this way we have raised $15 ($301.50 today) for which you will find a money order enclosed."[29] On September 17, Joseph H. Oppenheim of Savannah sent $25 to Mayor Courtenay, noting that it came from the sale of a satin tidy [embroidered cover for the arms or back of a chair] made and raffled by his daughters.[30] And in Saccarappa, Maine, the children of the "Wide Awake Club" of the local Congregational church held an evening of entertainment that raised $46.70 ($938.67 today) for the victims. The director of the event was Mrs. Alfred Sweetser, whose husband was a Union army veteran and an active supporter of the event. The note that accompanied the donation expressed the hope that the donation would "help to bring the glad day when there shall be no 'North' or 'South' in our land, but *one* single sentiment of love *in* all for all & *unto* all."[31]

Every leader who leads the recovery from a mass casualty disaster has to deal not only with the huge number of suffering people but also with an occasional lunatic. On September 26, Mayor Courtenay received a raging complaint from an alleged minister of Buffalo, New York. He had sent in a subscription of $5.00 for the relief effort, collected from "Jesus Christ, $1.00; Apostel Peter, $1.00; Apostel Judas Iscariot, $1.00; Bell Ashworth, $1.00; and Heaven's New Born King George Frederick Trostel." He received a prompt acknowledgement, but deemed it "utterly unintelligible," and went into a rant. He replied to Mayor Courtenay, "Now You Southern Hot Head, I will wake up your ideas, I am Jesus Christ the Son of God.... If the Bombardment during The Late Rebellion, And the more Recent Earthquake has not learned You that There is a God that Governs this Universe I will Teach you better." He continued that if he did not receive a satisfactory response, he would have Courtenay "Hung Untill Dead, Your Carcass will be more useful than open Rebellion."[32]

The single most amazing feature to emerge from the national fundraising relief effort was not the prominent role of Northerners—although that alone was worthy of serious note—but rather the personal involvement of former Union soldiers. On September 3, the deputy commander of the headquarters department, Grand Army of the Republic (G.A.R.), issued a call to all of his states' posts to take up charity relief subscriptions for the sufferers of Charleston. Formed in 1866, the G.A.R. was composed of Union soldiers and sailors who had survived the war and gone home, yet missed the friendships and camaraderie they had shared during the war. That same day, C. E. Goodrich, a G.A.R. post commander in Nyack, New York, wrote Mayor Courtenay to notify him that his post had started a subscription for relief funds among his own comrades and in the village itself. Goodrich had been a member of a Union army unit that had occupied Charleston at the end of the war. He closed his letter by stating, "Those of us whose privilege it was to spend the spring and early summer of 1865 in your city cherish with pleasure the recollection of those few months, and the friendships for me there which have continued to the present time, and our hearts go out now in sympathy to friends and strangers alike in this hour of their affliction and suffering."[33]

On September 13, ex-Governor (and former brigadier general) Lucius Fairchild of Wisconsin, commander-in-chief of the G.A.R., and Col. E. B. Gray arrived in Charleston. They spent the day examining the effects of the earthquake and consulting with Mayor Courtenay and the E.R.C. After Fairchild's visit, he urged every G.A.R. post in the nation to contribute to the relief and rebuilding of Charleston, noting that the local authorities had already made exceptional efforts to shelter and feed the needy. In Charleston, the *News and Courier* told its readers, "What we want right now is to hold up to the Union the glorious fact—equally credible to them that offer and to them that take—that, in God's providence, the city whose very name was synonymous with disunion and strife—however justifiable—is through its poignant misfortunes made the means of demonstrating to the Republic, and to the civilized world, that all Americans are kin, and that this is indeed, and in very truth, one country, one people." The result of Fairchild's appeal was that G.A.R. posts throughout the United States raised tens of thousands of dollars for the rebuilding of Charleston. The John G. Dix Post alone, in San Jose, California, sent $1,000, noting that the community of 15,000 residents had already contributed $323.

The churches representing of all of Charleston's religious denominations sustained serious physical damage, and all of the white church groups quickly launched direct appeals to their fellow churchmen and women nationwide for aid to rebuild. On September 6, the First Baptist Church and the Citadel Square Baptist Church requested that Baptist newspapers nationwide publish an appeal for funds for Charleston's Baptists. In addition, they sent out an emissary to visit Baltimore and other cities and solicit funds for both churches. A week later, the Baptist conference in New York instructed its colleagues in Charleston to have appeals printed and sent to New York and other major cities.[34]

Catholics wasted no time raising funds. On September 2, The Reverend Patrick L. Duffy telegraphed the editor of the *Brooklyn Examiner*, "The Catholic churches, orphanages, infirmary, schools, and pastoral residences have been seriously damaged. Miraculous were the escapes of the priests, sisters, and orphans, who are now camping out. I am erecting a temporary altar in the cathedral grounds. Our loss and suffering are great."[35] On September 10, the members of St. Patrick's Benevolent Society in Charleston sent out appeals to their fellow parishioners throughout the United States and Canada, asking that donations be sent to Monsignor Daniel J. Quigley, vicar general of the Diocese of Charleston, or The Reverend Patrick L. Duffy, the chancellor. The Archdiocese of Baltimore contributed $7,000; the Archdiocese of Philadelphia, $15,000; and in all, the Catholic community nationwide contributed $103,207 to the rebuilding effort.[36]

William B. Howe, bishop of the Episcopal Diocese of South Carolina, made a nationwide appeal via the *News and Courier* on September 7, estimating the diocese's losses at "at least $100,000." By the end of September, he had received $13,744.15. The Reverend Anthony Toomer Porter had his own financial connections. When he told the Hon. Fred A. White, a friend in London, that the Church of the Holy Communion and its school would need $20,000 to be repaired, White said that his firm would immediately wire Porter $3,800. A few days later, Porter received a telegram from industrialist J. Pierpont Morgan: "Intense sympathy; draw on me for $5,000."[37]

The congregation of Charleston's French Huguenot [Protestant] Church was especially challenged. The church had suffered damage in from the cyclone of 1885 and now found itself "nearly shaken down and its means of restoration apparently beyond reach" due to the earthquake. Pastor Charles Vedder, wrote that the church and its congregation were saved by the generosity of Charles Lanier of New York, "who furnished the means by which the church edifice was rehabilitated" after Vedder made a personal trip to New York to petition the Huguenot faithful there for help.[38]

The Jewish community bore losses to its places of worship, its houses, and its business buildings. Kahal Kadosh Beth Elohim, the handsome Greek Revival synagogue on Hasell

Street, was not only the birthplace of Reform Judaism in the United States but also the oldest surviving Reform synagogue in the world. It suffered cracks and interior damage, but its main structure and the Greek Doric portico survived. Appeals to the Jewish community nationwide resulted in over 200 contributions amounting to more than $5,000, which were applied to the $7,000 repair cost.[39] Services resumed on September 18. Many Jewish-owned businesses were located on King Street between Broad and Calhoun streets, a zone where almost every building suffered damage. Many of them, like their neighbors, received rebuilding assistance from the E.R.C.

The Lutherans (mostly of German descent) sent out their own appeal letter and called on "our brethren in the United States to help us in this strait."[40] They estimated their losses to be $15,000. On September 12, Mr. O. G. Marjenhoff of Charleston received a check for $50 from the Grocer's Association of Jersey City, New Jersey, which he forwarded to Mayor Courtenay "for the relief of needy Germans."[41]

The Methodist ministers of Charleston appealed to their members nationwide after a meeting on September 8. Trinity, Bethel, St. James (Spring Street), and the Cumberland Street churches all were damaged. The ministers also asked for aid on behalf of the Methodist Church in Summerville. Their chief ally was The Reverend W. D. Kirkland, a Methodist minister in Summerville and editor of the Charleston-based *Southern Christian Advocate*, which he used to encourage Methodists nationwide to assist their stricken churchmen in Charleston. South Carolina Methodist Bishop W. W. Duncan and several of his ministers also personally canvassed Methodist congregations in the South and North for funds.[42]

About September 4, the members of Charleston's First (Scots) Presbyterian Church unanimously resolved to make a general appeal for aid to Presbyterian churches throughout the country. The *Atlanta Constitution* pleaded their case, stating "Their beautiful building is a total ruin and the congregation is impoverished. Contributions can be forwarded to Rev. W. T. Thompson, D.D., pastor." On September 11, the appeal was quickly extended to the other two white Presbyterian churches in Charleston; the Second (Flynn's) Church, and to Westminster Church. The three churches, all located on Meeting Street, estimated their combined repair costs to be about $61,000.

In many cases, it is difficult to distinguish between the donations that originated as a result of appeals and those that came from spontaneous impulses of charity. From Dublin, Ireland, American actress Mary Anderson donated $2,500 to the Charleston sufferers. Born in Sacramento in 1859, she was three years old when her father, a Confederate officer, was killed, leaving her an orphan. Raised by Ursuline nuns in Louisville, Kentucky, she later became a successful actress. Her gift was the net proceeds from performances on the Dublin stage during the second week of October 1886.[43] Aboard the Cunard Line steamer, R.M.S. *Umbria*, the identical sister ship of the *Etruria*, which brought Mayor Courtenay home to the horrible news of his shattered city, Captain McMickham held a concert on board and raised $194 for the victims.[44]

Blacks in Charleston faced special hardships. Because many were poor, they constituted the largest pool of aid recipients. And because they were black, they were the least empowered and most restricted when it came to soliciting relief on their own. Soon after the quake, The Reverend W. M. Thomas, of Laurens, South Carolina, sent a contribution and a note via Captain Dawson. "Reading of the distress of the home of my boyhood, I tender my heartfelt sympathy, and, in behalf of Bethel A.M.E. Church congregation, you will find enclosed $12 for the benefit of the earthquake sufferers." The congregation of Bethel African Methodist Episcopal Church in Beaufort made a similar contribution of $16.02. Dawson's newspaper took special notice of these non-race-based donations from blacks, stating "These contributions were made for the relief of the sufferers of the earthquake, without restriction as to color, and deserve more than a passing mention. In proportion to their means, perhaps, no body of charitable people in the country have given more generously; and the fact that the gift was made by these poor colored people for the benefit of whites and blacks alike is another manifestation of that spirit of charity and sympathy between the races which is among the brightest tokens that have followed the great calamity of last week."

On September 10, the newspaper noted that The Reverend William Henry Harrison Heard, pastor of Mt. Zion African Methodist Episcopal Church, had received a contribution of $26 "for the relief of the colored sufferers of the earthquake" from The Reverend T. G. Stewart, pastor of the Metropolitan A.M.E. Church of Washington, D.C. Similar black-to-black donations followed. This kind of aid did not sit well with the city fathers, who wanted complete control over all the relief funds expended. Although the E.R.C. had no problem with white church groups soliciting separate aid, they were not at all disposed to cede any authority as financial caretakers of the community to blacks.

On September 6, the Charleston Knights of Labor had petitioned the committee for funds to assist its destitute members. Founded in Philadelphia in 1869, the trade union, which was open to blacks and women (and in Charleston, totally shunned by whites), initially maintained extreme secrecy for its own protection. The E.R.C., exercising its plan to maintain central control of all relief operations, advised the Knights that it would "gladly receive from them, as from any other body of citizens, any information that would aid them in ascertaining and relieving the distress in the city, and that any information furnished would be promptly and cheerfully acted upon to the extent of the committee's ability," but that it had no authority to transfer to any other body the funds entrusted to it. Authority or not, the E.R.C. chose to overlook their rule on making block grants to religious or ethnic groups on October 1, 1886. On that day, it resolved to pay Morris Israel, one of the E.R.C.'s hard-working members—and also president of the congregation of Kahal Kadosh Beth Elohim Synagogue—the sum of $1,000 "to be used for the relief of needy Hebrew sufferers."[45]

After seeing that high-quality U.S. government tents were being distributed chiefly to whites, Charleston's black citizens were justly skeptical that they would receive a proportionate share of the relief aid that was starting to flow into the city. A plan to solicit relief aid on its own, apart from the E.R.C., grew out of a meeting held at Mt. Zion A.M.E. Church on September 7. Pastor Heard, stated that he had been led to call the meeting "by the receipt of letters from the various parts of the country, asking to whom they might send contributions for the benefit of the colored sufferers of this city." When he stepped forward to carry the torch for the city's black sufferers, Heard became one of the city's earliest civil rights leaders.

The next day, Samuel W. McKinley, a member of the committee formed at Mt. Zion the previous night, sent out circulars to black newspapers nationwide. The E.R.C. viewed this as a highly dangerous loss of control, and city leaders and some of the South's leading newspapers began to actively undermine the efforts of Charleston's blacks to seek aid by themselves. White Charlestonians had a special distaste for Heard's church. When the Union army occupied Charleston in the spring of 1865, Mt. Zion had welcomed the black troops of the 54th Massachusetts Volunteer Infantry Regiment, the first black regiment from the North during the Civil War. The 54th was particularly detested, as it had led the attack against Confederate troops at Battery Wagner on Morris Island, at the mouth of Charleston Harbor, on July 18, 1863. From April 25 to August 17, 1865, they were part of the Union army force that occupied Georgetown and Charleston. They were mustered out of service at Mt. Pleasant on August 20, 1865.

The Rev. William H. H. Heard

To describe William Heard as a self-made man would be axiomatic, for no black man in the South after The War had any other way to advance himself. Heard was born a slave on June 25, 1850, during the corn-plowing time in rural Elbert County, Georgia.[46] His father was reputed to be Thomas Heard, a wealthy white planter, and his mother was Parthenia Galloway, a slave owned by the Galloway estate. William lived with his mother and her other children, fathered by George Heard, a talented blacksmith, wheelwright, and carpenter owned by Thomas Heard. Because George and Parthenia were owned by different people, they were unable to live together. George was given a pass from his master that permitted him to visit his wife and children on Wednesday nights after he completed his work, but he had to be back at the Heard plantation by Thursday morning. He was also permitted to visit his family between Saturday night and Monday morning.

When the Galloway estate was probated, Parthenia and her three children, including William, were put on the auction block and sold to the highest bidder. Their buyer was Lindsay Smith, a large planter of Elbert County who owned nearly one hundred slaves. Several years later, John A. Trenchard, principal of a high school in Elbert County, bought Parthenia and her children (now numbering four) from Trenchard and set her to work as his cook. The two elder children, including William, became house servants. When the slaves were emancipated in 1865, William was hired by a farmer. A bright young man, he quickly gained for himself an education that included reading, writing, spelling, and arithmetic during six-week school sessions in the summer.

Soon he had become a teacher himself and found work as a three-month-a-year teacher in the black schools, earning $1 per month per pupil. He ended the year with $300—a handsome sum in that time. In 1873 he moved to teach in South Carolina, then under the control of the Radical Republicans, who opened the University of South Carolina to blacks. When blacks—including Heard—were allowed in, virtually all the white professors and students fled the classrooms. Nevertheless, Heard was able to complete just over a year of college education before the end of Reconstruction in 1876. That year, without the yoke imposed by the carpetbag regime and the Union soldiers, local authorities reclaimed their power, re-established white rule, and turned out all the black students. Undaunted, Heard returned to Georgia, taught school, and attended Clarke University and Atlanta University, both black colleges. He later moved to Philadelphia, entered the Reformed Episcopal Seminary there, and continued his studies. In 1882 he entered the A.M.E. ministry, and by 1886, was serving as the minister of the Mt. Zion A.M.E. Church, Charleston's largest black congregation, on Glebe Street.

In his previous preaching assignments, Heard had demonstrated a great talent for recruiting new members, fundraising, and paying off the debt of impoverished congregations. By the time he reached Charleston, he was an acknowledged leader, and perhaps the best-qualified black man in Charleston to address the desperate needs of his race in a segregated city whose only unifying force in 1886 was shared catastrophe.

On September 8, the day after the first organizing committee met, the *News and Courier* published an editorial headed "No Sectional Line—No Color Line." It criticized the action of the black clergy of the city for organizing for the purpose of fundraising and demanded that they cease issuing appeals for aid—even though the black ministers had expressly stated that the funds received would be turned over to the white-run E.R.C. In Washington, the Associated Press published the appeal of Heard and McKinley's committee, but attempted to portray it as a forgery or a financial hoax. The *News and Courier* reprinted the description, which read, "The signatures are not autographs. The document appears to be a copy. It is written upon what was a letter heading from which all but the date line has been cut. The penmanship is business-like and the matter is given above without a particle of editing. The note of enclosure is brief and direct." No similar scrutiny and criticism was made of white Baptists, Catholics, Episcopalians, Huguenots, Jews, Lutherans, Methodists, or Presbyterians that were soliciting funds from their own groups for their own exclusive use, and none of them was required to funnel the resulting receipts through the E.R.C. As opposed to blacks, white religious and ethnic groups were all presumed by the city's leaders to be both ethical and competent.

The *Atlanta Constitution* was indignant that Charleston's blacks might fear being neglected by the white minority. "The appeal of the colored ministers to the people of the United States for means of relief for the colored people in distress, was wholly unnecessary," it said. "The relief committee treats the funds at their command as subscribed for the benefit of the whole community, and as it was given without distinction of race, color or condition, so will it be expended. Up to this time more accommodations have been provided for colored refugees than for the whites, and in issuing rations the committee are no respecter of persons. This has been demonstrated, and the expectation is that the colored ministers will be willing to co-operate with the general committee, instead of acting on their own account."[47]

On September 9, at a meeting held at Mt. Zion A.M.E. Church, the black clergy and their congregants organized an executive committee and handed the minutes of their proceedings to the *News and Courie*r so that the city leaders would understand their aims. The newspaper refused to publish them.

The next day, Heard and his committee refuted the newspaper's accusations of two days earlier, noting that no black person had been asked or permitted to serve on the E.R.C. They also stated that the all-white E.R.C. could not possibly know the needs of the black community, noting, "There are among us, as among the whites, those who are needy, but have too much self-respect to ask for aid and who would prefer to do without rather than to submit to impertinent questioning. A committee of their own race would open the way for relief without humiliation." The *News and Courier* finally capitulated and published a sanitized version of Heard's mission statement.

One of the flyers sent out from Mt. Zion on September 8 reached Harry C. Smith, the publisher of the *Cleveland* [Ohio] *Gazette*, a progressive and outspoken black weekly newspaper. In his appeal to Smith, Reverend McKinley wrote, "Enclosed you will find a copy of the proceedings of a meeting held by the colored people of this city, appointing a committee to appeal to the American citizens for aid in this, our time of distress. Hundreds are homeless and without the means to obtain food or clothing. The papers haven't exaggerated in the least. The situation is appalling. Please publish our proceedings and do all you can to help us. Yours for the race."[48]

Smith took up the cause. He made it clear that blacks outside the South knew what would likely happen if they did not support their black Southern brethren. Under the headline, "Help the Colored Folk in Charleston," Smith wrote, "We call the attention of our many readers to the appeal elsewhere in this paper. Many of our people were sufferers from the earthquake in Charleston, S. C., and colored ministers without exception should see to it that this matter is at once properly placed before their congregations, so that prompt and effectual assistance may be rendered. The large amount of funds and clothing being sent daily to Charleston by the *whites* will reach the *suffering whites,* and unless we take hold at once and assist our own suffering they will continue to be in dire need. Every one of us can do *something* for this cause."[49] He went on to urge his readers, "If you can contribute anything to the Charleston sufferers, do so at once.... Tell your minister to give this immediate attention, *because it is his duty* if he is a servant of God."[50]

Undaunted by the rebukes of the E.R.C. and the scorn of the newspapers, Heard left Charleston and conducted intensive fundraising efforts in the fall of 1886, visiting New York; Baltimore; Philadelphia, Bethlehem, and Pittsburgh, Pennsylvania; Newport, Rhode Island; and Norfolk and Portsmouth, Virginia, to solicit funds for black disaster relief and for repairs to black churches.[51]

By October 5, 1886, thirty-five days after the earthquake and only eighteen days after Mayor Courtenay made his nationwide appeal for relief donations, the E.R.C. made a shocking announcement to the city council. They intended to publicly end earthquake relief fundraising because, in their opinion, sufficient funds had been received or pledged so that no further aid was necessary

This confounded some members of the city council. For the most part, only the most basic emergency repairs had been completed. The vast majority of the rebuilding had yet to even be started. Indeed, there were not nearly enough workers available in the city to undertake the work, even though a flood of contractors and workmen had descended on Charleston in the previous two weeks. But the committee was adamant. It was time to end the fundraising.

First, they reported that those needing emergency shelter, food, and immediate financial aid had been cared for, "and in their judgment there has been such a resumption of the normal life of this city as to justify the discontinuance of this system at an early day." They stipulated that they could not accurately assess the cost of rebuilding the businesses and homes of the city, but wanted "the good givers of this universal charity" to know the limit of the city's actual needs. After reviewing their funds at hand, which on October 5 amounted to $497,309.43, the E.R.C. "placed the committee in a position to substantially help the *needy* sufferers, for whose relief *only* this committee was formed."[52]

The committee then stated that it had never been its role to solicit funds for the use of private citizens and businesses who had access to financial resources that would enable them to rebuild their own buildings. The E.R.C. was determined, at all cost, not to take advantage of the generosity of the American public. It concluded, "As the grateful recipient of the generosity and sympathy of those who have stretched out their helping hand to us, this committee have felt that they should be guided by facts and estimates of the actual loss of needy sufferers, and if on an untrodden path they might err, they preferred to err, by limiting the estimate rather than placing it too high."

The decision, based on the E.R.C.'s rigid interpretation of "needy," provoked controversy. None of the council members wanted to milk the public's generosity, but a number felt that cutting off all access to aid at this early date was rash and premature. They accepted the recommendation as information and tabled it for discussion until their next meeting.

On October 5, the city council resumed its deliberations. Alderman James Adger Smyth moved that the mayor be authorized to withdraw the proclamation of September 17, which asked for aid. Alderman E. F. Sweegan, a merchant, argued against the proposal as, in his opinion, the amount of the contributions received was not nearly sufficient to supply the relief for which the appeal was made. After considerable discussion, the motion was put to the vote. The mayor and nine aldermen voted to rescind the aid appeal; six aldermen opposed it. The motion passed. The city would not seek a penny more than absolutely necessary. By the time all the donations came in, the E.R.C. had received a total of $646,109.90—just under $13 million in present-day funds. It had fed tens of thousands of people during their days of greatest need, granted over two thousand applications enabling citizens to replace destroyed clothing and blankets and purchase lost tools of their trades. It had paid to transport hundreds of indigent citizens to lodging with friends and relatives outside the city. Thanks to the E.R.C.'s heroic efforts, and the charity of a loving nation, the people of Charleston could now stand on their own feet, take a deep breath, and rebuild their beautiful, historic City by the Sea.

16

Bricks, Mortar, and Earthquake Bolts

Gangs of workmen were found at all the buildings occupied or owned by the City.... the sounds of the hammer and saw with the click of the mason's trowel were heard and seen on every hand.
—Ned Cantwell to his cousin, Marie

Charleston, South Carolina
Wednesday, September 8, 1886

With aftershocks still toppling damaged buildings and the ground still shaking under their feet, Charlestonians threw themselves into rebuilding. No one waited for anyone's permission or promise of help. Encouraged by mayors Huger and Courtenay, and urged on daily through the pages of the *News and Courier*, the merchants and businessmen of Charleston resolved to take on all the business they could get as soon as possible. They knew that in the long term, the city's salvation lay in the resumption of its industry. Whatever financial aid came in as charity would one day come to an end. It was only through commerce that the city could generate the cash flow it needed to regain its former status and rebuild its shattered buildings.

On Friday, September 3, the *News and Courier's* lead editorial set the tone for the entire recovery process. "Charleston is ready for business despite the Earthquake and its ravages," it said. "The warehouses give ample accommodation; the wharves are in excellent condition; the [cotton] compresses are fully up to their work; the merchants and factors, undaunted by misfortune, have girded up their loins anew for the battle for commercial life. Charleston, we say, is as well able as ever to transact any business that can be obtained, and is as ready as ever to give to customers, from every part of the country, the commercial facilities they require and the attention and good faith they have a right to expect."

The *News and Courier* did not skirt the gravity of the damage. "The full consciousness of the loss is here. It is understood fully that far more than the earnings of a twelve-month were swept away in less than a minute. But for all of this, and because of this, Charleston is only the more determined to maintain the commercial position it has won."

Charleston's trading partners throughout the country provided assistance in a number of ways. They voiced their support often and publicly through the nation's newspapers, spreading the word that Charleston was open for business and had a working port and harbor, a full workforce, and adequate warehouse space. They urged the nation to send its cotton export trade and orders for goods. The city's leaders directly solicited the nation's larger banking operations, merchant exchanges, and businesses for direct aid and asked for "leniency toward Charleston merchants, some of whom were in arrears in their payments because of the earthquake."[1] Because Charleston's merchants were known to be trustworthy, these efforts were largely successful.

Charleston's businessmen were as plucky as Courtenay had painted them. Henry Steitz, the wholesale fruit merchant whose building at the corner of Meeting and Market streets had lost its entire front face, declared, "Building damaged but business going on as usual. Filling all orders from country customers. We hope for a good fall and winter business and can meet all demands that come."

The owner of the Cameron and Barkley Company at 162 Meeting Street, which sold and repaired steam engines and mining machinery, told the newspaper, "The earthquake has damaged our building to the extent of not less than $25,000 ($502,500 today). The cyclone [of 1885] only cost us $4,000. Our stock is not badly hurt, and will not be unless it rains. We are in position to fill all orders, and have the best stock of bar iron, belting, nails and spikes in the South. Our facilities for handling our stock and filling orders have not been lessened by the earthquake. 'In fact,' said Mr. Barkley with a comprehensive sweep of his hand, 'I don't know but what our facilities for moving engines and heavy machinery have been increased. We can run them out now without difficulty.'" He was probably making a joke at his expense because there were now gaping holes in some of his walls.

East Bay Street, which was built on reclaimed land, was one of Charleston's most heavily damaged areas. A set of three, three-story buildings built with adjoining walls exemplified the capricious nature of the quake's destructive power. These three buildings housed William M. Bird & Co., a major wholesale dealer in oil, paints, and glass; William E. Holmes, a wholesale oil dealer; and Chaffee & O'Brien, a grocery.

The earthquake spared some and destroyed others

On the night of August 31, William M. Bird & Co. saw its three-story brick building at 205 East Bay Street reduced to wooden sticks and brick rubble by the earthquake. The edifice was literally shaken to pieces. The top two stories came crashing down onto the ground floor, spilling wreckage halfway across

William M. Bird & Co. after the earthquake

Cumberland Street. Miraculously, the cast iron, four-column colonnade of Corinthian pilasters on the ground floor and the attractive city gaslight on the street corner, survived the catastrophe. One door to the north, William E. Holmes' business at number 203 lost its top floor, which collapsed onto the second floor, while the next neighbor to the north, Chaffee & O'Brien's grocery store, lost only its cornice. Photographs of the three adjacent buildings, with one, two, and three stories left standing, were published all over the world, and the wreck of the William Bird & Co. store became the best-known graphic icon of the earthquake. William Bird's own handsome residence at 17 Meeting Street was also "very much shattered," and one of its piazzas was completely destroyed. On September 7, Bird told a reporter,

> We have been hit as hard as any one we know, but if badly disfigured, we are still in the ring. We are not disheartened. At 5 o'clock the morning after our store was wrecked we had a gang of men at work, and the store we are occupying

> (175 East Bay) was rented fifteen minutes afterwards. We feel grateful that we got out of it as well as we did. At no other time could the building have fallen without injury to our employees. What tends to soften our losses is the sympathy which comes from every quarter. We are ready for business and are doing it and will work with redoubled energy. In four months we hope to occupy our new building on the old corner.

The company's advertisement stated, "We have relocated at 175 East Bay street during the rebuilding of our store demolished by the earthquake. A full stock now in store and to arrive. We solicit a continuance of the patronage of our friends, and those in need of any articles in our line—Oils, Paints, Window Glass, &c., &c., &c. All orders will be promptly attended to."

Other brave businesses emerged from the ruins. Richard J. Morris, the tinner whose roofing shop had burned down during the lower King Street fire, advertised on September 7 that tin roofing, guttering, and "repairs to same will be promptly attended to on leaving orders with R. J. Morris, formerly in King Street, near Broad. Now at G. W. Egan's, Meeting near Circular Church."

Walker, Evans, and Cogswell advertised as early as September 4, "We are happy to Announce that we are prepared for business in our Paper and Stationery departments, are partially working today in our Printing Office and Bindery and unless prevented by unforeseen circumstances, will in a very few days be as ready for all business as ever."

The South Carolina Railway Company's extensive docks on the Cooper River had not been affected, and only one of its warehouses was severely damaged. To demonstrate its confidence in Charleston's quick economic recovery, the company announced on September 10 that it was constructing a new 60-foot-by-400-foot warehouse shed on the wharf and was also building another large new wharf with a 250-foot frontage on the river. Work would begin on September 13 and would be completed as rapidly as possible, so there would be no lack of facilities for handling all business that might be offered.

When the Federal Engineering Commission, led by William E. Speir, completed its work in late September, it left the E.R.C. a thorough survey of the damage, which gave the city's leaders the information they needed to carry out the rebuilding stage. Yet even as the federal engineers were departing, an even more comprehensive survey was getting underway. Between September 28 and December 11, 1886, a three-member private board of inspectors, commissioned by a consortium of insurance companies doing business in Charleston, conducted an exhaustive building-by-building survey to determine the liability exposure their companies faced. Insurance agent Harry C. Stockdell of Atlanta served as chairman of the board and was joined by James A. Thomas and Hutson Lee. The only structures that were not inspected within the city limits were between five hundred and eight hundred small one- and two-story frame houses in the extreme northern and western portions of the city. These were mostly owned by poor blacks. Because they could not be identified by house numbers and had suffered relatively little structural damage, they were not listed individually in the report.

The "Stockdell Report," as it came to be known, described the condition of nearly eight thousand structures. For the first 6,956 buildings, the report listed street address; names of owners and occupiers; structural material; roof material; length, width, and height; damage to each of the four main walls (by compass orientation: north, south, east, and west); condition of chimneys and flues; action needed to make the buildings safe; estimated cost of repairs; and comments. The report was the most comprehensive and sophisticated mass-catastrophe damage database ever created to that date. It found that "not more than one hundred out of 14,000 [chimneys] escaped injury, and ninety-five percent of these 14,000 were broken off at the roof and went to the ground,"[2] and concluded:

> The greatest damage to the city has been principally in the business portion, on the wharves, and on the Battery. Ninety percent of the brick buildings have been injured more or less, while frame buildings have only suffered from falling chimneys, cracking the plastering and injuring the foundations. The factories and mills have all suffered more or less, and one in particular (Bennett's Rice Mill) was well nigh ruined. The churches, as a class, have suffered more than any other buildings. This is due to the fact that the walls are generally very thick and heavy; and there being no inner or cross walls to bind the exterior walls, allowed them to be easily moved by the rocking or heaving motion of the earthquake. The window and door openings are large, and in almost every instance gave way at the corners and below, which weakened the walls above and gave way to the waving, jarring sensation that was so plainly felt on the night of the 31st of August.[3]

The Stockdell Report's extreme level of detail revealed extraordinary damage patterns, all of which supported a concept that is now well-known to structural engineers: earthquakes don't kill people; buildings do. A 1983 analysis of the report established two important factors that determined how vulnerable a building would be to an earthquake. It was strikingly obvious that one factor chiefly determined how susceptible any building was to damage: whether it was made of brick or wood.

In 1886, about seventy percent of Charleston's structures were made of wood; the rest were brick. As noted in the later analysis, "the heaviest concentration of brick buildings was in the commercial district of Charleston, roughly corresponding to the oldest parts of the city. The mixed brick and frame areas covered the rest of the southeastern portion of the peninsula and the higher ground in the peninsula north of the commercial district. This included the dock areas, many of the churches and public buildings, and much of the industry in Charleston in 1886."[4]

While the Stockdell Report listed the post-earthquake fate of 6,956 structures on the Charleston peninsula, the data it provided was not analyzed in detail until 1983. Sufficiently complete information was available on 6,444 of the structures to code them for severity of damage. This information was then used to analyze the degree of damage resulting from the type of construction material used and from the type of soil (solid ground vs. made land) on which buildings were built.

Of the structures analyzed, 1,965 (thirty percent) were constructed of brick, and 4,479 (seventy percent) were of wood-frame construction. Of the 1,965 brick buildings, 1,288 (sixty-six percent) were severely damaged. Of the 4,479 wood-frame buildings, 306 (seven percent) were severely damaged. Of the total of 1,594 damaged buildings, eighty-one percent were made of brick, and nineteen percent were made of wood.[5]

Nearly sixty-six percent of all brick structures failed, versus about seven percent of wooden buildings. In other words, brick structures failed at nearly ten times the rate of wooden ones. Even though brick structures made up only about thirty percent of the city's buildings, they produced almost eighty-one percent of Charleston's heaviest damage.

Charleston's Structures by Construction Material

Brick Structures	1,965
Percentage of Total Structures	30.5
Wooden Structures	4,479
Percentage of Total Structures	69.5
Total Structures	6,444

DAMAGED STRUCTURES BY CONSTRUCTION MATERIAL

Brick Structures Damaged	1,288 of 1,965
Percentage of Brick Structures Damaged	65.6
Wooden Structures Damaged	306 of 4,479
Percentage of Wooden Structures Damaged	6.8
Total Structures Damaged	1,594

The enormous brick-versus-wood disparity was obvious in Charleston and was pointed out a few weeks after the quake by Kenneth McDonald, a Nashville, Tennessee, architect who spent several days inspecting the damage. In an interview with the *News and Courier*, he noted, "Every brick house in the city has a cracked wall, while every frame house suffered only in the cracking of the plastering." When struck by earthquake shocks, wood-frame structures, being far more flexible and elastic than

Inside an earthquake-shattered house

brick, could easily be bent out of their original shape, but they could also more readily snap back, rather than break.

On close inspection, the high failure rate of the masonry structures was found to have several causes besides the inherent inflexibility of masonry. Buildings that were generally well constructed cracked little and fared well. Those that were most severely damaged were frequently found to be the victims of inferior materials, substandard workmanship, or even outright building contractor fraud. In 1886, after the earthquake, one local expert was instrumental in bringing these problems to light.

While USGS scientist William McGee was in Charleston, he sought out the expertise and opinions of a wide array of local advisors. Few proved more valuable than a wiry, battle-scarred, fifty-three-year-old, French-speaking ex-Confederate officer who taught at the College of Charleston. Gabriel Edward Manigault, M.D., was born in Charleston in 1833. Descended from generations of wealthy, slave-owing rice planters, he was a man who, since childhood, had been touched by wanderlust, enraptured by foreign lands, and mesmerized by art, architecture, ornithology, zoology, geology, natural history. He was consumed by a passionate interest in all things scientific. Because of his vast knowledge of a dizzying range of esoteric subjects, Manigault provided the earthquake team with some of its most valuable information through his systematic and insightful study of two fundamental subjects:

Gabriel Manigault, M.D.

Charleston's techniques for erecting buildings and the nature of the land the structures were built upon.

Growing up in South Carolina's architecturally magnificent plantation houses and living in its finest urban dwellings gave Manigault an intimate understanding of the careful planning and precise workmanship that went into the best of them. Furthermore, he had sawdust in his blood. Gabriel Manigault was the great-nephew of Charleston's legendary "gentleman architect," Joseph Manigault, who designed Charleston's City Hall and the magnificent three-story Neoclassical suburban villa he built at the corner of Meeting and John streets.

Dr. Gabriel Manigault pointed out how well-designed, well-constructed buildings, such as the Miles Brewton House, built between 1765 and 1769 at 27 King Street, and the Blake Tenements, built between 1760 and 1762 at 4 Courthouse Square, had withstood even the most violent forces of nature. These houses, he noted, had thick walls, and the bricks were laid in patterns known as Flemish bond, on the front face, and English bond, on the less-public sides and rear. While reporters described lower King Street as a "death trap" after the earthquake, the Miles Brewton House suffered only fallen chimneys, the loss of some slate tiles from the roof, and slight cracks in the east wall. Miss Susan Pringle, who lived there during the earthquake, wrote, "The foundation and lower storeys stood like a rock."[6] The Blake Tenements still showcase the exquisite beaded mortar and masonry workmanship that is as strong and elegant today as it was before the American Revolution.

Bricks laid in Flemish bond at the Blake Tenements

There are exceptions to every rule, however, and the inherent inflexibility and brittleness of equally well-built brick public structures, such as the Old Powder Magazine (1713), the oldest government building in the city; majestic St. Michael's Episcopal Church (1752-1761); and the Old Exchange Building (1767-1771), were evidenced by the damage they suffered. Manigault admitted, "The violence of the earth movement will be the better realized when it is explained that some of these walls, which rival the masonry of old Rome in their solidity, were forced to yield to the severity of the great shock."[7]

Two of the major factors in determining the strength of a brick building are the bricks themselves and the mortar that holds them together. The finest masonry work in the Charleston area can be found in buildings built before 1838. The strongest walls are made of thick, hand-shaped, plantation-made Carolina grey brick, laid in Flemish or English bond, held together with mortar made from burnt oyster-shell lime. The brick, which is not grey but rather a mottled, dark rusty brown, does not resemble the smooth, machine-pressed brick of today. It has an irregular surface that mortar can adhere to with great tenacity. The manufacture of Carolina grey brick peaked during the colonial period and had ceased by the 1850s.

The pattern in which the bricks were laid was also a critical element in a building's survival during the earthquake. When laid with its long side facing out, a brick is referred to as a "stretcher." If laid with the short end facing out, it is known as a "header." When laid in Flemish bond, each course (row) of bricks is placed with a header next to a stretcher. On the course above, the spacing is staggered so that a header is placed above the center of a stretcher. In English bond, one course of bricks is laid with stretchers end-to-end, and the row above is laid with headers side-by-side. Either pattern assures that the long axis of one brick is superimposed over the short axis of an adjacent one, locking all of them into a strong, lattice pattern.

In his report to McGee, Manigault pointed out that as excellent as the Carolina grey brick was, "the durable character of the walls was probably more due to the lime from which the mortar was made, than to the bricks themselves."[8] The mortar used in Charleston's eighteenth- and early nineteenth-century buildings was a mixture of water, coarse building sand, crushed oyster shells, and lime made from burned oyster-shells. As the mortar dried, it (like other mortars) bonded the bricks together—but with unsurpassed strength. The manufacture of oyster-shell lime was an important industry in early South Carolina and continued until cheaper, stone limes from the northern states were introduced about 1838.[9] After 1838, the original high-quality construction materials and methods continued to be available to those who could afford them until the Civil War reduced the South to poverty.

Many of the building failures during the 1886 earthquake had their roots a half-century earlier in the aftermath of the great Ansonborough fire of 1838. The fire consumed more than one thousand structures in a single day and led to a law requiring that new structures in the burned area had to be constructed of brick. This requirement resulted in a demand that immediately overwhelmed the local supply of experienced masonry contractors and skilled masons. The huge demand also forced wages up, which resulted in a flood of contractors and masons from outside the state, particularly from northern cities. This in itself was not a problem, but the enormity of the demand and the scarcity of supply tempted many contractors to rush through one job to profit from others. This led to shaving the quality of the workmanship and sometimes to outright contractor fraud. This post-disaster scenario, where a catastrophe—earthquake, windstorm, flood, or fire—overwhelms the local supply of high-quality labor and leads to substandard rebuilding—has plagued humankind for millennia.

The Northern masons brought their own bricklaying patterns and materials. Their preferred bond, known as American bond, had become a standard in the North. This pattern consisted of five courses of stretchers followed by a row of headers, which was repeated until the desired building height was reached. Manigault conceded that "it is a very strong bond," and that "experts consider that a wall built with this new bond is less likely to crack than one built with the Flemish bond," adding, however, that "it is next to impossible to split a wall built with the Flemish bond," chiefly because all of the bricks were interlaced.[10] Walls laid in Flemish or English bond had an additional advantage over the northern pattern. Every course of brick had to be finished before the next one was laid, virtually ensuring that every brick was completely surrounded with mortar. Walls laid in American bond were more open to sloppy workmanship.

All the problems that stemmed from poor post-1838 workmanship and materials soon became apparent when thousands of damaged and fallen buildings were examined after the earthquake. In one case, the inner face of an exterior wall showed nineteen courses of stretcher bricks, with not a single row of headers to link it with the outer course. In addition, some of the brickwork from the post-1838 era was "rough-cast" brick. In this masonry technique, walls were built of a mixture of brick chunks and a good deal of mortar, all covered with a coat of stucco. The use of cheap, broken bricks and large amounts of mortar, all concealed by the stucco, looked fine but was structurally weak, and little of it withstood the earthquake. After the earthquake, Inspectors also noted that many buildings were being "repaired" simply by stuffing the cracks with cement mortar and then covering the whole surface with one or two coats of yellow or white lime wash. "This, you will observe, presents the wall with a clean looking surface, but has not strengthened the wall one particle."[11]

Proper mortar is a cement that hardens with age and exposure, and with time, becomes an artificial stone that is even harder and firmer than the brick. Therefore, the quality of the mortar is as important as the quality of the brick. A New York architect and builder who visited Charleston shortly after the earthquake pointed out serious construction defects in many of the buildings that had been destroyed.

> It seems to me that a good deal of the destruction in Charleston is due to careless and imperfect work. I have noticed in very many instances that the mortar used in the construction of dwellings here is of a very inferior kind.... I have noticed

> several brick buildings standing, the walls of which are put up with that inferior kind of mortar. You can tell them at a glance. You will notice that the mortar between the bricks has worked out. It has the appearance of being gouged out or washed out. That is the kind of mortar which will not hold bricks together, and that is the kind I suspect that has been used in many of the buildings that were shattered. I notice a good deal of this peculiar kind of mortar all over the city in these heaps of debris. I don't mean to say that houses built with proper mortar would have stood the shock of the earthquake you had in Charleston, but I am certain that the destruction would not have been as great.

The defective mortar consisted largely of sand, which was cheap, with virtually no lime, which was more expensive. This mortar had little adhesive capacity and was barely able to hold the bricks in place. A finger scratched over this mortar will rub it off, leading to its infamous nickname, "finger mortar." Federal building inspector H. C. Speir referred to the mortar as being "little better than common mud. It soon dries and commences to crumble.... In all cases I advise the people to use gravel in their mortar, and to have it clean, sharp, and angular. No sand should be used at all."

The aftermath of "Buddensiek mortar"

Finger mortar was also called "Buddensiek mortar" in honor of unscrupulous New York building contractor Charles Buddensiek, who specialized in constructing substandard, thrown-together tenement houses and used every conceivable fraudulent building shortcut possible to save money. His use of mortar consisting almost exclusively of sand without sufficient lime to bind it resulted in brick walls that literally fell apart. In 1885, Buddensiek mortar was responsible for the simultaneous collapse of eight tenement buildings under construction in Brooklyn, killing several workmen. "The houses were built to sell," wrote Jacob Riis, a social reformer. "That they killed the tenants was no concern of the builder's."[12] Buddensiek was ultimately convicted of manslaughter and sentenced to ten years in New York's infamous Sing Sing prison. His name lives on as a synonym for building contractor fraud.

A reporter confirmed Speir's concerns about inferior rebuilding practices in Charleston. When he examined a dozen or more mortar piles in various parts of the city where repairs were in progress, he noted the absence of the yellow-clay sandy gravel that had been customarily used in Charleston to make good mortar. This traditional gravel had rough, irregular edges, which promoted a solid bond. The sand he found at the construction sites was similar to common sea sand, which is worn by friction into a rounded form, making it nearly useless for building purposes. The same inferior, post-1838 mortar that caused so much damage when the earthquake hit was now being used to rebuild the city. On September 30, an investigative reporter wrote, "A specimen of Buddensieck mortar, of which Charleston is full, can be seen on Meeting street, between George and Calhoun streets. The workmen who were engaged in taking down the walls yesterday simply used a broom handle to push off the bricks."

Gabriel Manigault, too, was quick to point out the value of using quality workmanship and construction materials. "Upon examining the interior parts of these walls [of a pre-1838 building], which have been exposed for purposes of repair," he wrote, "it can easily be seen how thorough the work was; there being complete adhesion between the bricks and the mortar in which they were laid. The entire walls, when completed, were put together compactly, and have stood for over a century as monuments of the substantial and honest work of the time."[13]

Another observer noted that where proper brick and mortar had been used in the older houses, "crowbars and chisels are required to dislodge and separate the bricks, and it is almost impossible to free them entirely of the strong material that holds them so firmly together. Compare this cement with the crumbling masses that are rapidly disintegrating into heaps of sand and mud everywhere along the streets, after a few days' exposure to the air and rains, and the difference between good and bad mortar need not be otherwise explained to one who makes the comparison." It was a humbling experience for Charlestonians to hear, chiefly from Northern building experts, that "so considerable a portion of the damage occasioned to their property by the earthquake was due and rapidly traceable to faults of construction and to the use of improper material."

Just after the earthquake, the newspaper called for the city to establish quality control oversight of the contractors and masons, but no such ordinances or regulations were ever adopted, and the shoddy methods and materials used to rebuild Charleston after the 1886 earthquake may be seen in the city to this day.[14]

While the type and quality of construction materials used were the primary reasons why buildings did or did not withstand the earthquake, the composition of the land a building sat on—original firm ground or filled-in creek beds and marshes, known as "made land"—proved to be another important factor.

Original (light) vs. made (dark) land in 1886

The Charleston peninsula had originally been a narrow strip of land between two rivers, intersected by numerous tidal creeks. The original high land consists of a layer of sandy soil, ten to sixty feet deep, which rests on a layer of marl, a fairly dense composition of clay, sand, and calcium carbonate, that lies forty to sixty feet below. In lower lying areas such as the tip of South Battery, some of the city's most elaborate and historic houses are only one or two feet above the high tide mark, and the average dwelling on the peninsula lies between eight and fifteen feet above sea level.

As the city grew, new land was created by filling in the creek beds and marshes with building refuse, dry garbage, and sand. By 1886, approximately forty percent of the city's buildings sat on made land. To achieve a solid construction base when building on made land, pilings must be sunk deep enough into the ground to anchor them in the marl that underlies the sandy soil of the coast. However, in the interest of economy, this step was often ignored.

Made land magnified the effects of the earthquake because of the geophysical phenomenon known as liquefaction. Filling in the creeks and marshes with trash and sand did not prevent the subsurface intrusion of seawater, leaving the made land saturated with a high percentage of it. When the shock waves of an earthquake strike water-saturated sand, the sand liquefies, behaving more like a thick soup than a solid. This amplifies the power of the shock waves, which exert greater force on a building above the liquefacted area than they would on a building sitting on heavy soil or rock.

Charlestonians did not know about liquefaction in 1886, but they were quite aware that shocks felt strongly in one part of the city could pass unnoticed in another. A month after the first shock, a reporter noted that a tremor had been clearly felt in the vegetable market on Market Street (built on made land over a former creek bed) but not elsewhere. The reporter speculated that "the ground in that place being 'made ground' is more susceptible to the subterranean movements than that in other parts of the city."

Structures Damaged on Firm Ground

Brick Structures	765 of 1,214
Percentage Damaged	63.0
Wooden Structures	12 of 2,420
Percentage Damaged	0.5
Total Damaged Structures on Firm Ground	777

Damaged Structures on Made Land

Brick Structures	523 of 751
Percentage Damaged	69.6
Wooden Structures	294 of 2,059
Percentage Damaged	14.3
Total Damaged Structures on Made Land	817

The type of construction materials used also had an effect on the amount of damage suffered by a building on firm ground or made land. Sixty-three percent of brick buildings on firm land were damaged, versus nearly seventy percent on made land. For wooden buildings, the effect was quite different. About one-half of one percent of the wooden buildings on firm land were damaged, versus fourteen percent on made land. This shows that although the added damage caused by the liquefaction of the made land was significant, the determining factor in whether a structure survived or not was still the building material used: brick vs. wood.

The inspections after the earthquake revealed a number of hazards that had to be coped with or avoided when rebuilding. The first of these hazards was the frequent practice of erecting front walls much higher than the roof of the building, which left the top of the front wall without any substantial support—and rendered it easy prey for the shaking and twisting of the earthquake. The peaked gable ends of brick buildings proved especially vulnerable, as they had no support from the walls. When the upper part of the gable walls fell, they frequently ripped out much of the wall below them, thus weakening the side walls, many of which also came down or had to be pulled down later and rebuilt. "The hip roofs of the older dwellings and stores in the city were probably not so generally adopted by our forefathers without reason," the newspaper chided. Hip (or Mansard) roofs, which have two slopes on each of their four sides, were recommended as the model for rebuilding and new construction in Charleston after the quake. They had been introduced to the city in the 1850s but did not gain significant popularity until the 1870s.[15]

A similar hazard was Charleston's infatuation with fancy, decorative parapets and heavy, ornate cornices, which projected out over the street. The weight of these decorations, which was poorly supported by the front walls of the buildings to which they were attached, made them dangerous. Chimneys, poorly supported front walls, and heavy cornices had been the chief killers during the first two shocks of the disaster, and their debris produced most of the wreckage in the streets. Throughout September and October, the *News and Courier*, citing numerous authorities, pointed out the dangers of these features and urged Charlestonians not to incorporate this kind of dangerous, top-heavy ornamentation into their repair and rebuilding plans.

Homeowners were also warned to be careful of lighting fires in fireplaces whose chimneys had not been professionally inspected. "Quite a number of wooden houses have been shaken so violently that their chimneys have been broken off in some cases, level with, and in other cases, below the roof," the newspaper said. "In such residences there is absolutely no protection against the sparks communicating with the exposed shingles, and without the utmost care it is not only possible but probable that frequent fires will be the result."

Federal building engineer Speir offered his advice for rebuilding the city's fourteen thousand fallen chimneys.

> As most of your city will have to be rebuilt, the matter of chimney-tops becomes an important one. I suppose about one-half of the people killed here owed their deaths to bricks from the chimney-tops. At least I judge so, from what I have seen of the destruction of chimney-tops and the way in which they were thrown down. . . . When your houses are rebuilt, for Heaven's sake abolish the old ponderous chimney-top. Build your chimneys flush with the roof and then top them off with light terra cotta pipes or pots, as they are called. These are perfectly safe. If they are blown down they don't crush in your roofs, nor the heads of passers-by, and then the cost of replacing them is trifling.

By September 11, the E.R.C. had the city's most critical needs—emergency food and shelter—sufficiently under control that they could start to address the issue of obtaining rebuilding funds. Mayor Courtenay told the committee that contributions had already passed the $150,000 mark and "would probably reach $200,000 or $250,000." With the knowledge that at least a portion of the needed funds would be available, the committee debated how to determine who got assistance. By September 13, it had adopted an application form and ordered that five thousand copies be printed.[16] The form was distributed from the city recorder's office in City Hall. Only homeowners could apply. Businesspeople had to rebuild on their own, as did the churches.

The vestry of St. Michael's Church had only recently completed extensive repairs after the 1885 cyclone. After the earthquake, building inspectors declared that the church's handsome Tuscan portico, built between 1752 and 1761, should be torn down because it was a safety hazard. It had separated from the bell tower when the tower sank eight inches into the ground. The vestry did not have the money for the demolition and asked the city council for a loan to carry out the work. The council ruled that "it is unable to comply with the request for assistance . . . as all church edifices are more or less injured and would be equally entitled to monetary aid, which Council is not in a position to grant."[17] Fortunately, fate stepped in and saved the magnificent structure from being permanently disfigured. The portico and the rest of the church were ultimately repaired with private funds, and St. Michael's still stands tall and majestic on the corner of Meeting and Broad streets.

To ensure that rebuilding aid only went to the needy homeowners, the application form called for detailed information. The committee stated unequivocally that if all details were not provided, the application would be returned to the applicant and that "married women whose husbands are living must not make applications themselves." The newspaper reported:

> The applicant is then required to give his residence at the time of the earthquake and at the present time, and to show what was his occupation then and what it is now; also what was his income, by salary or otherwise, at the time of the earthquake and what it is now. Besides, he is required to set forth the character of the injury inflicted by the earthquake, showing what part or parts of the building, occupied and owned by him, have been injured, and also what general loss in addition has been sustained. The estimated amount of the losses must be given. The applicant must show whether he has any other real or personal property, and what it is, the questions in the case of a widow or unmarried woman being the same as in the case of a man who applies. Finally, the applicant is required to say whether he can, out of his own

> means, pay for the necessary repairs of his dwelling and make good the losses without the assistance of the committee.

Four days after the rules were announced, the "daughter of an uninterested father" wrote a letter to Mayor Courtenay, pointing out that if married women whose husbands are alive were not permitted to apply, "it will exclude a very needy class of women who will be left to suffer." The newspaper noted that the committee had made the rule to ensure that only one claim was made for each piece of property, "but at the same time, where for any reason the husband has abdicated his position as the head of the family, the application may be made by the deserted or neglected wife. A statement of the circumstances which necessitate the matter of making of the appeal by a lady whose husband is living should, of course, accompany her application, in order that it may go directly to the sympathetic side of the committee." As the result of the E.R.C.'s flexibility on the matter, numerous women were able to apply on their own behalf.

The applications for rebuilding incorporated all of the E.R.C.'s safeguards against fraud. When someone applied, building experts were sent to inspect the property unannounced. There they examined the damage and calculated the estimated repair costs, all without knowing how much money the homeowner had applied for. Mayor Courtenay said with pride, "As far as this cross-examination has proceeded, we find a majority of the applications differing very little from the revised figures of the experts, showing that, as to most of those who have applied, there has been neither exaggeration nor misstatement of the facts as far as they knew them." In short, Charlestonians proved that they were honest and had no intention of taking unfair advantage of the funds made available to them by the generous people of the United States.

Having established the mechanics of the application process, the committee next had to determine which people from which residential areas were authorized to apply, and where the dividing lines were to be drawn. The earthquake had affected not only the city of Charleston, whose corporate limits extended only a mile or so north of Calhoun Street, but also the whole state. Nevertheless, without any known funding at the start, The E.R.C. prudently confined its initial activities to within the city limits of Charleston.

Charleston was the seat of Charleston County, which covered a strip of land along the coast with the city at the approximate center. In reality, Charleston was a city-state unto itself, and it dwarfed the rest of the county in both population and wealth. Hence, the Charleston County government, focused as it was on the city, was no help—but the neighboring towns and villages clamored for aid from the E.R.C. Some allocation of relief money had to be made to the adjacent communities, but the committee did not know how to do this.

The mayors of Summerville and Mt. Pleasant realized that no matter what they did independently, the bulk of relief funds would pour directly into Charleston. As a result, they immediately—and wisely—sought to forge strategic alliances with the City by the Sea.

Mayor John Averill had already taken strong measures to care for the residents of his village. On September 5, he called a special meeting of the Summerville town council to devise a relief plan. After deliberating, they passed two resolutions. The first was to establish a committee to request and receive aid. The second stated that "it was judged inexpedient at this time

Army tents shelter Summerville victims

to take any steps toward a publication of a call for assistance for Summerville that will conflict with Charleston, but that steps be taken to see what proportion of the joint relief fund as given to the sufferers of Charleston will be allotted to Summerville as called for by injuries caused by the earthquake."

With the consent of the council, Averill then appointed nine members to serve on the Summerville relief committee: five ministers and four laymen. Representing the clergy were The Reverends J. M. Pike; W. H. Taylor of the Summerville Presbyterian Church; L. F. Guerry of St. Paul's Episcopal Church; Joseph A. Sasportas; and Anthony A. Alston of the First Baptist Church. The laymen were A. W. Taylor; F. M. Sires; W. R. Dehon, a local leader in the Temperance movement; and the ubiquitous John B. Gadsden. The Summerville committee had a far more progressive makeup than Charleston's E.R.C.—Sasportas and Alston were black. The *News and Courier* reported, "They are men of sound integrity, and the white and colored people have equal confidence in their endorsement of cases of distress."

By September 6, the committee reported that urgent cases were receiving aid but that no tents were available, although some ship's sails had been rented and erected as shelter from the elements. The Summerville committee quickly concluded that nearby rural settlements also desperately needed help, and they took the hamlets of Lincolnville, Stallsville, Germantown, Sweatsville, Knightsville, Brownsville, and other places between Ten-Mile Hill and Ridgeville under their care.

To expedite his town's badly needed assistance, Colonel Averill began to draw on his close relationships with Charleston's business and community leaders. This action paid immediate benefits. On September 8, the E.R.C. voted to send $500 to Summerville, along with fifteen of the tents received from North Carolina. Five of those tents were forwarded to Lincolnville. A few days later, the E.R.C. approved Summerville's plan to care for the nearby settlements, including parts of St. Andrew's Parish on the Ashley River. On October 11, John B. Gadsden wrote of Summerville's ordeal:

> We have all moved back into our houses now, that is, all who have houses in which they are able to live. Being a member of the relief committee, I can testify that the destitution of the greater part of our people for some three or four days after the shock, and until we could organize in some way to help, was very great. Right nobly have the people of this our common country, and even foreign countries, come to our help, and unless we have another shock like the one of August 31st I think in two or three months we shall be on our feet again as strong as ever. I say another shock like the one of August 31st, for we look upon the small after shocks, which we have heard and continue to hear every day, as the settling down of convulsed nature, and trust that Charleston and her daughter, Summerville, having passed through fires, war, cyclones, and now earthquakes, may be mercifully spared in the future, and obtain a much needed rest.[18]

Colonel Averill shared Mayor Courtenay's professional devotion to keeping accurate, publicly disclosed records of the funds entrusted to him, and he published a report showing all monies received and disbursed. It stated that the Summerville committee received $20,713.11 ($416,331 today) for its relief expenses from Charleston's E.R.C. and $8,788.52 from other sources, totaling $29,501.63. Most of the money—$21,159.00—was spent to repair 93 houses in Summerville and 237 more in the surrounding villages.[19] The Summerville committee completed all of its relief operations and closed its books on December 10, 1886.

Mt. Pleasant's city elders also quickly appointed a committee to survey the damage. By September 8, the committee had completed its work and reported its findings to Charleston's E.R.C. They found losses from the "falling of chimneys, breaking down of plastering, &c., amounting to not less than five thousand dollars." The rectory of St. Andrew's Episcopal Church, a three-story wooden residence with a brick foundation, was seriously damaged. Mrs. Sarah Fell's two-and-a-half-story wooden house was shaken from its foundation and was left in a dangerous condition. Mr. R. Magwood's wooden house was badly wrecked and dangerous. The three-story brick residence belonging to Mrs. Thomas Shingler and occupied by Probate Judge Kirk and his family was cracked "through and through." It was the only brick residence in the village and had been used as headquarters for Confederate General Roswell S. Ripley during the Civil War. The Mt. Pleasant Presbyterian Church had been severely wrenched, and the partial failure of one of its brick supporting piers left the building sagging five inches on one side. The village elders went on to note that a large number of Mt. Pleasant's residents were unable to repair their damaged buildings without outside aid, as they had not yet recovered from the effects of the 1885 hurricane. Following the example of Summerville a few days earlier, Mt. Pleasant requested to be added to the regional pool of sufferers for which Charleston's E.R.C. was soliciting and distributing aid.

Unfortunately, the civic leaders of Mt. Pleasant did not have Summerville's high-level connections with Mayor Courtenay, and they had to work harder to be incorporated in the relief distribution system. Mt. Pleasant newspaperman Louis A. Beaty wrote that the village's application "was taken to Charleston and presented to the Mayor at his office, but that gentleman's conduct toward the Mount Pleasant men was such that they left almost immediately, and without accomplishing anything. The memorial was then taken to Capt. Dawson of the *News and Courier*, another member of the committee, who kindly consented to bring the matter before that body at the next meeting."[20] On September 18, the E.R.C. voted to include Mt. Pleasant and its surrounding villages under its umbrella with the same status as Summerville.[21] Ultimately, the E.R.C generously paid out damage claims for rebuilding houses far outside the city and the Summerville and Mt. Pleasant areas.

Aid was provided to most of the epicentral tract around Summerville; northeast to include Mt. Pleasant and Sullivan's Island; north to St. James', Goose Creek and Plantersville; to homes fifteen miles west of the Ashley River; and south to include James Island, Johns Island, Edisto Island, and Wadmalaw Island.

The first stage in the rebuilding process was to clear the main streets of rubble so that streetcars could run again and horse carts could haul in building material. This was accomplished in the first three days after the initial shock. Hauling off debris from the thousands of damaged or fallen buildings consumed the rest of the fall. Most of the wood and brick wreckage was dumped into the former rice mill ponds and tidal creeks and marshes in the western and southwestern parts of the city. Within a decade, this made land was being used as building lots. The next steps were to shore up damaged walls with wooden braces until they could be rebuilt and to frame in gaping holes in ceilings with lightweight lumber and canvas to protect the owners and the insides of their buildings from the elements.

Holes in the ceiling of the Unitarian Church

The Unitarian Church on Archdale Street was typical of buildings that had suffered severe structural damage. The Gothic tower of the church had fallen through the roof, destroying the elaborate fan-vaulted ceiling and nearly slicing the building in two. Pastor E. C. L. Brown, noted that his congregation had only recently made its utmost effort to recover from the damage inflicted by the hurricane a year earlier, and "supposed the work done for a generation at least." "Now," he said, "all is undone by damages four times greater, but I have not heard a syllable of surrender or suspension. Rather all seem nerved to determination and effort. Our church is literally cut in two. But it seems to be a clean fracture and there is vitality in both parts. They will grow together again and we shall be as strong as ever." And as for Charleston itself, he proclaimed, "She is not going to surrender and the world will not let her die."

By September 7, there had been a "praiseworthy endeavor" on the part of building owners to remove anything overhead that could threaten the safety of those below, such as crumbling walls, broken chimneys, and smashed copings and cornices. Broken roofs and windows had to be repaired as quickly as possible, or covered with wood if time did not permit. The lessons learned from the 1885 hurricane were put into practice, and old tarpaulins and coverings used after that storm were brought out again. On King Street, awnings from storefronts were used to cover the roofs of the damaged buildings.

While the bracing was going on, heavily damaged walls were being torn down, and the rubble was picked over by laborers who reclaimed usable brick by chipping off broken fragments of mortar. The rebuilding effort was not hampered by a lack of materials—just the money to pay for them. The quake had not affected the supplies already on hand, and new shipments started to come in within two weeks. On September 16, for example, Montague E. Grimke, owner of Builder's Depot at 118 Church Street, advertised the arrival by steamer of twenty-five thousand pieces of Bangor slate for roofing use. In addition, his on-hand supply of materials included one thousand barrels of

Rockport lime, five hundred barrels of Rosendale cement, five hundred barrels of Portland cement, three hundred barrels of plaster of Paris, and fifty tons of land plaster, as well as bricks, fire bricks, drain pipe, and sand gravel.

With the enormous amount of rebuilding necessary, the need for construction workers quintupled. Because of the limited number of workers available, the law of supply and demand immediately started driving up the cost of labor. On September 7, the newspaper published a table of increased daily pay rates that had been endorsed the previous night by a mass meeting of the Knights of Labor, the black tradesmen's union. The new rates included a fifty-cent per diem increase over the previous rates. First-class bricklayers would now command $3.50 ($70.35) a day; less-skilled second-class bricklayers would get $3. First-class carpenters were entitled to $3; second-class, $2.50; and third-class, $2. Pay grades for skilled painters were $3, $2.50, and $2, and plasterers received $3 and $2.50; tinners (roofers), $3 and $2.50; and laborers, $2 and $1.50.

Many Charlestonians considered these pay raises to be tragedy-based extortion. One member of the E.R.C. was said to have stated that every workman who had signed the pay raise demand "ought to be hung to a lamp post." Those who were not offered the higher rates refused to work—an act considered intolerable by Charleston's white citizens. James B. Smith, an experienced New York builder, suggested that Charleston defer as much work as possible until the winter, when more Northern labor would be available. However, because of the coming drop in temperatures and the increased threat of hurricanes in the fall, this was not feasible.

The enormity of the need for labor quickly superseded all other considerations, and a week after the new rates were published, bricklayers of all ability levels were able to find ample work at $5 a day—and in many cases, higher. Most of the workers who had contracted to do plastering at the Citadel for $4.50 a day soon walked off the job to take better-paying work at other sites. On September 19, the newspaper reported that plasterers were demanding $7 to $9 a day. The city sent out notices nationwide to entice workers to come to Charleston, with the hope that increased supply would lower prices.

By September 5, Charleston received news from Yates County, New York, that several bricklayers and plasterers were ready to leave for Charleston at a moment's notice. Already aware of the increased labor prices, the *News and Courier* commented, "It is hoped that if they come they will not follow the example of extortion set for them by our own people."

By the end of September, when money to rebuild houses had become readily available through the E.R.C., Mayor Courtenay laid out the labor and rebuilding issues for the Charleston correspondent of the *New York Herald*.

> The progress of building is simply limited by the number of workmen. One thousand bricklayers, slaters, and plasterers and some carpenters could get employment in the next two, three, or four months. The scarcity has put up [raised] the price of skilled labor and many people have blocked up their houses temporarily with timbers to keep them from falling, and have pretty much made up their minds to wait until next spring and stand an occasional leak as preferable to paying exorbitant wages. In fact, a great many people pride themselves on living in cracked houses into which the rain drips pleasantly during a shower, and all because they haven't the money to pay high prices current for workmen who don't do very much work.

Nevertheless, labor rates did not return to anything near pre-earthquake levels until a year later, and the massive influx of poorly skilled workers led to a great amount of shoddy masonry work and outright contractor fraud, just as it had in the wake of the 1838 fire.

As the repairs progressed, one feature began to appear on almost every pre-1886 masonry building in Charleston:

The earthquake bolt gib plates came in many shapes and sizes

"earthquake bolts." Known to builders throughout Europe and America simply as reinforcing rods, they earned their local name when used by the thousands to reinforce damaged masonry buildings—and a few wooden structures—after the earthquake.

Although some people believe that the use of earthquake bolts originated in Charleston, they were, in fact, at least two-hundred-year-old building technology at the time the earthquake struck. The central marketplace of Bad Münstereifel, Germany, is a good example. It is ringed with three-and-a-half-story masonry buildings, virtually all of which employ iron reinforcing rods and highly visible exterior gib plates as part of their original construction. Most of the gib plates are formed in the shape of an "S," but one set of plates spells out the German building's construction date: 1674.[22] The use of these rods was based on a plan "explained to and adopted by the French Academy of Sciences to straighten out old stone buildings many years ago," said Henry Kittridge, a New York carpenter and machinist, in 1886.

An earthquake bolt, gib plate, and nut at Middleburg Plantation

In South Carolina, this technique for reinforcing walls dates back to at least 1714, when iron reinforcing rods were installed on the gable ends of Mulberry Plantation's main house as part of its original construction.[23] Mulberry is the third-oldest plantation house in South Carolina and one of the oldest brick dwellings to survive in the South. It was severely shaken and damaged during the 1886 earthquake and now also bears the quake's well-known signature: additional earthquake bolts,

Reinforcing rods from 1714 and 1886 at Mulberry Plantation

capped with nuts and Charleston's familiar round, nine-inch iron gib plates. This plantation house is believed to be the only structure in South Carolina to employ both types of reinforcing bolts: the original 1714 set used for additional strength and the 1886 set installed to repair earthquake damage. In Charleston, another colonial-era example of reinforcing rods can be seen at the Old Powder Magazine on Chalmers Street, whose cast-iron tie rods capped with enormous, X-shaped gib plates were installed in the 1740s.[24]

Reinforcing rods were used to brace the U. S. Mint in New Orleans in 1840, and many others were used in that city's French Quarter and commercial district. In the 1840s and 1850s, architect Edward Brickell White used rods concealed in masonry in Charleston's Grace Episcopal Church and in the east wing of the Citadel, which may have reduced the amount of damage sustained by both buildings during the 1886 quake.

On January 4, 1843, a severe earthquake (intensity VIII) affected Memphis and other places in western Tennessee. The Wheatlands, a house constructed of handmade brick in 1825 in Sevierville, Tennessee, was damaged by an earthquake in 1843 and repaired using earthquake bolts.[25] In New Knoxville, Ohio, a quake in 1875 damaged many brick buildings, which were repaired with iron rods and star-shaped gib plates identical to those in Charleston.

One week after the 1886 earthquake, Henry Kittridge proposed to Mayor Courtenay that iron reinforcing rods be used in Charleston. "Most of the damaged houses in your city can again be drawn into shape and made as strong and reliable as ever, and that, too, without disfigurement, by placing heavy angle iron on the corners, and connecting rods with nuts on the end in such a way as by turning them the displacement will be remedied. These strong rods can remain hidden from sight by ornamental heads to cover them."[26]

When the Federal Engineering Commission made its daily rounds in early September, it agreed with Kittridge and recommended the use of reinforcing rods and gib plates in about half of the damaged masonry houses it encountered. On September

18, for example, commission members examined forty-nine buildings, mostly residences, in Charleston's first and second wards, near the tip of the peninsula. Eight were declared safe. Ten were declared dangerous wrecks to be taken down immediately. Of the remaining thirty-one houses, it was recommended that fifteen could be rendered safe after the damaged walls were secured by rods and gib plates.

The main pumping station of the City of Charleston Water Works Company was also a candidate for reinforcement. It had sustained massive damage, breaking the main line and cutting off water to the city's fire hydrants after the first shock. The walls were so badly cracked and sprung that reinforcing rods were needed for each story, and seventy-four earthquake bolts and straps and eighty-four gib plates, requiring in all about sixteen thousand pounds of iron, were installed to secure the building. The cost of the reinforcement ran to about $2,200 ($44,200 today).[27]

Charleston's earthquake bolts consist of long, iron rods about an inch in diameter and threaded on each end, which were inserted into holes drilled in both sides or both ends of a building. On the outside of each wall, a gib plate—a large iron washer with a hole in the center—would be placed over the bolt, next to the wall. Then a nut would be screwed onto the end of the rod and tightened, forcing the gib plate against the wall. As the nuts were further tightened, the gib plates forced the walls closer together. In a variation of the same principle, the same nuts and gib plates would be used, but two rods would run to the center of the building. There, the threaded ends were inserted into a turnbuckle, which, when turned, drew both rods in toward the center, accomplishing the same effect as tightening the end nuts.

By the end of the month, virtually every badly damaged masonry building in the city was being repaired with earthquake bolts. Even today, the bolts can be seen in almost every old brick building in downtown Charleston. Their use was not confined to the city, however. The ubiquitous bolts can be found in the former Berkeley County courthouse (now the Darby

Building) in Mt. Pleasant; the Beaufort Arsenal; the Horry County courthouse in Conway; the Lancaster County courthouse, in Lancaster; the Gen. Francis Marion Bamberg House in Bamberg, Barnwell County; the Yorkville Enquirer building and the Latta House, both in York; the Bethesda Presbyterian Church in Camden; and throughout the Horseshoe on the campus of the University of South Carolina in Columbia. Georgia and North Carolina also used their share of earthquake bolts. Examples include a building on York Street in Savannah and two structures in Charlotte, North Carolina: the Sugaw Creek Schoolhouse and the Litaker Insurance Building, where earthquake bolts were installed as protection against future tremors when the building was built in 1895.

Although earthquake bolts were primarily used for masonry buildings, there are two notable examples of their use in repairing and re-squaring large wooden frame structures in South Carolina in 1886. Mt. Pleasant Presbyterian Church had two earthquake bolts inserted between the front and back walls of the church, parallel to the twin balconies. These were capped with star-shaped gib plates and nuts. The bolts were removed during a major structural restoration and expansion of the building in 1982, and steel beams were inserted to re-straighten and reinforce the side walls.[28] By far the most visible use of the bolts in a South Carolina wooden building is at St. Paul's Episcopal Church in Summerville (see page 247). There, three bolts run beneath the floor joists, while three more pierce the sanctuary itself, twelve feet above the heads of the worshippers.

Earthquake bolts continued to be used to reinforce damaged structures long after the 1886 tremors died away. In the 1930s, the Work Projects Administration enlarged and repaired the Beaufort Arsenal, installing numerous earthquake bolts with turnbuckles. At the Phillips-Yates-Snowden House, built about 1842 at 15 Church Street in Charleston, nine additional bolts were installed in 1992 to further stabilize the building.[29]

When the Federal Engineering Commission finished its recommendations, and various methods of remedying the damage had been suggested, the repair work could begin in earnest. With money flowing into the city, rebuilding took off like a skyrocket. By September 24, more than nine hundred people had applied for rebuilding aid. By October 3, the number had toped 1,550, and nearly 100 vouchers for work had been issued. Julian Bacot noted in his journal entry for September 24, "Business on Broad Street very dull; the bay however looks pretty bright and lively as cotton has started to come in pretty steadily and there are several large English Steamers lying in the docks to receive it as fast as it comes in. A feeling of confidence is beginning to make itself felt throughout the entire community; repairs are rapidly progressing and old Charleston bids fair to rise triumphant from the wreck and ruin of her recent and fearful calamity."[30]

Despite the high pay rates, there was a mad scramble for labor, and homeowners who found workers watched them closely so that someone else did not lure them off. Augustine T. "Gus" Smythe wrote to his wife on September 6, "Another most busy day. I was up a little after 5 o'clock to see and meet the carpenters at my office at 6 o'c. They have worked hard with the tinners all day and night. I have at least a roof over my office [at 7 Broad Street]. But the dust and dirt is fearful. To-morrow I will start them at the roofs at home & if the rain only keeps off a day or so longer I will be at least weatherproof."[31] On September 11, Smythe wrote to his daughter, who had evacuated the city and was living in the Upstate, "Our yard is in worse confusion than ever. Piles of brick all around. Workmen going all about. Everything and everybody stirring like so many ants. The Two Miss Gibbes [whose Legaré Street house had burned] & their brother live on our front steps or by the clock (just inside the front door), Matilda & her family sleep in the entry on mattrasses, the Gibbes' servants sleep around loose.... The dogs are well and your little play house served for a shelter for some people."[32]

When it came to living conditions during the rebuilding, Charlestonians threw pride out the window. If their houses were unsafe and they had a carriage house, they slept with the horses and changed their clothing behind a bed sheet.

Whatever it took to get their houses livable, they did, and the result was an explosion of do-it-yourselfers who filled in for the lack of skilled laborers and craftsmen.

The huge demand for housing could have fueled enormous increases in rent, but once again, Charlestonians showed their civility. The newspaper noted, "One of the most gratifying circumstances connected with the recent disasters is that there has been no advantage taken of the people in the way of advancing rent rates for the more eligible places. Many of these were untenanted, and such of them were comparatively uninjured were of course in great demand. With the real estate agents there was no speculation in the people's misfortune, and homes were at once provided for them at the usual rates."

As the rebuilding got underway, some of the recommendations made by the building inspectors were incorporated into the repair work. In an uncharacteristic moment of humor, Mayor Courtenay offered a few facetious suggestions for Charleston's building codes: "If earthquakes were to be our normal condition hereafter, it might become necessary to adapt our architecture to the erection of structures similar to those of Japan, where shocks are felt every hour, and where the houses were of bamboo, one story high, the foundation of which were round balls or gliders, on which the occasion of earthquakes moved forward and backward gently without disturbance to the inmates, and it is reported produce pleasant sensations."[33]

Being traditional by nature, Charlestonians did not adopt every new recommendation, but some residents did take heed. Although Mansard roofs did not become the universal standard after 1886, many were erected after the earthquake, solving, to some extent, the problem of collapsing gable ends. The same was the case with chimneys. A substantial number were rebuilt much shorter than before and used the recommended lightweight terra-cotta tops. However, by the time William M. Bird & Co. completed its rebuilding work in early 1887, its three-story store was even more ornate than the original. Bird added a wraparound, eyebrow coping molding over the ground floor storefront, which had survived, and placed a heavier, matching

William M. Bird & Co. as rebuilt in 1887

molding atop the third floor, which wrapped around the front and one side of the building, overhanging East Bay and Chalmers streets. The top of the East Bay and Chalmers Street walls both rose several feet above the roof, and the front wall was capped with pediment featuring a large eagle. The new building was a defiant challenge to future earthquakes because it had exactly the type of decoration that the building inspectors had denounced as being hazardous.

For a while it looked as if government officials were going to tear down the heavily damaged post office (the Old Exchange Building), built when South Carolina was still a royal province. The proposed demolition of the historic structure was unthinkable to the city's residents. In a letter to the newspaper, a writer known only as "A Charlestonian" asked Chamber of Commerce President Samuel Y. Tupper, a man known for his appreciation of history and fine architecture, to protest the destruction of "one of the few ante-revolutionary and historic buildings spared to us by the destructive fire of 1861." The writer noted that although the building, with all its wooden bracing in place, looked fatally flawed, it was, in fact, structurally sound. "There is no evidence of injury to the foundation," he wrote. "The building ought be saved." Fortunately, the fates smiled, and this irreplaceable gem of American colonial architecture was preserved for generations to come.

The damage to the Confederate Home and College, a double tenement built about 1800 at 60-64 Broad Street, turned out to be far more extensive than originally thought. The building had been badly injured during the 1885 cyclone, and repairs had just been finished when the earthquake hit. The roof had gaping holes, and the east and west gable ends and the oversized central pediment had collapsed. Two Confederate veteran's groups raised $170 to help, but the home's true salvation came from a very unlikely source, far to the north of the Mason-Dixon Line.

On or about September 10, former Union soldier, Samuel D. Paine, now a Methodist minister and a representative of the Grand Army of the Republic (G.A.R.), visited Charleston to see the damage for himself. General Thomas A. Huguenin, of the E.R.C., took him to the Confederate Home, where he met with its director, Amarinthea Yates Snowden. After learning about the home's mission and achievements, and of the plight of its 211 mothers and widows of Confederate soldiers and their 52 dependents, Paine was impressed, and left with the promise that he would bring the home's plight to the attention of the G.A.R. nationwide.

With Paine's encouragement, former Union soldiers raised $1,123.39 ($22,761 today) to rebuild the Confederate Home, with an additional $2,500 donated by Boston's relief committee. Funds also came in from England and Scotland. So grateful were the residents of the home that they had a plaque made and placed it prominently over the door. Possibly the only memorial to the Union ever installed in a former Confederate state, it reads, "Ruined by the Earthquake 1886. Restored by the People of the Union 1887."

17

SUCCESS AND CELEBRATION

The ruins have so nearly disappeared that their vestiges must be sought diligently, and with the aid of a guide, in order to find them, and to recognize them when found.

—Charleston *News and Courier*, August 31, 1887

Charleston, South Carolina
Fall 1886

As if showing some small bit of remorse for what it had wrought on August 31, fate bestowed exceedingly gentle weather on the Lowcountry for the rest of 1886. Temperatures were normal. The hurricane season, which runs from June until the end of November, produced no storms at all. Indeed, the highest winds recorded in the last four months of 1886 reached only thirty-six miles per hour. The heavy rains that accompany tropical gales and hurricanes were totally absent. The U.S. Signal Service reported that the average rainfall in Charleston for the last four months of the year was normally 18.54 inches. The actual rainfall for that time in 1886 amounted to only 5.16 inches, an enormous blessing for people who were still living in improvised quarters and damaged homes.[1]

By October 1, unemployment was zero for any able-bodied worker, and men were being paid twice the rate they had commanded only a month earlier. Fatally damaged buildings were being pulled down and replaced at a fever pitch. The burnt district on King Street had been cleared of its wreckage and new construction was underway. The tent cities were gone from the public parks and squares, and the city had cleaned and packed the tents and shipped most of them back to their owners. The mayor had directed work crews to thoroughly clean the public spaces and replant flowers. Most of the refugees who had evacuated the Lowcountry in early September had returned. Foot, carriage, and trolley traffic along King, Meeting, and Market streets was brisk, and people had resumed buying in numbers almost equal to those before the quake. There was an upbeat feeling in the air.

On September 20, 1886, in response to the emotional devastation caused by the tragedies of the previous two months, the members of the Charleston's Preacher's Union made a petition to Mayor Courtenay and the city council to set aside a day for prayer, fasting, and remembrance of the earthquake's victims. The group consisted of The Reverend William Heard of Mt. Zion A.M.E. Church and seven of his fellow black ministers in Charleston. The council "received it as information," and never acted on it.

While residents concentrated on looking to the future and rebuilding their shattered homes, businesses, and churches, earthquakes continued to stalk the Lowcountry. Tremors were an almost everyday event during the fall. Noticeable aftershocks occurred every week or ten days, and several severe shocks from the Summerville area caused great alarm and additional damage.

The psyches of Charleston and Summerville—and the rest of the nation, for that matter—had been altered. Prior to August 31, the general public in the Southeast lived in blissful ignorance of earthquakes. A month later, not a person within five hundred miles of Charleston had escaped the terror they wrought. "The Great Charleston Earthquake," as it was now known, had not only been front-page news everywhere in the country on September 1, but was also an ongoing topic of dinner-table conversation for months to follow. In just a few days, the entire nation had become earthquake conscious. Now, it seemed, anyone might be at risk, and a disturbing truth had become evident: the shocks showed no sign of ceasing.

On September 17, an intensity VI aftershock shifted the pillars of a Summerville hotel about two-and-a-half inches, cracking most of them in the process.[2] Four days later, another intensity VI shock toppled the heavily injured South Carolina Medical College building, which had been severely weakened by previous shocks. Worse was yet to come. Two of the most severe 1886 aftershocks were felt in Charleston on October 22. An intensity VI shock at 5:20 a.m. cracked the west wing of the Custom House and caused the wall supporting the roof on the west to "give away" slightly. That afternoon at 2:45 p.m., an intensity VII shock—the most severe since August 31—threw down several chimneys and cracked several others in Summerville. The tremor knocked bricks from chimneys, shook down plaster in Charleston and Columbia, and was felt as far away as Washington, D.C.[3] The next severe shock, with an intensity of VI, arrived at 12:20 p.m. on November 5, 1886. It shattered several chimneys in Charleston and Summerville and was also felt in the nation's capital.

As the year headed toward its conclusion, the city council voted to sell the site of the destroyed Main Guard House and build a new police station elsewhere in the city. The federal engineers, having declared the Old Exchange unsafe, were searching for a new home for the post office and federal court-rooms, then "most inconveniently situated at the Custom House." The city council agreed that the Guard House property was an appropriate place for these federal offices, and approved the sale. They proposed to build a new Main Guard House on the site of the former west wing of the Citadel, which had burned.

On December 14, the city council moved to deposit $100,000 of E.R.C. collections, along with any future receipts, into a fund to be used to repair and rebuild the city's hospitals and schools. Alderman James Adger Smyth said that "in his opinion no better disposition could be made of that amount of the relief funds, as the money was intended to reach the poor." The motion passed unanimously.[4] The E.R.C. then notified the *News and Courier* that since applications for rebuilding had been accepted from the middle of September to the middle of December, no further requests would be processed.

On Christmas Eve day, a newspaper headline cried, "Everybody Buying Something Good to Eat To-Morrow." It went on to say that, "The markets, both up and down town, will be the resort of thousands today, and there never was a time in the history of Charleston when the stalls of the vendors, in all kinds, were so loaded with delectable material as present." The store windows were filled with toys; the jewelry stores were loaded with diamond and gold jewelry and fine watches; and the food stores were overflowing with luscious fruits and nuts for holiday baking and enjoyment. Stores offered raisins, citrons, Persian dates and figs, French prunes, mincemeats, crystallized fruit, almonds, pecans, filbert and Brazil nuts, walnuts, brandied peaches and pears, and chocolate creams and bonbons. The beverage purveyors had full stocks of Maryland A1 rye whiskey, Amontillado sherry, and "old Oporto" port wine at $3 per gallon. Regular port, sherry, and Catawba (local) grape wines went for $1 per gallon, and "fine old Walhalla [South Carolina] corn whiskey" brought $1.75 per gallon ($35.18, or about $7 per present-day bottle). In the stalls of Market Street, there was not a single variety of domestic food animal or wild game that was not available in abundance. After four months of tragedy and channeling all their resources into rebuilding, Charleston's time to splurge and celebrate had arrived.

Churches and homes were decorated in fragrant pine garlands, and candles burned in the windows at night. Walker, Evans, and Cogswell, the large Charleston printing firm, held a drawing for prizes on Christmas Eve day, calling out the winning ticket numbers held by their December customers. The same day, Mr. E. J. Kanapaux, the foreman of the carpenters at work on St. Patrick's Church, was presented with a beautiful watch and chain by his employees, both as a Christmas gift and as "a testimonial of the friendly regard and esteem in which Mr. Kanapaux is held by the gentlemen whose work he supervises."

The rooms of the Young Men's Christian Association were open from 9 a.m. until 10 p.m. as usual on Christmas Day "thus giving the members and their friends a place for social gathering." At the armory of the Sumter Guards on Hudson Street, the guards' eighty-seven members were invited to share lemonade at 11 a.m. The City Railway Company closed down operations for Christmas Day, so that the overworked drivers could have a day of rest. A correspondent wrote the newspaper in their support, noting that they were "about the hardest-worked men in the city," and suggesting that a public subscription for donations be held between Christmas and New Year's Day for their benefit.

Christmas day fell on a Saturday in 1886. The holiday was celebrated quietly in Charleston, as was the general custom. Services were held at every church with white congregants in the city. The most festive service was held at St. Luke's Episcopal Church, where the music was promised to be "unusually fine, following an elaborate programme having been arranged for the service." The order of service included both sacred music and traditional carols, including "Hark, the Herald Angels Sing," "Adeste Fideles," and "Gloria in Excelsis." According to the *News and Courier*, the black churches followed a different tradition. It noted, "With the exception of Emanuel A.M.E. Church, [they] will make no special observation of Christmas Day. The young men of Centenary Church will sit upon Christmas Eve night until day dawns. This is something novel, though it is the universal custom of the colored congregations to sit up and see the New Year in."

At the November 23 meeting, Mayor Courtenay had proposed that the city council authorize "making a proper acknowledgement to the very large number of persons who had contributed money for the earthquake sufferers, and that a formal letter of thanks be sent to every individual who had contributed as of the end of the year." On December 28, the

City of Charleston,
Executive Department Dec. 31st 1886

To the Contributors to the Relief
of the Earthquake Sufferers.

At the close of this year, memorable by our sudden and terrible affliction from the mysterious visitation of Earthquake, the Corporation of the City of Charleston, in grateful memory of the blessed bounty that quickly came from all parts of this land and even from beyond the seas, for the needy sufferers in this stricken City, sends to each and all these generous givers, this City's heartfelt gratitude and glad greetings in the new hopes of the New Year.

The families in more than two thousand homes restored, the multitude of the poor fed and sheltered, the orphan, the sick, the aged and the needy in their rebuilt hospital and places of refuge will always cherish this gracious and universal giving and in the heart of this City the memory of it shall be precious as a charity the fruit of which is on earth and its growth in heaven.

To all our brothers far and near, who have been touched by our woe, our heartfelt wish is that as unto us in this human springtime of peace and goodwill, so unto them may come in God's good providence "The charities that soothe and heal and bless."

Voted unanimously in City Council 28th December 1886.

W. W. Simons, Clerk. Wm A Courtenay, Mayor

Mayor Courtenay's letter of thanks

council adopted the text to be sent to approximately two thousand donors to the relief fund. The handsome five-inch-by-eight-inch document, printed on heavy, buff-colored stock, bore the city's official seal and was engraved in a beautiful formal script. It was dated December 31, 1886; bore the signatures of both the mayor and W. W. Simons, the city clerk; and read:

> To the Contributors to the Relief
> of the Earthquake Sufferers.
>
> At the close of this year, memorable by our sudden and terrible affliction from the mysterious visitation of Earthquake, the Corporation of the City of Charleston, in grateful memory of the blessed bounty that came from all parts of this land, and even beyond the seas, for the needy sufferers in this stricken City, send to each and all these generous givers this City's heartfelt gratitude and glad greetings in the new hopes of the New Year.
>
> The families in more than two thousand homes restored, the multitude of the poor fed and sheltered, the orphan, the sick, the aged and the needy in their rebuilt hospital and places of refuge, will always cherish this gracious and universal giving, and in the heart of this City the memory of it shall be precious as a charity, the fruit of which is on earth, and its growth in heaven.
>
> To all our brothers, far and near, who have been touched by our woe, our heartfelt wish is that even as unto us in this human springtime of peace and good-will, so unto them may come in God's good providence, 'The charities that soothe and heal and bless.'
>
> Voted unanimously in City Council 28th December 1886.

The city tried to ensure that its selfless heroes were formally acknowledged. The firemen and police force had already been thanked, but two firemen received special notice. Julius O. Goutvenier and Mr. Mackin, foreman and callman respectively, of Engine No. 2, were each presented with testimonials. Both had been injured fighting fires on the night of August 31. On February 24, 1887, at the annual stockholder meeting of the City of Charleston Water Works Company, three more heroes were publicly honored. They were Joseph P. Chapman, the chief engineer of the works, and his two assistants, H. C. Vogt and T. Masserlerno. During the great shock on August 31, the six-inch water valve, which leads from the standpipe [a water reservoir] to the fire hydrant at the water works, was wrenched out, flooding the premises, and threatening to wash away the foundations of the building. "The stream of water was so great that no one could withstand it, but by the time the stand pipe was nearly emptied, these heroic men, having remained at their posts in the tottering building, were ready with a 6" plug, which they at once inserted, and then, putting both pumps at work, in thirty minutes they had the standpipe full again, with a full pressure upon the water mains." The men's quick actions brought back the water pressure in the city's fire hydrants, enabling the firemen to bring the King Street fire under control before it could burn its way to the residences on Broad Street and further down King Street.[5]

The Summerville relief committee had closed down all operations and balanced its books as of December 10, 1886. On March 21, 1887, the Charleston E.R.C formally concluded its operations and held its final meeting. It ruled on the last unresolved claims for house repairs and plastering, paid out those that were approved, and made small final payments to applicants whose claims had been forwarded by the Summerville and Mt. Pleasant Relief Committees. With the E.R.C.'s work at an end, the mayor thanked all its members for their outstanding work and devotion during the ordeal and congratulated them on the successful completion of their labors. After the E.R.C had housed and fed tens of thousands of people, cared for the sick, issued grants to repair over two thousand houses, and carefully accounted for the enormous influx of tents and cash donations, Chairman Barnwell rapped his gavel for the last time, and the committee passed out of existence, exactly two hundred days after it was created.

The E.R.C.'s final financial report was filed with the city council on July 25, the last act of the body that had performed so nobly in the face of the catastrophe. In addition to the $646,109.90 the E.R.C. raised, private societies and lodges raised about $50,000; individuals, $25,000; and churches, $288,000; for a total of approximately $1,009,109.90 ($20,283,090 in today's funds). Each dollar of E.R.C. income, when invested in food, medicine, or building repairs, generated $4 to $5 in total trade, as the money was spent by the recipient on food, clothing, bricks, wood, and labor. Then the providers of these goods and services each took their part and re-spent it on more of the same or other things. As a result, the $1 million of relief income generated $4 to $5 million in actual trade for the city ($80 to $100 million today)—an immense boon and stimulus to the economy. In commemoration of the unstinting service of the E.R.C.'s ten members, the city council authorized the creation of "simple but elegant" bronze memorial plaques, mounted on ebony panels, for each member.

Most of the earthquake drama was over by the first of the year, but the rebuilding was on hiatus, waiting to spring back into action as soon as the cold, rainy winter came to an end in February. As soon as the crocuses pushed their heads up through the dirt, construction work was back in full swing. Much of the structural work had been completed the previous fall, facilitating an early start on the inside work, especially the plastering, which was not as dependent on good weather as was masonry and roof tiling.

Although life continued to return to normal in Charleston and Summerville in the early months of 1887, the aftershocks continued. On the morning of March 17, a violent intensity VI shock widened existing cracks in buildings and knocked down more plaster in Charleston. It was described as more severe in Summerville. Since August 31, everyone had prayed that the earthquake was a freak, once-in-a-lifetime event. But as the months passed by, and the earthquake shocks continued, a fundamental paradigm shift took place, and a new reality quickly evolved. Six months earlier, everyone hoped the most recent quake would be the last. But by the spring of 1887, the continuing recurrence of shocks had become an expectation. South Carolinians now lived with the assumption that earthquakes, like hurricanes, were an inevitable part of normal life. The shift was clearly demonstrated in the daily diaries of several Lowcountry residents. When the first shock hit on August 31, numerous diarists had recorded the event with alarm, bewilderment, awe, in page after page of details. But after experiencing and recording numerous shocks over a period of months, the diarists regarded them as routine, and even a strong shaking only merited four words in Thomas Porcher Ravenel's journal. On January 10, 1887, all he wrote was "Shock about 8 P.M."[6] It took about twenty years—a full generation—for most of the aftershocks to die out, and for people to again start being complacent about the threat of earthquakes.

By August 31, 1887, one year after the killer shock, visitors from out of town were having a hard time finding "the earthquake damage." The relief income had invigorated and revitalized the city in ways no one could have imagined. The *News and Courier* wrote on August 31, 1887, "How faint and far in the distance the great calamity seems, when we think and speak of it. It is hard to believe that it occurred only a year ago. It is harder to make the stranger who visits the changed scenes believe that it was a calamity, indeed, and a most terrible one. The ruins have so nearly disappeared that their vestiges must be sought diligently, and with the aid of a guide, in order to find them, and to recognize them when found."

Where buildings had crumbled and fallen, new ones now took their place, "brighter and better and stronger than the old. The seams in the rent walls have been closed as by the healing power of nature and have left few scars. The shaking of the city appears, truly, to have settled it upon a firmer foundation, and to have awaked subtle forces of life and growth which lay dormant before. The trampled grass in the public squares grows greener than before, and bright-hued flowers, everywhere, raise their heads and look around, as if wondering to see the marvelous transformation that has been effected around them while they slept under the winter's sod."

Few in Charleston wanted to lose Mayor Courtenay's leadership, yet withstanding considerable pressure from all sides, he declined to run for a third term. Barring another cataclysmic catastrophe (though he might well have felt jinxed, having been saddled with the cyclone in 1885 and the earthquake in 1886), a third term would have allowed him to continue to pursue the perfection of Charleston's civic resources and launch a renaissance of economic growth. Nevertheless, he decided to serve out his second term, retire in December, and return to the world of business he loved as much as he did the Holy City.

After he and his fellow Charlestonians had experienced so much death, destruction, privation, and misery, Courtenay wanted to see the last line in the disaster recovery story marked with a bold exclamation point. To give Charlestonians true closure on their years of disaster and recovery, he conceived a massive, thrilling, colorful extravaganza of spectacle, fun, and excitement, the likes of which had never been staged in the South. He named it Gala Week, and decided that it would run from Monday, October 31 to Saturday, November 5.

The whole city was abuzz with excitement for two months in advance of the extravaganza. The Gala Week Association, consisting of a board of eighty managers, announced ever more thrilling details every week over a period of two months before the event. No one could imagine anything so large, complex, exotic, or fantastic. Parents pleaded to have all schoolchildren excused from attending school for the event. The school board never agreed, but parentally approved truancy was rampant.

King Street decorated for Gala Week

As opening day neared, all railroads connecting with Charleston scheduled extra trains to handle the influx of at least 60,000 visitors for the week. Every hotel, inn, boarding house, and available private residence prepared to welcome throngs of visitors.

Arriving visitors saw six triumphal arches spanning the principal streets, all decked out with American flags and illuminated by thousands of Japanese lanterns. Merchants decorated their stores to the maximum extent. A reporter for the *New York Times* noted, "Charleston is wrapped up in the stars and stripes. Fifty thousand American flags have been sold here during the past week, and every one of them is floating in the air tonight. Horses and mules, coachmen and car drivers have been wearing the National colors all day. In the whole city there is only one old Confederate flag, bullet-riddled and moth-eaten, displayed, and the State flag, with its palmetto tree cannot be found.... The decorations of all the houses on the principal streets are bewildering in their beauty... and to-night the city looks like a scene from fairyland.[7]

The festivities got underway on Monday afternoon at Colonial Lake, where boat processions, boat races, and tub races were held in front of bleachers erected to accommodate 20,000 spectators. The evening's main event, admired by 60,000 onlookers, featured a dozen brass bands and a spectacular fireworks show with Chinese mortar shells. The rest of the week was filled with an all-encompassing array of attractions, including hot air balloon ascensions and a sham battle between wooden "battleships." The Chicago Giants and St. Louis Browns baseball teams arrived to play two games in town. Horse races were held every day, weather permitting, and foot races were also part of the fun, all for substantial cash prizes. Rowing, sailing, and tugboat races took place in the harbor. All the night activities were illuminated by electric lights. The Electric Light Company had rented every light available and was running its electrical production plant at full capacity. These affairs were scheduled to permit viewers to see the events and still have time to attend the evening opera performances, which were held at the Academy of Music on King Street at nine o'clock.

The biggest hit of the week was the evening display on Colonial Lake. All the houses around the lake were decorated with tier after tier of candle lanterns, giving the lake "the appearance of being enclosed in walls of living, iridescent light." The German Artillery Band performed, their brilliant uniforms illuminated by electric lights. In the lake, a pageant such as had never been seen in Charleston played out every night. "Every boat, from stem to stern, was fitted up with simple arches, on which were hung hundreds of Chinese lanterns, and each boat had on board a night's supply of special marine fireworks, Belgian lights, Bengal lights, and other material. A full-rigged ship, decorated as Cleopatra's barge, was so brilliantly illuminated that it seemed to burn on the water." As the show got underway, the ships circled the harbor to the tunes of the military band, with fireworks punctuating the intervals between songs. The *pièce de résistance* of each night was a mock artillery battle between two federal monitor gun ships and a mockup of Fort Sumter, which the crowd found enthralling.

The festival provided such an enormous menu of activities for people of all ages that no one could fail to find something interesting going on at any given time. One event, perhaps not even noted as unusual by the vast horde of attendees, will certainly never be repeated in Charleston. The occasions on which St. Michael's bells are tolled are regulated by a strict and solemn list of protocols. These ancient rules had been enforced and expanded by the vestry of the church since the bells first tolled for the funeral of Mrs. Martha Grimke on September 22, 1764. It can only be said that Mrs. Grimke, along with the deceased vestrymen and wardens of the church, must have been sent spinning in their graves when, on the opening day of Gala Week, the initial aria on St. Michael's chimes was the old-time Southern folk song,

Oh, Susie Anna
Don't you cry for me,
Kase I'm jest from Alabama
Wid de gold dust on my knee.

After an unprecedented week of fun, entertainment, food, music, opera, fireworks, shooting contests, horse races, foot races, pageants, parades, bands, wrestling matches, baseball games, minstrel shows, light displays and harbor tours, the festival came to an end on Saturday night, November 5, a stunning success. Until the very last moment of the last program, the entire city remained as beautifully and brightly illuminated as it had on opening day. Everyone who attended had a wonderful time.

Refreshed, relaxed, and roundly thanked by the hundreds of people he encountered during the festival, Mayor Courtenay could now spend the last few weeks of his administration preparing for his departure from office. He chaired his final meeting of the city council on Monday, December 12, 1887. After the night's normal business had been concluded, he asked for the privilege of saying a few words before he vacated his post. He did not review the accomplishments achieved over the past eight years, modestly noting that all the city's affairs and improvements had been made a matter of public record and published in the city *Yearbooks* which he had inaugurated. He thanked all the aldermen and city officials, past and present, with whom he had worked, and closed his remarks with these words:

> In parting officially with you let me say how deeply and gratefully I feel as the recipient, during so many years in public office, of such continued personal kindness, good will and support from my associates here, and such unfailing sympathy and encouragement from all classes of my fellow-citizens. These recollections will be pleasant memories all through my life, and a sufficient reward and satisfaction under all circumstances. There are times in the experience of most men when they feel the poverty of human speech to express the emotions which struggle for utterance, and in taking leave of you let me from a full heart ask you to realize my deep sense of obligation to each and all of you, for your personal kindness and official co-operation in my public duties, and to wish each of you health, happiness, and prosperity through life.[8]

At that point, Alderman J. Adger Smyth moved that the mayor's farewell address be entered into the minutes and that the mayor vacate the chair to Alderman Christopher S. Gadsden, mayor *pro tem*, so that a special committee might make its report. Courtenay left the council chamber, and Alderman Smythe then read to the council a loving tribute to the mayor's service and a recital of his numerous accomplishments. The council moved that the tribute be entered into the official minutes and that a copy be prepared for presentation the mayor.

Courtenay was then requested to rejoin the meeting and take his seat as mayor for the last time. Alderman Gadsden said to him, "We have been profoundly impressed by this record of the magnitude and value of your services to this city. The recital furnishes an instructive lesson as to how much can be accomplished by one man of energy and sagacity, in so brief a period, for the benefit of the community in which he lives. You have richly earned the repose you seek, but such endowment of administrative talent must not rust out in private life. I trust, sir, that ere long, you will illustrate on a wider field of duty and responsibility the activities and energies you so eminently possess."

Courtenay thanked the members of council for what he felt was a "too partial expression of their regard." With that, the council adjourned, and two eras came to an end: the eight-year mayoral career of fifty-six-year-old William Ashmead Courtenay and the official conclusion of the city's recovery from the Great Charleston Earthquake of 1886.

One huge question remained unanswered: why had the earthquake happened in the first place? Mayor Courtenay may have successfully rebuilt Charleston in fourteen months, but the seismic Holy Grail—the answer to the question, "What triggered the 1886 earthquake?"—remained elusive. The Grail Quest would ultimately go on for more than a century after geologist William McGee arrived in Charleston on September 2, 1886, but the elusive secret would eventually reveal itself. The unwrapping of that enigma was to become one of the most fascinating stories in the history of American science.

18

HOW THE GRAIL WAS FOUND

It is a bitter and humiliating thing to see works which have cost man so much time and labor overthrown in one minute. —Charles Darwin

While Charleston was digging out from under its debris and starting to rebuild, scientists all over the world were clamoring for hard information about the earthquake. It was no wonder. The quake's first important attribute was the high level of mortality and suffering it inflicted: eighty-three fatalities and countless injuries known at the time. Then there was its immense power. The 1886 earthquake was the most violent seismic event in the recorded history of the United States since the New Madrid quakes in 1811-1812. The physical damage in Charleston was appalling. The shocks of August 31 left the city looking like the scene of a long, large-scale enemy artillery barrage. Finally, it was the best-reported U.S. mass-casualty disaster since the Civil War. Not only was the story covered for months in every newspaper and news magazine in America, it also appeared in publications throughout Europe and Latin America.

While news reports were detailed and frequent, virtually no dependable geological or seismic information about the earthquake was available to the scientific community until April 19, 1887. On that day, Capt. Clarence Edward Dutton and Ensign Edward Everett Hayden, both USGS scientists, presented their paper, "Abstract of the Results of the Investigation of the Charleston Earthquake" to a packed hall at the National Academy of Sciences in Washington, D.C.. The paper was published in the magazine *Science* a month later.[1] At about the same time, Mayor Courtenay's 1886 *Yearbook*, which contained a ninety-seven-page report of the earthquake, illustrated with dramatic photographs of damaged buildings, was published. The two reports revealed an extraordinary amount of scientific and sociological information about earthquakes.

The authors of the scientific study were well qualified for the task. Born in Connecticut in 1841, Dutton, a Yale graduate, had served in the Union army and spent sixteen years with the USGS. Most of his early research was focused on the Rocky Mountains, and he had a special interest in the geological history of the Grand Canyon. His extensive, first-hand geological experience and fluid writing style made him a natural choice to co-author the initial abstract and later write the extensive final report on the earthquake. Ensign Hayden, twenty-eight years old at the time of the quake, had lost his leg in a climbing accident while on duty with the USGS in the West and had to resign his commission as an assistant geologist. He worked for the Navy as a specialist in marine meteorology for the rest of his life. His commission was later reinstated and he ultimately retired as a rear admiral.

Dutton and Hayden collected an unprecedented amount of information about the quake's effects. The data came from more than 1,600 reports solicited by the USGS via extensive questionnaires, which were published in newspapers throughout the country. These questionnaires enabled the USGS to determine the speed at which the earthquake's shock waves traveled. Spontaneous contributions of information from individuals and scientific collaborators nationwide added to that data.

Isoseismal map of the earthquake from the Dutton Report, 1889

The USGS had already compiled an isoseismal map of the earthquake. Those who studied the map were stunned to see the immense area affected by the earthquake—a circle with a radius of about one thousand miles, embracing "somewhere between two and one half and three million square miles."[2] Within 250 miles of the epicenters—throughout the states of North Carolina, South Carolina, and Georgia and in northeastern Florida—the energy of the shocks was enormous. Dutton and Hayden explained that the available data would show only the relative damage suffered on the earth's surface at a given distance from the epicenters and would only generally indicate the magnitude of the earthquake at its hypocenters. Nevertheless, the area of significant structural damage extended for one hundred twenty miles, and the area of moderate damage extended for three hundred miles. The earthquake was felt as far as Bermuda, 800 miles from Charleston, and Cuba, 750 miles away.

In the epicentral area, the shocks were sharp and intense because the thrust was nearly vertical. Farther away, the shocks came from a more horizontal angle, producing the visible rolling waves seen in Charleston. This was the first time that the wave motion of earthquakes had been clearly explained. These waves were described by Dr. F. L. Parker, who was standing on Tradd Street in Charleston when the first shock hit. His observations were recorded in the 1886 *Yearbook*. First, he heard a roaring sound from the southwest, then another from the northwest. Then he started to feel the vibrations of the earthquake.

> The street was well lighted, having three gas lamps within a distance of two hundred feet, and I could see the earthwaves as they passed, as distinctly as I had a thousand times seen the waves roll along Sullivan's Island beach. The first wave came from the southwest, and as I attempted to make my way towards my house, about one hundred yards off, I was borne irresistibly across from the south side to the north

> side of the street. The waves seemed then to come from both the southwest and northwest and crossed the street diagonally, intersecting each other, and lifting me up and letting me down as if I were standing on a chop sea. I could see perfectly and made careful observations, and I estimate that the waves were at least two feet in height.[3]

The statement that the 9:51 p.m. shock consisted of energy waves from distinctly different directions was confirmed in multiple reports, although the direction that the waves traveled varied from reporter to reporter.

The confirmation of the wave motion helped to answer the question posed by the editor of the *Palmetto Post* in Port Royal, South Carolina. A week after the earthquake, he pleaded, "Will someone clothed in the robes of science tell our readers why it is after an earthquake shock one is nauseated?"[4] Dutton and Hayden had the answer: "Six hundred miles from the origin the long swaying motion was felt, and was often sufficient to produce seasickness."[5] The investigators conservatively estimated the wave height in Charleston to have been in the ten-to-twelve-inch range. However, they made the significant notation that "In Charleston it appears to have been greatest in the 'made ground,' where the ravines and sloughs [creek beds and marshlands] were filled up in the early years of the city's history. The structures on higher ground, though severely shaken, did not suffer so much injury."[6]

The Dutton-Hayden paper presented another amazing conclusion: given the level of damage that Charleston suffered, and its distance from the epicenters, "the intensity of the shock at Charleston was only three-tenths what it must have been at the epicentrum, and about one-third the intensity [felt] at Summerville." The authors went on to make a chilling statement: "Had the seismic centre been ten miles nearer to Charleston, the calamity would have been incomparably greater than it was, and the loss of life would probably have been appalling."[7]

Because of incomplete and conflicting data, the federal earthquake commission had problems in determining the speed of the shockwaves through the earth. The two most challenging issues were how to determine, to the second, when the shock arrived at a given point, and how to tell which of the various shocks—or shock components—was arriving at that time. Not only did the initial shock at 9:51 p.m. have two discrete components (a shock at 9:51:15 and another 9:51:48, each produced by a different hypocenter), but the shocks traveled faster through some types of rock than others. By spring 1887, despite the difficulties of the task and ambiguities of the data, the commission reached the following conclusions:

- The earthquake originated in an oval epicentral tract approximately eighteen miles wide and twenty-six miles long, running on a northeast to southwest axis whose approximate center was Middleton Place or Dorchester on the Ashley River. The closest edge of the tract lay approximately ten miles northwest of Charleston; the center of the tract lay about twenty miles northwest.
- The earthquake had at least two, and possibly three epicenters: the most powerful near Woodstock Station; the next most powerful near Rantowles; and possibly a third, the least powerful, near Middleton Place and Dorchester.
- The 9:51 p.m. first shock had two separate components, thirty-three seconds apart, each emanating from a different hypocenter.
- Structures built on made land suffered more damage than those on firm ground, and those of brick more than those of wood.
- The force of the earthquake was transmitted by waves of energy whose intensity and arrival times were influenced chiefly by their distance from the hypocenter and the density of the material they traveled through.

- The shock waves traveled through the earth at rates between about 3.02 to 3.74 miles per second.
- The earthquake and its aftershocks had no connection to weather conditions, tidal conditions, or subsurface or undersea landslides.
- It did not generate any tsunamis or tidal waves because the source of the shocks was onshore.

Dutton and Hayden concluded their presentation by addressing the single subject with which scientists were obsessed: what caused the earthquake? What specific set of conditions came together and triggered it at that time and place? Unlike some of their colleagues a year earlier, they exercised great caution in what they said. They were totally forthright in stating that they could not prove any theory of its origins and said, "we prefer to remain silent for the present, fearing that if we commit ourselves here to any preference for a particular view, we may find ourselves encumbered with a bias arising from the intensely human propensity to defend, through thick and thin, utterances which have once been formally given."[8]

The scientific community was tantalized by the information in Dutton and Hayden's paper, and clamored for more data. Scientists got what they asked for just over a year later. The mother lode of the 1886 earthquake information collected and analyzed by the USGS was published in the *Ninth Annual Report of the Director of the U.S. Geological Survey for 1887-1888.* Edited by Dutton, the Charleston earthquake section consisted of 323 pages of text, photographs, illustrations, diagrams, maps, and drawings and was the largest, most intense investigation of any earthquake ever undertaken in the history of American seismic research. Dutton explicitly (and wisely) designed his report "for a larger class of readers than those who have made earthquakes a special subject of investigation."[9] For this reason, and to this day, the Dutton Report remains one of the primary sources of information about the earthquake.

Dutton's report began with a description of the research team that the USGS had assembled for the project. Special

Capt. Clarence Edward Dutton

accolades were given to the work of Earle Sloan, whose relentless fieldwork provided the bulk of the report's data and conclusions about the size of the epicentral tract, the epicenters, and the force and directions of the wave motions they generated. Dutton summed up his opinion of Sloan by saying, "It is impossible to bestow higher praise upon an observer."

The report continued with a detailed description of the first week's events in Charleston provided by Carl McKinley, whose extraordinary reporting of the event appeared in the *News and Courier*. Then Dutton presented the findings of Dr. Gabriel Manigault, describing the building construction methods and materials used in Charleston and their important role in the destruction caused by the shocks. Dutton included comprehensive accounts of the damage in Charleston, Summerville, and the rest of the epicentral tract, which drew heavily from the work of Earle Sloan and William McGee. Dutton's extensive use of photographs, diagrams, and precise illustrations made the esoteric subject matter easier for the layperson to understand.

A separate chapter was devoted to the geometry and science used to determine the depth of the hypocenters at Woodstock and Rantowles. Dutton confirmed Sloan's estimated depth of twelve miles, plus or minus two miles, as the depth at the former and eight miles for the latter. As for Sloan's inferred third epicenter near Dorchester and Middleton Place on the Ashley River, Dutton did not find sufficient evidence to support the theory.[10]

More chapters followed on the damage to the rest of the country, the mechanism of the wave motions, and the speed of the shocks. The last subject was the product of mathematician Simon Newcomb. Then, Dutton closed the report with an 118-page appendix containing a state-by-state digest of the several thousand questionnaires and other responses received by the USGS in response to its appeals.

After presenting his massive accumulation of well-organized evidence, Dutton, as the foreman of the scientific jury, had to face the judge to declare their verdict: what caused the earthquake? With the same caution he exercised when co-writing the 1887 paper for *Science*, Dutton concluded his report by saying, "After the most careful and prolonged study of the data at hand, nothing has been disclosed which seems to bring us any nearer to the precise nature of the forces which generated the great disturbance. Severe labor has been expended for many months in the endeavor to extract from them some indications respecting this question, but in vain. This problem remains where it was before. Having nothing to contribute towards its solutions, I have carefully refrained from all discussion of speculations regarding the causes of earthquakes."[11] In short, the scientists had been able to precisely determine how much and what kinds of damage the earthquake did and where it came from—but not what caused it to happen in the first place. They knew where it was located and the exact nature of its powers—but they had still not found their Holy Grail—the triggering mechanism that unleashed its power. Not yet, that is.

With the release of the Dutton Report, the curtain went down on the Charleston earthquake show. All the now-unemployed actors in the Charleston geological opera left the theater and went their various ways, off to new geological studies, back to their old careers, or on to other challenges. As the aftershocks diminished, so did both the scientific and general interest in the subject.

People started to forget the earthquake, but the earthquake never forgot its comfortable home near Summerville. Aftershocks continued to be felt long after May 2, 1888, when Ada M. Trotter, an Englishwoman visiting Summerville, recorded in her personal journal, "Lots of detonations last eve, and a very long shake during the night."[12] In fact, the aftershocks and new small earthquakes never stopped. They just faded again from the public's consciousness, as they had between the New Madrid shocks of 1811-1812 and the big one in Charleston and Summerville in 1886.

One by one, the main participants in the great experience of 1886-1887 went their separate ways. Acting Mayor William Elliott Huger's quick wits, sound judgment, and tireless work had enabled the city to establish full-fledged shelter and emergency food distribution operations in the first six days after the earthquake. A lifetime businessman, he owned and managed cotton and manufacturing businesses. His second term as an alderman for Charleston's First Ward expired at the same time as Courtenay's term as mayor, and like Courtenay, he returned to the world of business. He died in his seventy-eighth year at Flat Rock, North Carolina, in 1925. Survived by his wife, Elizabeth, he was interred in Magnolia Cemetery, Charleston.

William J. McGee, the candid, good-natured USGS geologist and friend of John Wesley Powell, worked with the Survey until 1893. Then he followed Powell to take a post under him as an ethnologist at the Smithsonian Institution's Bureau of American Ethnology, where he spent another ten years. In 1888, he married Anita Newcomb, daughter of brilliant astronomer and member of the earthquake commission Simon Newcomb. His remarkable wife was a star in her own right. A physician when women in the field were rare, she became an acting assistant surgeon in the U.S. Army and founded the U.S. Army Nurse Corps.[13] A talented, intelligent man, McGee held responsible government and private positions in the fields of ethnology, anthropology, museum leadership, hydrology, natural resources conservation, and soil erosion. He edited the *National Geographic Magazine* and also served as the society's president. McGee died in 1912.

After making his earthquake observations for the USGS in Charleston, physicist Thomas C. Mendenhall became an educator and president of the Rose Polytechnic Institute, now the Rose-Hulman Institute of Technology, in Terre Haute, Indiana, from 1886 to 1889. He served as director of the U.S. Coast and Geodetic Survey from 1889 to 1894, and from 1894 to 1901, he was president of the Worcester Polytechnic Institute in Massachusetts. During his prolific career, he made numerous discoveries and published significant papers in the fields of measurement, electricity, gravity, and magnetism. He died in 1924 in his eighty-second year.

Earle Sloan resolutely slogged through swamps, thickets, and rattlesnake-infested timberlands to map and locate the earthquake's epicentral tract. After his final report to the USGS on October 16, 1886, he returned to private life, but the University of Virginia was so impressed with his work that they awarded him an honorary doctorate degree. He continued his career in the fields of chemical and mining engineering, focusing on the phosphate industry. In 1901, he became the State Geologist of South Carolina, and held the ill-paying position for ten years out of love for the work. He died in Charleston in 1926 and was interred in Magnolia Cemetery.

Clarence E. Dutton's comprehensive, balanced record of the Charleston earthquake remains in print to this day. As head of the USGS Division of Volcanic Geology, he studied volcanic activity in Hawaii, California, and Oregon. Prior to writing the famous earthquake report, he had authored another geological classic, *Hawaiian Volcanoes*, in 1883. Dutton died in 1912 in New Jersey at the age of seventy.

Captain Francis W. Dawson, the British-born, ex-Confederate seaman-turned-journalistic flag bearer for the economic recovery of his adopted city, remained at the helm of the *News and Courier* until his bizarre death in 1889 at the age of forty-eight. One morning that year, Dawson learned from a policeman that his family's attractive, twenty-two-year-old governess, Helene Burdayron, was being stalked by Dr. Thomas B. McDow, who lived in Dawson's neighborhood. Dawson went to McDow's home to confront him—and was never seen alive again. Dawson, knighted by the Pope for ending dueling in South Carolina, was found in a pool of blood, shot dead—from behind. McDow claimed self-defense. After only two hours of deliberation, the jury found McDow not guilty.

Carlyle McKinley, whose brave, non-stop journalism produced the definitive account of the first days of the earthquake, ably served the *News and Courier* as an editor for twenty-five years. The self-effacing poet and scholar was described by W. W. Ball, editor of the Charleston *Evening Post,* as one "who nearly crossed the borderline of genius."[14] However, his genius had a dark side. In his 1889 book, *An Appeal to Pharaoh. The Negro Problem, and its Radical Solution,* McKinley contended that blacks were inherently inferior and savage, advocated induced emigration, and, if necessary, the forcible deportation of the entire black population from the United States. He died in his Mt. Pleasant home in 1904.

The Reverend William H. H. Heard served Mt. Zion A.M.E. Church in Charleston for three years. A fundraiser of exceptional skill, he took over its pulpit with the church $14,850 in debt. After paying for a new $2,500 organ and $1,800 in hurricane repairs, paying down $5,000 of the church's long-term debt, and adding 1,100 members to the church, he left by transfer to the Philadelphia Annual Conference of the A.M.E Church. In 1895, he was appointed U.S. Minister to Liberia and, in 1908, was elected a bishop in the A.M.E. Church. He died in 1937.

Colonel John H. Averill, superintendent of the South Carolina Railway and Summerville's mayor and director of its relief committee, continued his business career. He achieved further prominence when he was named secretary of the executive committee that planned The South Carolina Interstate and West Indian Exposition, hosted by Charleston from December 1901 to May 1902. A financial disappointment, this exposition was the only world's fair ever held in South Carolina and was attended by 675,000 people, despite unusually cold weather. Averill died in 1922.

The Edward Valentine bust of William A. Courtenay

After two four-year terms as mayor, and after nursing his beloved city through two catastrophic disasters, Mayor William Ashmead Courtenay left public office and returned to the business world. In gratitude for his dedicated public service, and over his strenuous objections, a group of Charlestonians insisted on commissioning a marble bust of him to be placed in the city council chambers. The work was executed by sculptor Edward Valentine and unveiled in 1888. The base bore the inscription: "As Chief Magistrate, he administered the government with firmness, impartiality, and success, even amid the disasters of cyclone and earthquake, signally illustrating the safe maxim, that 'public office is a public trust.'"

For several years after he left office, Courtenay was rumored to be a candidate for the governorship, and announced publicly that "although he had not sought the governorship, he would not decline to serve if he were called by a majority."[15] The active campaigning and ultimate election of Benjamin R. Tillman snuffed out the hopes of Courtenay's supporters, and put a final period at the end of the story of his political career. In 1889, he became president of the Bessemer Land and Improvement Company, a steel mill, and moved to Bessemer, Alabama. A year later he was back in the Holy City.

By 1890, when he applied for membership in New York's prestigious Grolier Club, America's oldest and largest society for bibliophiles and enthusiasts in the graphic arts, a sponsor valued his assets as between $75,000 and $100,000 ($1.5 to $2 million in present funds).[16] His final business vision led him to found the Courtenay Manufacturing Company, a large cotton mill, at Newry, Oconee County, South Carolina, in the extreme northwest part of the state, in 1894. His enduring support of the state's cultural affairs was often acknowledged throughout his lifetime and culminated with the award of an honorary Doctor of Laws degree from the University of Nashville in 1900.

After twelve years at Newry, he moved to the state capital in Columbia, where he could better pursue his interests in history, art, and literature. In Columbia, he was one of the founding commissioners of the Historical Commission of the State of South Carolina, the predecessor of the South Carolina Department of Archives and History. In his seventy-eighth year, his health began a rapid decline, and he died in his Columbia home on March 17, 1908, survived by his wife, Julia. Eulogized and mourned throughout the South, Charleston's indomitable former mayor was laid to rest in Magnolia Cemetery.[17]

In 1886, the earthquake provoked an enormous amount of research that helped redefine the basic concepts of seismology. But by the time that America became involved in World War I, most of the principal investigators of the grand event had retired or died, and the Charleston earthquake was, with few exceptions, a closed chapter in the history of geology for nearly fifty years.

An exception was the work of Stephen W. Taber, who presented a major paper on the Charleston earthquake to the LeConte Scientific Society at the University of South Carolina in 1914. He noted that between 1893 and 1913, 395 earthquake shocks were felt on 257 days in the Summerville-Charleston area, an average rate of one shock every two months. On June 12, 1912, a shock rated VII on the ten-point Rossi-Forel Scale, was the most severe felt since the shocks of 1886-1887. In agreement with Dutton's 1889 report and a book Dutton had written in 1904, Taber concluded that "most of the earthquakes felt in the South Carolina Coastal Plain during the last thirty years were probably to be attributed to readjustments taking place along a plane of faulting, located in the crystalline basement underlying the Coastal Plain sediments, not far from Woodstock, and extending in a generally northeast-southwest direction."[18] Taber also conducted exhaustive analyses of earthquakes as related to rainfall levels, tide levels, sea levels, groundwater levels, frequency by month, barometric pressure, and lunar and solar periodicities. He mercifully lay to rest all assertions that earthquakes were tied to any of these conditions.

It was not until seventy years after "de quake," after great strides had been made in the earth sciences, that the case of the Charleston earthquake was reopened. By the 1960s, the theory of plate tectonics had begun to play a crucial role in the explanation of earthquakes. Scientists had discovered that the earth's brittle surface—the lithosphere—was broken into a dozen or more thin (ten to one hundred twenty miles thick), rigid shells known as plates, which float and slowly move around the planet, creating or filling oceans and dividing or uniting continents. The constant interaction of these plates is known as tectonics, and as the plates grind past, collide, or shove their edges under or over each other, enormous strains are produced and released at the plate boundaries. These plate boundary interactions—such as at California's well-known San Andreas Fault—are the prime (but not the only) triggers of earthquakes.

When scientists mapped earthquakes that had occurred between 1961 and 1967, they found that these seismic events took place in narrow belts at the boundaries of the large tectonic plates.[19] These belts spawned volcanoes and interplate earthquakes, generated by the direct interaction of one plate with another. Yet the tectonic plate theory—well-founded and comprehensive as it was— did not seem to explain events like those at New Madrid or Charleston, for those were *intra*plate earthquakes—geological events that took place *within* tectonic plates, hundreds of miles from any known plate boundaries. The tectonic plate theory was a major milestone in geological science, but it would take a great deal more research to determine what triggered the Charleston earthquake.

Between 1886 and the earthquake's centennial anniversary in 1986, South Carolina experienced continuing seismic activity. The earthquakes felt between September 1886 and the end of 1889 were initially believed to be aftershocks of the August 31 event, that is, geological readjustments at the same hypocenter that produced the first and second shocks on August 31, rather than new earthquakes from different hypocenters in the same seismic zone. Whichever they were, the shocks continued. A 1977 USGS report cataloged three hundred shocks in the first thirty-five years following 1886.[20] Geologists soon concluded that the post-1889 shocks were not aftershocks. They were new earthquakes, and inevitably, more would follow. The only real questions were how many, how often, and how big?

Notable earthquakes were felt in the Summerville-Charleston area in 1912. In Summerville, chimneys cracked and plaster fell in several houses; beds moved several inches in Charleston; and shocks were felt throughout Georgia, South Carolina, and North Carolina. In 1945, a strong shock cracked brick walls and chimneys at Camden and Chester, and it was felt in various towns in four states. A 1959 quake threw down a chimney and cracked plaster at Summerville, Charleston, and Wadmalaw Island and was felt in Georgia. In the northeast part of South Carolina, where earthquake activity had been a rarity, a 1959 earthquake cracked plaster and sheetrock at Chesterfield, and the shock was felt in North Carolina. In the village built by William A. Courtenay at Newry, in Oconee County, an earthquake cracked a chimney and displaced furniture in 1971. That tremor was also felt in Georgia.

In 1974, a magnitude 4.3 shock cracked the brick veneer, driveway, and sidewalk of a house in North Charleston and separated a brick wall from the rest of the house. Throughout Charleston, cracks formed in driveways, sidewalks, and plaster. At Summerville, the same types of damage prevailed, and at Ladson, a few miles away, a five-hundred-ton machine tool "jumped around on its bed" at the General Electric manufacturing plant. It was the largest earthquake to date since the 1886 event. Back in Oconee County, a 1979 shock cracked drywall and concrete floors and was also felt in Georgia and North Carolina. The University of South Carolina reported about twenty aftershocks.[21] Clearly, earthquakes were not finished with South Carolina.

In the 1970s, the level of interest in the Charleston earthquake continued to rise, as new technology made detection and reporting more accurate. Seismic monitoring equipment was first used in South Carolina in 1971, when Gilbert A. Bollinger conducted observations in the Summerville area. In 1973, a five-station reconnaissance array was established. That fall, the USGS established a seismic network as part of a multidisciplinary study.

In 1977, the USGS published a series of studies related to the Charleston earthquake. Editor Douglas W. Rankin set the quake in its proper historical perspective when he wrote, "The seismic history of the southeastern United States is dominated by the 1886 earthquake near Charleston, S.C. An understanding of the specific source and the uniqueness of the neotectonic setting of this large earthquake is essential to properly assess seismic hazards in the southeastern United States."[22]

Rankin confirmed that the epicentral tract for the 1886 event was an elliptical area, twenty miles wide and thirty miles long, the center of which was Middleton Place, and confirmed Earle Sloan's computation that the 1886 quake's main shock had a hypocenter approximately twelve miles below the surface. In 1981, the seismic zone that Sloan had identified, and others had more precisely defined, was officially designated as the Middleton Place-Summerville Seismic Zone (MPSSZ).[23] The fault that causes the earthquakes in the MPSSZ lies buried beneath hundreds of meters of sediment and rock, which makes it difficult to pinpoint, although the exact locations of the hundreds of earthquakes it has produced in recent decades can be determined with great precision. Dr. Pradeep Talwani, professor of geophysics at the University of South Carolina and director of the South Carolina Seismic Network, named the Woodstock Fault in 1982, and believes that it is part of a northeast-southwest trending East Coast fault system that runs up the coast through Maryland and Virginia and perhaps beyond.[24]

Rankin also stated that the epicentral area is compressed by the sides of the fault and noted that earthquakes could be caused by the buildup of strain from these opposing forces.[25] His report suggested that "the present seismicity is originating at depths of one to eight kilometers [0.6 to 5 miles], mostly in the crystalline basement beneath the sedimentary rocks of the Coastal Plain."[26] Yet despite the scientific advances made in identifying the trigger mechanism of the 1886 earthquake, he admitted, "The cause of the Charleston-Summerville seismicity is still not determined."[27] Again, the Grail remained as slippery to grasp as before—a fact that did nothing to soften Rankin's ominous statement that "seismic activity is continuing today in the Middleton Place-Summerville area at a higher level than prior to 1886."[28]

As the centennial anniversary of the earthquake approached, research interest increased, providing fascinating new insights into the triggering mechanism and rekindling the search for the Holy Grail. In 1982, Dr. Talwani's investigation of the epicentral tract confirmed Earle Sloan's deduction that the 1886 earthquake consisted of multiple components. He also identified a new fault, the Ashley River Fault, along a portion of the Ashley River near the center of the MPSSZ.[29]

A USGS national workshop held in Charleston in 1983 focused on the 1886 earthquake and its implications for the future. It attracted 125 participants and showcased an extraordinary array of studies of the earthquake itself and of seismic activity along the East Coast in general. Leland T. Lang, of the Georgia Institute of Technology, explained why the Charleston event was so baffling when he admitted, "Few events present so frustrating an enigma as does the occurrence of the 1886 Charleston, South Carolina, earthquake. As a demonstration of potential seismic risk, the Charleston earthquake is one of a very few scattered exceptions in the Eastern United States."[30]

The first consensus reached was that the current earthquakes emanating from the MPSSZ were not aftershocks of the 1886 event but that "there is a local [geological] structure at Charleston that is the source of the continued seismic activity."[31] The second consensus was that the causative fault for the 1886 event had not yet been unequivocally identified.[32]

One of the most striking pieces of scientific information to come out of the conference was proof of the enormous displacement of the earth by the 1886 earthquake. This was clearly demonstrated by the distortion and buckling of the railroad tracks, which was most severe near the Woodstock and Rantowles epicenters of the Woodstock Fault. This displacement was intensively studied by Leonardo Seeber and John G. Armbruster of the Lamont-Dougherty Geological Observatory of Columbia University. They determined that seventeen feet of track in the Charleston-Summerville area had to be cut out of the three railroad lines to straighten out the rails. This was *prima facie* evidence that the earth had been severely compressed in this area and that the compression had been released by the 1886 earthquake—which buckled the tracks in the process. But it still did not answer the question of what specific geological action associated with the fault triggered the quake. The scientists were closing in on the answer, but they could not yet declare victory.[33]

The workshop participants agreed that "With only few exceptions, the majority of buildings, facilities, and lifelines built in Charleston and surrounding areas since 1886 are still without adequate seismic resistance. Accordingly, if a repeat of the 1886 event occurs, the disaster it caused would be repeated on a magnified scale."[34] They noted that South Carolina "is on the long climb up the hill of policy innovation," but that awareness of the magnitude of earthquake threats throughout the state was being aggressively pushed by two entrepreneurial pioneers. The first was Dr. Joyce Bagwell, a member of the Department of Geology and Chemistry at the Baptist College of Charleston (now Charleston Southern University). She gave frequent talks to citizens' groups and schools and effectively used the media to spread the word of the dangers of earthquakes. Her colleague, Dr.

Charles Lindbergh, an engineer at The Citadel, worked directly at the government level, seeking to influence officials to adopt earthquake preparedness policies.

The workshop's study group on increasing awareness and preparedness presented the results of a simulation of what would happen if an earthquake the size of the 1886 event occurred in Charleston in daytime in 1983. The simulation used the 1886 damage levels, interpreted after studying actual damage sustained in two major recent earthquakes. The results indicated that eighty percent of the schools, fifteen percent of frame residences not on stilts, eight-five percent of frame residences on stilts, and seventy percent of masonry residences in the epicentral area would be damaged beyond the point of safe occupancy. The study estimated that between five hundred and one thousand people would die, of which seventy percent would be students, killed in their structurally unreinforced schools. Dr. Otto W. Nuttli, of Saint Louis University, concluded, "The impact of the next large Charleston earthquake on the Eastern United States is going to be severe."[35]

The group agreed that the scientific community ought to encourage families to set aside emergency supplies, educate school officials and PTAs as to the benefits of having earthquake drills, and emphasize to builders the need to design and construct schools, hospitals, and other critical facilities to withstand damaging earthquakes. It warned that preparations for a large earthquake in the Charleston area are not being addressed adequately. In addition to strengthening buildings and enacting stronger building codes, the study group noted, "the public should be prepared to react to a damaging earthquake when it is happening because there will likely be no warning."[36]

In a discussion that assessed earthquake hazard awareness, the panel put people in two categories and drew its conclusions. Group I was made up of city and county officials in Berkeley, Charleston, and Dorchester counties; Summerville-area residents who had experienced minor earthquakes or seismic sonic booms; beach residents who had felt seismic sonic booms; and lifelong Charleston residents or residents of more than five years who had been reminded periodically through the news media of the 1886 earthquake. The level of awareness of earthquake hazards in this category was called "moderate" and the level of preparation was "none." Group II consisted of new residents of the Charleston-Summerville area and residents outside of the Lowcountry in central and upper South Carolina. It was rated as having "zero" level of awareness, and the level of preparation was "none."

After the conference, at least one group took a more active interest in Charleston's seismic events. In July 1986, during the construction of Charleston Place, a large hotel and retail complex at the corner of Market and Meeting streets in Charleston, the USGS installed instrumentation to record the building's response to the next damaging earthquake that strikes downtown Charleston. This instrumentation was the first strong-motion detection system installed in Charleston. It consists of twelve motion sensors, installed on the ground, sixth, and eighth floors. Different sensors detect motion in north-south, east-west, and up-down directions. The cables from the sensors drop down to a central control panel and recorder located behind the bell captain's stand on the ground floor.[37]

Studies of the earthquake continued as a result of the heightened awareness caused by the conference. In 1993, Dr. Talwani analyzed all the available data for the area around Fort Dorchester on the Ashley River and came to an important conclusion: the northeast-southwest line of the Woodstock Fault had been cut in the middle and significantly offset by a separate fault. In 2004, he determined that this fault ran twenty degrees to the northwest, down the Sawmill Branch of the Ashley River, and past Fort Dorchester to Middleton Place. This portion of the Ashley River Fault he named the Sawmill Branch Fault. When he returned to re-examine the earthquake damage to the fort and to the bell tower of the parish church of St. George, Dorchester, Dr. Talwani found additional physical evidence to support the existence of the Sawmill Branch Fault. At the fort, two large cracks at opposite ends showed significant offset displacements parallel with the proposed fault. In

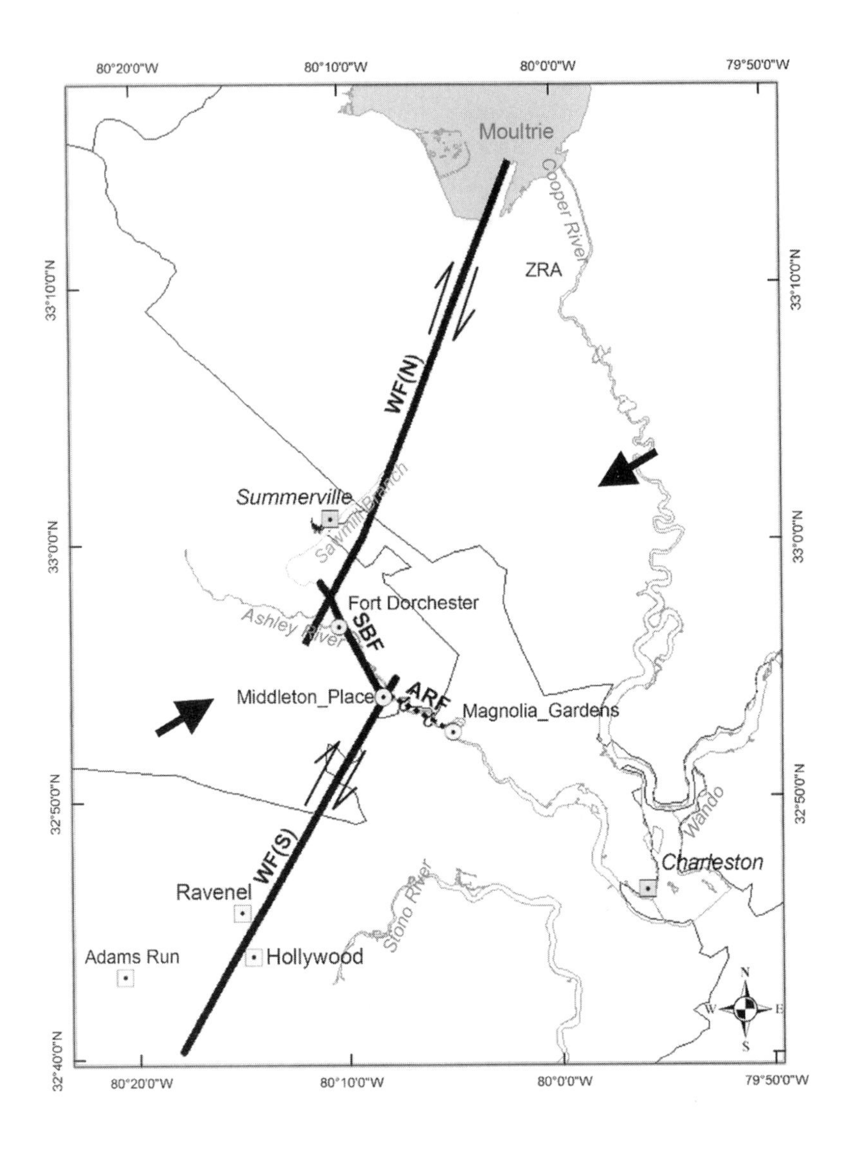

The Woodstock, Ashley River, and Sawmill Branch Faults near Summerville

The land to the west of the Woodstock Branch Fault is moving northeast. The land to the east is moving southwest. The bold arrows indicate the maximum horizontal stress in the region. Most of the current seismic activity lies in a zone where the two fault lines intersect.

addition, large blocks of the tower had been thrown to the northeast and southwest—the same directions as the maximum horizontal compression of the moving land that built up the stresses that triggered the earthquakes.

To determine the rate at which strain was compressing the area near the Woodstock Fault, a twenty-station, high-precision global-positioning system (GPS) was put in place in the Summerville area and operated from 1993 to 1994. Dr. Talwani and Robert A. Trenkamp analyzed the GPS data and determined that the area was being compressed at the rate of about one centimeter (0.4 inches) per kilometer in a northeast-southwest direction along the fault line, which is approximately fifty kilometers (thirty miles) long. This rate of compression is similar to that being produced in the San Andreas Fault in California.[38] The compression produces stress, and when the level of stress passes the critical point, the stored energy is released as an earthquake. The release of accumulated compression and stress along the Woodstock Fault was the elusive trigger mechanism geologists had been seeking for over a century. The reason for one part of the region—that near Summerville—to be moving in a different direction and speed than the rest of the surrounding area is still unknown.

In 2001, Dr. Talwani and William T. Schaeffer finished an analysis of fifteen years of paleoseismological work exploring prehistoric earthquake liquefaction sites in South Carolina. These were ancient sand blows, exactly the same as William McGee, Thomas Mendenhall, and Earle Sloan had seen in the hundreds during their 1886 fieldwork. The analysis found evidence that the magnitude 7.3 Charleston earthquake in 1886 was not the first of that size in the area. In fact, the study identified seven episodes of magnitude 7.0+ earthquakes over the past six thousand years. This evidence shows that the 1886 earthquake was not a freak event but simply the latest in an ongoing series. From their findings, the authors concluded that magnitude 7.0+ earthquakes occur about every five hundred years in the Charleston area.[39] Dr. Talwani refers to this chain of events—the increase in the rate of strain over the period of time between major earthquakes, which ultimately triggers an earthquake—as the MPSSZ "Earthquake Cycle."[40]

With the knowledge gathered through this intricate 120-year chain of events, with each successive link contributed by a different researcher, geologists now know what triggered the 1886 seismic event, and the several thousand small earthquakes that followed it to this day. The Great Charleston Earthquake of 1886 was caused by the release of accumulated compression along the Woodstock Fault.

The Grail had has finally been found.

19

Lessons Learned

There is nothing, I believe, so trying to a healthy nervous system as a succession of earthquakes.
—Louis Housel, *An Earthquake Experience*, 1878

To this day, no disaster recovery operation in America has ever been as successful as Charleston's after the 1886 earthquake. The leaders and people of the city, with no financial assistance whatsoever from the federal or state governments, carried out the most efficient, humane, rapid, and financially responsible recovery from a mass-casualty disaster in all of U.S. history prior to 1886. The workforce of Charleston was back to nearly full employment within one month after the first shock, and Charleston and Summerville were rebuilt within fourteen months. Had the procedures and philosophies employed in the recovery and rebuilding of the city and the South Carolina Lowcountry been recorded, organized, and used in later disasters elsewhere, thousands of lives and billions of dollars could have been saved. In light of the nation's recurring disasters that require a rapid response, the successful Charleston Disaster Recovery Model is worth reviewing.

THE CHARLESTON DISASTER RECOVERY MODEL

Civic leaders acted immediately to organize the relief effort and set priorities. They did not sit and wait for any outside agency to tell them what to do or to bail them out.

When the earthquake struck on August 31, 1886, the city's chief executive, Mayor William A. Courtenay, was on a ship in the middle of the Atlantic Ocean, and there was no way to contact him until he reached New York five days later. In addition, with all telegraph and railroad lines out of operation, Charleston was enveloped in a complete communications blackout. There was no way to contact anyone more than ten miles away for help. Furthermore, no private, state, regional, or national organization had any legal obligation to provide aid. However, there was a designated interim chief executive, with the full power and authority of the elected mayor. Acting Mayor William E. Huger immediately accepted responsibility for emergency relief operations. The first day after the quake, the commissioners of the heavily damaged hospitals conferred with Huger and the available members of the city council and decided to move all patients from the ruined medical facilities to Agricultural Hall on Meeting Street. In partnership with a private transport company, which volunteered its services, medical personnel moved patients, equipment, and furniture, and within forty-eight hours, the city had a fully operational new public hospital. On the third day after the quake, Huger and the city council assembled and carried out the first city-wide damage assessment. They resolved to meet the following day, after they had the basic facts in hand, to decide on a plan of action.

Civic leaders quickly formed an emergency relief organization to deal with the most immediate needs: food and shelter.

On the fourth day after the earthquake, the acting mayor and city council formed a separate agency of city government to deal solely with the relief process: the relief committee, later renamed the Executive Relief Committee (E.R.C.). This group was made up of private citizens who were experts in their fields and were not politicians. Although the E.R.C. operated under the legal authority of Charleston's city council and reported regularly to it, the committee was empowered to take all necessary actions to save the lives and protect the safety of the citizens of Charleston. During the emergency, the city council took care of the routine matters of city governance, giving the E.R.C. full control of running the relief operation. To carry out their work, the E.R.C. initially formed two subcommittees: one to handle shelter and one to handle food and medicine. Each was headed by a community leader with executive and practical skills for the specific job at hand.

The E.R.C.'s leaders drew on their experiences from previous disasters.

E.R.C. members adopted successful survival and recovery strategies they had learned from past experiences dealing with numerous previous catastrophes, including damage, destruction, and casualties from prolonged artillery bombardment and occupation by two enemy armies; seven major fires; a tornado; and numerous Atlantic hurricanes. An example was the rapid establishment of a voucher-based food distribution system, which had been used twenty-five years earlier after the Great Fire of 1861.

The city council immediately diverted all non-critical financial assets to the relief efforts.

Until alternative funding was located, the city council diverted all non-critical city funds (such as those intended for capital improvement projects) to the relief efforts.

The E.R.C. focused on providing emergency services and meeting emergency needs with a minimum of red tape.

During the first weeks of the E.R.C.'s operation, immediate cash grants of up to $10 ($200 current value) were made to any person who could demonstrate a legitimate immediate need, such as loss of all their clothing or tools of their trade. No application paperwork was required. Any member of the

E.R.C. could authorize an immediate aid voucher on his own signature, as could "any person of good character known to the committee." A "person of good character" was generally construed to mean members of the city council, well-known businessmen, and any of the city's clergymen.

Shelter assistance from the outside was sought as soon as communication was possible, and the construction of emergency shelters began quickly.

In response to newspaper reports of Charleston's catastrophe, donations of tents poured in from public and private sources. First to respond was the federal government. On September 1, the day after the earthquake, at the request of Charleston officials, U.S. Congressman Samuel Dibble requested the U.S. Army to send an emergency shipment of army hospital tents to Charleston. The government acted immediately. One hundred tents arrived on September 6. This represented the entire stock of federal tents available on the East Coast. To ensure that the federal supplies were actually needed, the Quartermaster's Department sent an officer to Charleston on September 5 to evaluate the extent of the damage. Unlike food and medical aid, which was distributed in an impartial manner, the distribution of high-quality federal tents (and hundreds later from other places) was carried out along racial lines. The majority of the army tents went to whites. For blacks, hundreds of wooden shed shelters were constructed in public parks and squares in the first two weeks after the earthquake. However, all the tents and sheds only met the housing needs of about twenty percent of the city's homeless.

Verifiable eligibility criteria for free food and medicine were established and enforced.

Eligibility for free food distribution was limited to those incapable of purchasing or preparing food for themselves. This included those too young or old to work and the sick, physically disabled, and mentally incompetent. These persons and families were issued ration tickets, which could be obtained from the

E.R.C. members, city council members, the clergy (both black and white), or any person vouched for by a person of good character known to the committee. The people who were allowed to issue ration tickets were intimately familiar with the residents of their wards or congregations, and they knew who was truly needy and who was not. Able-bodied people who held paying jobs or who could afford to buy food were not eligible for free food, except at the E.R.C.'s soup kitchen.

At first, only citizens of the city of Charleston were eligible for food rations. After many people who lived outside the city limits found ways to obtain ration cards, the E.R.C. changed the rationing system. Rations were no longer given out at the central distribution center, and a cart delivery system was inaugurated. The carts only delivered food to those who lived inside the city limits. When relief funds began to come in from outside Charleston, the food distribution system was broadened to serve other areas, including Mt. Pleasant, Summerville, and surrounding small communities.

Procedures were established to ensure that the truly needy did not slip through the cracks.

To ensure that all needy residents of Charleston were accounted for—even if they were too sick or too proud to ask for help—the E.R.C. asked the clergymen of both races to find such people and issue them food ration cards. The E.R.C. also sent black and white canvassers into all neighborhoods to ensure that no one eligible for assistance slipped through the cracks. Although the eligibility rules were strict, they were not inflexible. Exceptions were made for infirm and elderly poor people who lived outside the Charleston-Summerville-Mt. Pleasant designated aid assistance areas.

Civic leaders mounted a massive national fundraising campaign, using the nation's newspapers as their chief public awareness and fundraising tool.

It was soon apparent that neither the federal nor the state government was going to provide any financial assistance to the earthquake victims. Consequently, Mayor Courtenay, who was widely known and respected in the South and had a high profile nationally and internationally, made direct requests for financial aid to the American people. His chief ally was *News and Courier* editor Francis W. Dawson, who used his newspaper as a "mighty megaphone" to spread the appeals for funds. Between the two men, they raised $646,109.90 (just under $13 million today) for the relief efforts and rebuilding of Charleston. Private individuals and organizations donated approximately another $300,000 directly to the beneficiaries of their choice, making the total for relief donations approximately $1 million ($20.1 million today).

The newspaper kept everyone in the city aware of the opportunities and rules in effect and informed the outside world about Charleston's true condition and needs.

The E.R.C. communicated critical relief data to the public through the newspapers. This information included procedures for obtaining ration tickets, when and where to obtain food, and ways to apply for shelter, emergency cash grants, grants to rebuild damaged residences, and financial aid to allow wives and children to temporarily leave the city. No grants were made to rebuild commercial property. The *News and Courier* labored ceaselessly to promote the vitality of the city, solicit funds, and quash rumors.

Civic leaders gave high priority to inspecting all residences so that people would be encouraged to vacate crowded, unsanitary refugee camps and return to their homes.

City leaders—most of whom were ex-soldiers and knew the deadly effects of unsanitary camp life—worked strenuously to encourage citizens to return to their homes if possible. To achieve this goal, they ordered rapid, yet thorough, house-to-house building inspections, the results of which were published in the daily newspaper. These inspections identified houses that were habitable (though damaged), as well as those too dangerous to re-enter or which had to be pulled down. Civic leaders pointed out that a damaged house, if safe for occupancy (even if it was not attractive or had some uninhabitable parts), was infinitely preferable as a residence to a tent during the coming hurricane season, with its rains, winds, and falling temperatures. Mayor Courtenay set the example on September 10 when he moved back into his damaged brick house. Through the strenuous efforts and incessant encouragement of civic leaders, the camps were quickly cleared, and the dark menace of disease epidemics was completely avoided.

All donated funds and disbursements were accounted for in a completely transparent and financially responsible manner.

Many donations came from well-wishers who sent money in care of individual officials and institutions. Except for these direct person-to-person or organization-to-organization donations, all funds were channeled through a single agency, the E.R.C., whose treasurer was the city treasurer. The financial handling of all public aid money was done in a totally transparent manner. All transactions were documented. All donations were logged and announced in the newspaper. Vouchers were issued and receipts obtained for every authorized expenditure. A comprehensive financial report was published and distributed on July 25, 1887, showing in detail all income and how it had been spent. The amount of assistance given to every recipient of rebuilding aid and the name of the recipient were listed in the report, which was widely distributed throughout the community and to all major donor organizations.

When sufficient funds had been received to carry out the relief goals, the mayor publicly announced that no further donations would be required.

Mayor Courtenay formally announced through the newspapers on October 5, 1886, that Charleston had received, or expected to receive, sufficient funds to meet its needs, and that no further donations were needed. This nearly unparalleled act of fiscal restraint was carried out despite the fact that money was rolling in at an unprecedented rate and the city could have reaped double the amount that it accepted.

This policy stands in stark contrast to a recent campaign undertaken by a South Carolina church to raise non-emergency funds. After the church had already reached 135 percent of its financial goal for capital improvements, it sent out a follow-up letter to its members, pointing out that, although more than $1 million had been donated, the money had come from less than thirty percent of the membership. The church then asked, "Please help us reach our goal of 100 percent participation by making your pledge today."

No member of the E.R.C. benefited financially from the relief funds.

None of the E.R.C.'s members applied for or received any financial benefit from the funds that were collected and disbursed. Indeed, the only city official who ever filed any kind of claim was W. W. Simms, the clerk of city council, who, like two thousand other homeowners, filed a request for funds to fix his damaged house. He asked for $545 and received $452.

The E.R.C. was disbanded as soon as its basic mission of providing emergency services was concluded.

Civic leaders had no interest in maintaining an unnecessary bureaucracy. As emergency needs declined, the E.R.C. was cut down proportionately. By November 1886, its staff had been reduced to half its initial size, and it was totally dismantled in the spring of 1887. As noted, it filed its financial report to the city council in 1887. The report was published and widely disseminated to demonstrate to donors that their money had been used responsibly.

Although the exact procedures developed and used in the wake of the Great Charleston Earthquake of 1886 cannot be applied today to every catastrophe in every locality, the Charleston examples of self-reliance; immediate response; use of the knowledge and skills of local civic leaders, such as the clergy; decisive action; willingness to adapt to changing conditions; fundraising integrity; financial accountability; and minimal red tape and bureaucracy are goals applicable to any disaster recovery operation at any time and any place.

20

THE NEXT BIG ONE

The absence of large-magnitude earthquakes in eastern North America since the Charleston, S.C., earthquake of 1886 has resulted in complacency, or perhaps unawareness on the part of the general populace of the existence of any earthquake threat to them.

—Dr. Otto Nuttli, 1973

In the one hundred twenty years since the Great Charleston Earthquake of 1886, there has been an information revolution of unprecedented magnitude. Knowledge about virtually any topic is available almost instantly via the Internet. High school freshmen can access scientific information that was not available even to the world's most knowledgeable professionals ten years ago. Yet in this age of instant access to global databases, almost nothing has changed when it comes to the earthquake awareness of the people who live near the epicentral zone of the 1886 event.

In 1886, the scientific causes of earthquakes were poorly understood. Many knowledgeable and highly respected scientists linked the seismic events with such diverse triggering mechanisms as the tides; variations in rainfall, temperature, or atmospheric pressure; certain months or seasons of the year;

or solar or lunar cycles. Within the scientific community, the theories that most earthquakes were triggered by volcanic activity or by geological faults in the earth's crust were just emerging.

But the lack of scientific explanations for earthquakes was not the cause of the pre-1886 complacency. It was lack of personal experience with seismic events. In 1886, few living South Carolinians had experienced an earthquake. Virtually no one—including the state's historians—knew that South Carolina had a long and substantial history of earthquakes. Certainly, no one had ever heard of an earthquake that had killed anyone. Even noted historian Robert Mills said of his native state in 1826, "from the fatal consequences of earthquakes, it is happily exempt."[1] Everyone knew that hurricanes, tornadoes, floods, and lightning were serious threats—just like yellow fever and malaria. But earthquakes? Those who knew anything about them believed that they were unthinkable east of the Mississippi.

Well over a century after the event, scientists now know what caused the 1886 Charleston earthquake: the buildup and sudden release of geological stress along the Woodstock Fault, the centerline of an oval area known as the Middleton Place-Summerville Seismic Zone (MPSSZ). Hundreds of earthquakes—mostly small, but several quite noticeable—have occurred in the MPSSZ since 1886. Furthermore, the rate of seismic activity in the zone is continuing at a higher level than it was before 1886.[2] A total of eight magnitude 7.0+ earthquakes have occurred in the MPSSZ in the last six thousand years, with the 1886 magnitude 7.3 earthquake being the most recent one.

Because of the loss of life and the huge amount of property damage from "de quake," you might think that today's residents of Summerville and Charleston would be aware of the risks and prepared to face them. The opposite is true. With few exceptions, people who live within one hundred miles of the epicentral tract have no significant knowledge of its earthquake history, of the likelihood of a recurrence, or of ways to prepare for a repeat of the disaster. Geologist Dr. Otto Nuttli, who made extensive studies of the 1886 seismic event, clearly identified why this lack of awareness exists. In 1973, he wrote, "The absence of large-magnitude earthquakes in eastern North America since the Charleston, S.C., earthquake of 1886 has resulted in complacency, or perhaps unawareness on the part of the general populace of the existence of any earthquake threat to them."[3] He hit the nail squarely on the head. Unless we have personally experienced something, we don't tend to think of it as "real."

On the other hand, exposure to small earthquakes does markedly stimulate awareness and earthquake preparedness education activities. Everyone in California is at least aware that earthquakes pose a serious risk to their safety. However, being aware is not the same thing as being prepared. In Summerville, South Carolina, which lies directly over the Woodstock Fault, I gave an earthquake awareness talk in February 2006 to a group of about one hundred adults. Before I began my presentation, I asked those in the audience to raise their hands if they had personally felt an earthquake tremor in Summerville. More than ninety percent of the audience members raised their hands. However, in private conversations with members of the same group, when I asked if they had taken steps to prepare their homes to withstand the effects of earthquakes or had prepared earthquake emergency evacuation kits, they all said no. Making people aware of earthquake hazards is the first step in helping them prepare for and minimize the effects of an earthquake.

Earthquake awareness training is crucial. If everyone knew the dangers of the massive earthquakes that, for example, the MPSSZ is capable of generating, people might take them more seriously, prepare their homes and businesses better, and be ready to survive the disaster.

Unfortunately, the bulk of information that Americans get about earthquakes is derived from grossly exaggerated, scientifically impossible scenarios in disaster-thriller movies. In "Aftershock" (1999), a fault unknown to scientists destroys New York. In "10.5," (2004), as a catastrophic series of earthquakes threatens to slice off California and sink it in the Pacific Ocean, heroes try to glue the fault line together using nuclear bombs. And in the sequel, "10.5 Apocalypse" (2006), a "swarm" of ultraquakes wipes the presidential smiles off the faces of Mount Rushmore, sucks Las Vegas into a sinkhole, fires up the Mount St. Helens volcano, washes Hawaii away with a two-hundred-foot tsunami, and threatens to split the country in two. Movies of this type are good for selling popcorn, entertaining thriller-loving adults, and scaring young children, but they have no positive (and much negative) information value. They all give the same message: when it comes to earthquakes, we are all helpless victims. This concept is not only totally false, it is extremely dangerous because it instills the idea that nothing can be done to protect individuals, families, homes, or public buildings from the effects of an earthquake.

When it comes to obtaining practical information about earthquakes and how to prepare for them, high-quality material is readily available to anyone with access to a computer. For those without computer access (with California as the notable exception), public earthquake education programs are virtually nonexistent in most parts of the country.

At the national level, accurate and useful earthquake information is available from or through the Federal Emergency Management Agency's (FEMA) website: http://www.fema.gov/hazard/earthquake/index.shtm. Many of FEMA's publications can be downloaded, and all of the printed materials are free and may be ordered through a toll-free telephone call or by mail. These publications describe what to do before, during, and after an earthquake and provide information for individuals and homeowners, teachers and children, community planning and public policy personnel, and building professionals and engineers.

The United States Geological Survey (USGS) is the nation's primary provider of scientific and technical information about earthquakes. Its website, http://www.usgs.gov/, provides links to the earthquake history and hazards of each state. Its publications warehouse (infotrek.er.usgs.gov/pubs/) enables you to search for any type of earthquake information, and its automated notification system (earthquake.usgs.gov/eqcenter/ens/) enables you to get reports by email whenever a measurable earthquake occurs in your region of the country.

In the Southeast, most of which felt the effects of the 1886 disaster, little earthquake awareness training is being carried out by public officials. In South Carolina, there is virtually none at all, despite the fact that the MPSSZ is a well-known lethal earthquake fault. As a result, the vast majority of the state's 4.1 million residents have little or no awareness of the state's seismic history and hazards.

On the main page of South Carolina's official Internet website, www.sc.gov, there are no links to any information about the state's natural hazards, hazard awareness, or hazard mitigation. Typing the term "earthquake" into the state website's search engine, however, brings up articles, information, and news releases about earthquakes in South Carolina.

The South Carolina Emergency Management Division (SCEMD) is the state agency charged with disaster preparedness. Its chief role is to develop state policies and procedures for dealing with disasters. It is obvious that the SCEMD has given deep and careful thought to preparing for major earthquakes. Its plans are recorded in two important public documents, both of which are downloadable from the SCEMD web page, www.scemd.org/Prepare/earthquakes.htm. The first document is the *South Carolina Earthquake Plan* (2004), which describes how each state and local government agency is to work together in case of a major earthquake. The second is a federally sponsored 2001 study of earthquake threats titled *Comprehensive Seismic Risk and Vulnerability Study for the State of South Carolina*. The enormous importance and implications of this 609-page study will be discussed later.

Although the SCEMD offers no disaster preparedness training programs for the general public, it does provide personal emergency planning and response information on its web page. Its downloadable educational materials for individuals and families include *Building a Family Energency Kit* and *Developing a Family Disaster Plan*. On its online library (www.scemd.org/

Library/index.htm), the SCEMD also offers a downloadable four-page brochure with technically accurate and useful information about preparing for and surviving earthquakes.

The South Carolina Department of Education is not actively involved in earthquake awareness education. On its website (http://www.myscschools.com/), there is no significant information about the earthquake-resistance status of the state's schools or how the state's approximately 700,000 students will be protected if an earthquake strikes. No South Carolina law requires the state's schools to develop earthquake emergency preparedness plans or practice earthquake safety drills. However, several South Carolina schools, on their own initiative, have chosen to adopt and practice an earthquake safety drill developed by FEMA. The information they used was contained in the *Guidebook for Developing a School Earthquake Safety Program* (FEMA Publication #8-511 FEMA 088), which FEMA no longer distributes, but it may be available through interlibrary loan.

On a local level, even Charleston, Berkeley, and Dorchester counties, whose approximately 569,000 citizens live in or within fifty miles of the MPSSZ and will probably bear the brunt of the damage from the next major earthquake, have no active county-sponsored earthquake awareness education programs. The same earthquake education vacuum applies to the rest of South Carolina's counties.

The lack of public earthquake awareness training in South Carolina can partially be blamed on another natural phenomenon—hurricanes. Hurricanes, which often spawn tornadoes and leave floods in their wake, strike the South Carolina coast almost every year, and frequently cause heavy damage. Major earthquakes—though they have enormous damage potential—are infrequent. Therefore, understandably, a large portion of the state's disaster preparedness funding goes into hurricane safety awareness and preparation, leaving little or no money for earthquake hazard education. As the inevitable result, few citizens of South Carolina will be prepared or know what to do when the next major earthquake strikes.

Will there be another 1886-grade earthquake in the South Carolina Lowcountry? When will it strike? How will it differ from the 1886 event? How bad will it be? Once people become aware of the earthquake hazards in their area, these are the questions they want answered. The answers are complex, but enough knowledge about what causes earthquakes and, specifically, what caused the 1886 Charleston-Summerville earthquake, has been assembled so that some meaningful information is available.

Will there be another magnitude 7.0+ earthquake in the MPSSZ? The answer is yes. The reasons are fairly simple. Geologists know that this zone has generated at least eight magnitude 7.0+ earthquakes over the last six thousand years, with an average cycle (time between major earthquakes) of about five hundred years. They also know that hundreds of small earthquakes have been generated in the MPSSZ since 1886 and that these are occuring more frequently than they did before 1886. In short, the seismic activity level within the MPSSZ is increasing, not decreasing. From ongoing measurements, scientists also know that strain is being generated every day along the Woodstock/Sawmille Branch Fault lines, and when the accumulated strain exceeds a certain limit, it will be released as an earthquake. The sum of all this evidence suggests that there will be a recurrence of an 1886-grade earthquake centered in the MPSSZ. The next logical question is, when will it happen?

Determining *when* the next major earthquake will strike the Charleston-Summerville area is more complicated than figuring out *if* it will happen. Simply put, no group of scientists can predict with any great precision when an earthquake will strike in a given area. In the case of the MPSSZ, the average time between major seismic events is about five hundred years. The last large quake before the 1886 event happened in 1455 (+/- 17 years), or 431 years ago. The one before that happened in 1,000 (+/- 33 years), or 455 years before the later one. The next one further back struck in 353, or 647 years earlier, and so on back to 5,800 years BCE. But the fact that the average cycle is approximately five hundred years does not tell us exactly when the next earthquake will occur. At the present time, we are about one hundred twenty years into what appears to be a five-hundred-year cycle, but that does not mean that the next earthquake will strike in 2386. The quake will happen when the stress builds up to an uncontrollable level—and that could happen tomorrow or not for a thousand years.

Science does have the ability to create accurate models of what will probably happen if a magnitude 7.3 (1886-grade) earthquake strikes tomorrow. To create the models, scientists and engineers compare what types of structures (brick, wood, or a combination) survived the last earthquake and what types have been built on what kinds of land (solid ground vs.made land) since the 1886 quake. When the the increased population density in and near the MPSSZ, where most of the heavy damage will take place, and the increased earthquake resistance of the most modern structures are factored in, a picture of the likely aftermath emerges.

- The effects of liquefaction will probably be more severe in the next big earthquake than in 1886. According to Carl H. Simmons, who has been Charleston County's Director of Building Inspections for more than twenty years, "areas of prior surface faulting (such as sand blows or surface fissures) have a fifty percent greater probability of future faulting than firm land, as they have been transformed into weak points for the release of seismic energy to the earth's surface."[4] In other words, areas where faults and similar phenomena appeared in 1886 are weaker now, so the damage there will be greater when another earthquake occurs. This is especially true for structures built on made land where surface faults occurred during the 1886 event In addition, these buildings will suffer more damage than those on firm ground. In 1886, about forty percent of peninsular Charleston was made land. The amount of made land increased after 1886, as the city used the rubble from the 1886 earthquake to fill in more ponds and marsh.

- Pre-1838 brick structures will probably survive as well or as poorly as they did in the 1886 earthquake. Those that were well-built with good materials and workmanship are the likeliest to survive with the least damage.
- Post-1838 brick structures that were well-built with good materials and workmanship will probably also survive, but the substantial decline in the quality of masonry work after 1838 means that a higher percentage of the affected 1838-1886 buildings will fail.
- Post-1886 brick replacement structures, many of which were the product of substandard materials and workmanship, may not fare well. Following the 1886 earthquake, the widespread use of poor mortar to repair or replace buildings, plus many examples of brick laid in American bond with insufficient courses of header bricks to tie the walls safely together, will probably cause many masonry structures to fail. Examples of this poor workmanship are still being discovered today, as homeowners, businesses, and churches that own nineteenth-century buildings undertake what they thought would be routine maintenance and, instead, uncover serious structural and masonry defects traceable to the 1886 earthquake.
- According to Charleston County's Director of Building Inspections Carl H. Simmons, the thousands of earthquake bolts and gib plates installed after the 1886 earthquake will be "useless" in preventing damage during the next

major earthquake. "When the next major seismic event occurs, the surface waves of the earthquake will roll a building both vertically and laterally, just as in 1886," he said. "As the building is stressed first up and to one side, and then down and to the other side, the inflexible iron earthquake bolts will pull out the bricks behind them and those adjacent to the gib plates. This will, in many cases, cause the walls to fail. If that happens, the floors will be left with little significant vertical support, and they are likely to pancake down."[5] Although Charleston's legendary earthquake bolts have kept brick walls from bulging during normal times, it seems they will be of no value when tested by the next major earthquake.

- The 1886 earthquake demonstrated that while brick buildings often cracked, wooden structures generally flexed, but they usually sprang back into shape. Well-constructed wooden structures built before 1886, and properly repaired, and those built after 1886 using good design principles, materials, and craftsmanship should fare as well as wooden houses did in 1886, when only about seven percent of wood-frame buildings were seriously damaged.
- Structures constructed of all materials and built during the last ten years stand a much higher chance of survival than older ones because they were required to meet strong building codes designed with the area's earthquake and hurricane wind threats in mind.
- In Charleston County, there is no requirement to retrofit old or historic buildings to meet current building code standards, unless the cost of rebuilding a structure equals or exceeds fifty percent of its current value. The vast majority

of the county's historic properties have not been retrofitted to meet the current, more stringent, building codes for two reasons. First, to do so would require such extensive structural modification and disfigurement that the buildings would lose much of their original appearance, and thereby significantly diminish their value as models of how life was lived in the past. The second reason is the immense—and often impossibly high—cost of updating seventeenth-, eighteenth-, or nineteenth-century buildings to comply with current building codes.

- The vast majority of all private and public buildings in the Charleston-Berkeley-Dorchester County area are not earthquake-resistant and will probably be damaged to the same degree as they would have been in the 1886 earthquake.

In 2001, the SCEMD published its *Comprehensive Seismic Risk and Vulnerability Study of the State of South Carolina,* the most comprehensive analysis ever undertaken to determine what would happen during future earthquakes in South Carolina. Using a computer model developed by FEMA, the SCEMD simulated earthquake scenarios, using HAZARDS U.S. (HAZUS®), FEMA's state-of-the-art loss estimation model. It modeled four earthquakes: magnitude 7.3 (same as 1886), 6.3, and 5.3 events originating in the MPSSZ and a magnitude 5.0 earthquake in Columbia, the state capital.

Because the HAZUS® model was based largely on California buildings (which are extremely different from South Carolina's much older structures), the results may or may not accurately reflect the actual outcome of a repeat of the 1886 Charleston earthquake. In addition, the model does not take into consideration the influx of tourists into the state, especially during the summer months. If a large earthquake were to occur in the summer, the losses could be significantly higher. Nevertheless, the HAZUS® model is the best one available at this time. The results of its analysis of the 1886-level event did not appear to be seriously out of line with South Carolina's actual experiences, even though the state's masonry structures were built without regard to seismic design standards until 1993.

The results of SCEMD's study of a repeat of the magnitude 7.3 Charleston-Summerville earthquake during daytime hours are chilling:

- There would be approximately 45,069 casualties, including 891 fatalities (six times the 1886 known dead), 7,951 major injuries requiring hospitalization (sixty times the number of known injured from 1886), and 36,227 other injuries.
- About 70,000 households, or about 200,000 people, would be displaced, with an estimated 60,000 people requiring short-term shelter. In 1886, two-thirds of the population of Charleston (40,000 people) were too frightened to re-enter their homes.
- Approximately 255 fires would break out in the Tri-county region, compared with eight in 1886. The lack of operational firefighting equipment and a supply of water for fighting fires will probably be a major concern. In 1886, half the Charleston firefighters could not immediately respond to eight fires because their horses and equipment were unavailable, and their water supply was cut off for about an hour. If the current water supply were interrupted, the saltwater fire wells that helped to save many buildings in 1886 are not available.
- Approximately 404 schools would sustain major damages, causing significant casualties, because the majority of school buildings are not earthquake-resistant. School children could be the single largest source of fatalities.

- About 30 hospitals would sustain at least moderate damage.
- Approximately 216 fire stations would sustain at least moderate damage.
- All 8 of the area's private and commercial/military airports would sustain at least moderate damage. The region's largest civilian and military air terminal, Charleston International Airport (CHS) and Charleston Air Force Base, is located directly over one of the most geologically disturbed areas in the MPSSZ: the Ten-Mile Hill area, where violent earth disturbances caused serious railroad track distortion and a fatal railroad accident in 1886.
- Approximately 63 of the area's power faciities would sustain at least moderate damage.
- About 761 of the area's bridges would sustain at least moderate damage, bringing most traffic to a stop, and blocking the passage of emergency response venicles, such as ambulances and fire trucks.
- Approximately 172,144 buildings would sustain at least extensive damage, with another 331,700 more experiencing slight to moderate damage.

Even now, the MPPSZ continues to be the most active seismic area in South Carolina. Between 2001 and 2004, the South Carolina Seismic Network recorded eighty-two earthquakes in the zone, with maximum annual magnitudes in the 2.8 to 3.1 range. In addition, the state's first two offshore earthquakes were recorded in 2002. Centered approximately 15 miles off Seabrook Island, they registered 4.3 on November 11 and 3.8 on November 28.[6] Most people who have lived in Summerville for five years or more report that they have felt at least one earthquake, and long-time residents take the minor shocks for granted.

Twenty-two miles northwest of Charleston, the Middleton Place-Summerville Seismic Zone bides its time, building up stress and telegraphing its activity to the surface. A micro-earthquake occurred there at 2:02:20 p.m. (CST) on Saturday, November 19, 2005. The magnitude 2.6 event occurred about two miles southwest of Summerville and three miles below the earth's surface—almost exactly the same place that generated the Great Charleston Earthquake of 1886.

21

In Memoriam

This section is dedicated to the memory of those who died or were injured as a result of the Great Charleston Earthquake of 1886. No one knows—or will ever know—exactly how many casualties were sustained as a direct result of the earthquake and its countless aftershocks. Until now, no research-based tally of lethal and non-lethal casualties has ever been attempted. A grant from the Post and Courier Foundation, Charleston, made the following mortality and injury research possible.

We know that the toll starts with the eighty-three victims who were specifically enumerated in the official "Record of Deaths Within the City of Charleston" from August 31 to October 31, 1886.[1] The information in the "Record of Deaths" was listed in columns captioned as follows: Names / Sex / Color / Date of Death / Disease, or Cause of Death / Place of Residence / Age Y-M-D / Place of Nativity / Time of Residence Y-M-D / Attending Physician or Coroner / Place of Interment. This list of sixty-one blacks and twenty-two whites who died within the Charleston city limits in the first sixty days after the first main shock has been used as the "official" death count for the 1886 earthquake in numerous published studies. A summary from the "Record of Deaths" was published by the U. S. Geological Survey in 1887 as part of Clarence E. Dutton's *The Charleston Earthquake of August 31, 1886*. Many other writers guessed at the death count, citing figures as low as thirty and as high as one hundred ten, but all of these numbers were purely conjecture. Even though the Dutton Report's citation of the eighty-three certified deaths has been noted in numerous scientific papers since 1890, a scientific monograph published as recently as 1993 stated that "no events involving significant casualty have occurred in the Central and Eastern United States since the New Madrid events of 1811-1812."[2] While the New Madrid earthquake is believed to have produced one fatality, obviously, that statement is incorrect.

The city's precisely compiled record has built-in limitations that lead it to reveal only a fraction of the casualties caused by the earthquake. The lists of deaths and injuries presented here, which have been compiled from multiple sources, attempt to give a more accurate count of the casualties suffered as a direct cause of the 1886 Charleston earthquake. However, as is the case with most mass-casualty disasters, many of the earthquake victims' names were never recorded in any official public record, and may never be known.

The following two-part list includes the eighty-three earthquake-specific deaths recorded in Charleston's "Record of Deaths" and provides additional information derived from newspapers; church, synagogue, and cemetery records; and private documents on people, including those from outside of South Carolina, who could be identified as having been killed or injured by the earthquake. In part one of this list, the names of victims not found in Charleston's "Record of Deaths" have been marked with an asterisk, and the information sources are noted. In all cases, the wording of the information in the original records has been preserved. All addresses are in the city of Charleston, unless otherwise noted.

Part I

Earthquake Deaths, August 31 - November 30, 1886

1. *"[A colored woman] was killed also, in Pitt street." *News and Courier*, September 1, 1886.

2. *"A colored girl in the yard of G. W. Williams, Jr., [No. 2] Atlantic street, is seriously injured." *News and Courier*, September 1, 1886. George W. Williams states: "A twelve-year-old daughter of our cook was killed by a falling wall." *News and Courier*, October 2, 1886.

3. *"A colored woman in Beaufain street, killed." *News and Courier*, September 3, 1886.

4. *"A colored woman living in the yard of Capt. Smalls, in Bull street, was instantly killed." *News and Courier*, September 1, 1886.

5. *"It is reported that one of the factory girls living in a boarding-house at the corner of America and Blake streets was killed." *News and Courier*, September 3, 1886.

6 - 8. *Near Aiken, South Carolina. "An excited stranger rushed into the Constitution office tonight [Sept. 3, 1886], hurriedly, and told of a family of three negroes, man, wife, and daughter, having bound themselves together on Horse creek and jumped in the creek, all three drowning. The stranger's name is unknown." *Atlanta Constitution*, September 4, 1886.

9 - 10. *Near Aiken, South Carolina. Two black residents near [Horse Creek] were also reportedly drowned. *Atlanta Constitution*, September 2, 1886.

11. *"An old lady died of fright." *Savannah* (Georgia) *Morning News*, September 2, 1886, citing Augusta, Georgia, newspaper of September 1, 1886.

12. *"A woman drowned herself through fear at Bath [Georgia]" *Atlanta Constitution*, September 3, 1886.

13. *"A mulatto man was hit on the head with a brick at 110 Tradd street. Dead." *News and Courier*, September 1, 1886.

14. *"Hearing of this suicide (*see* Murphy, C.H.), a negro, name unknown, went to his home, told his wife that he, too, would end his anguish by death, and then locking himself inside his room, made attempts to hang himself, but a police officer burst into the room and pulled him down. He however will hardly live during the night." *Atlanta Constitution*, September 4, 1886.

15. *"Just at the first shock a wagon bearing two men was passing the corner of Reid and King streets, when the gable end of the store on that corner fell and covered the unfortunates with the debris, one of whom was killed. It was very late when the body was extricated and it was impossible to ascertain the person's name." *News and Courier*, September 1, 1886. "A wagon driver, killed by the fall of a building at the corner of Reid and King streets [lengthy description]." *News and Courier*, September 5, 1886.

16 - 17 *"A policeman said that he had seen two dead bodies in King Street, north of Broad street." *News and Courier*, September 3, 1886.

18 - 19 * Rural Georgia. "At least two deaths are reported from fright, and numbers of people have become insane." *Atlanta Constitution*, September 3, 1886.

20. *"A colored woman [of McClellanville] was buried yesterday [September 7, 1886] who died of fright. *Berkeley Gazette*, September 11, 1886.

21. *"[In Savannah] Revivals were in progress in two negro churches, and in the panic several women fainted and fell and were severely injured by the crowds who rushed over them. One negress died from fright." Dutton, 326-327, citing *New York Tribune* report of a Savannah newspaper.

22. Aarons, Mrs. Goldie. White female, died September 1, 1886, of "injuries received - earthquake," residence City Hospital, age 35, born in Germany, time of residence 3 years, A.H. Schwacke, Jr., buried in ~~Hanover Street~~ (struck over and replaced with [illegible]).

23. Aarons, Rachel. White female, died September 1, 1886, of "injuries received - earthquake," residence City Hospital,

age 6, born in Brooklyn, N.Y., time of residence 3 years, A.H. Schwacke, Jr., buried in ~~K.K. Beth Eloheim~~ (struck over and replaced with [illegible].

24. Agnew, Eugene. Colored male, died October 5, 1886, of "cold & exposure - earthquake," residence Cromwell's Alley, age 9 months, born in "City," time of residence "Life," J.P. DeVeaux, buried in Public cemetery.

25. Albrecht, Charles. White male, died August 31, 1886, of "accident - earthquake," residence 101 Tradd St., age 52, born in Germany, time of residence 13 years, C.H. Rivers, Deputy Coroner, buried Bethany Cemetery. *News and Courier,* September 3: "Charles Albrecht, a German cabinetmaker living at [illegible] King Street, was struck by a falling wall and instantly killed."

26. Alexander, Robert. White male, died August 31, 1886, of "accident - earthquake," residence 20 Meeting St., age 24, born London, England, time of residence 5 months, C.H. Rivers, Deputy Coroner, buried in Magnolia Cemetery. *News and Courier*, September 1: "At the [City] Hospital was...the body of Mr. Robert Alexander, the young English analytical chemist, was sent there horribly mangled;" also "Mr. R. Alexander, a young chemist, who was crushed to death at a boarding-house on Meeting Street. He had just bought a small steam pleasure yacht, and took his first and last trip in it last evening." *News and Courier*, September 6, 1886: "No. 34 [Meeting Street], the Huger mansion, has lost the cornice and is cracked in the front wall. It was while coming out of the door of this house that the lamented young chemist, Mr. Alexander, was killed on Tuesday night."

27. *Allen, John. "At the Nettles House [Summerville] two colored people, Thomas Ellis and John Allen, were killed by the falling ruins." *News and Courier*, September 3, 1886.

28. Barber, Mrs. Christine. White female, died August 31, 1886, "accident - earthquake," residence 10 Bull St., age 52, born in Germany, no time of residence listed, C.H. Rivers, Deputy Coroner, buried Bethany Cemetery. *News and Courier,* September 1, 1886: "Mrs. C. Barber, __ years, servant, of Germany, __ Bull St."

29. Barnwell, Marie. Colored female, died August 31, 1886, of "accident – earthquake," residence 36 Calhoun St., age 2 years, born in "City," time of residence "Life," C.H. Rivers, Deputy Coroner, buried James Island, S.C. *News and Courier*, September 3, 1886: "Colored baby, one year old, child of Moses Barnwell, 36 Calhoun street." *News and Courier*, September 5, 1886: "No. 36 Calhoun street, brick kitchen fell down, killing child of Moses Barnwell, colored."

30. Bell, James. Colored male, died September 11, 1886, of "scrofula, exposure, and injuries received - earthquake," residence George St., age 2 years, 1 month, born in "City," time of residence "Life," J.J. Edwards, buried in Public cemetery.

31. Bennett, Laura A., infant of. Colored female, died September 1, 1886, of "inattention" and "injuries received - earthquake," residence White Point Garden, age 1 hour, born in "City," time of residence "Life," John Forrest, buried in Public cemetery.

32. Berwick, Lavinia, infant of. Colored female, died October 4, 1886, of "cold & exposure - earthquake," residence Vanderhorst St., age 1 month, born in "City," time of residence "Life," J.P. DeVeaux, buried in Centennary cemetery.

33. Bright, V. D. Colored female, died September 7, 1886, of "exposure and injuries received - earthquake," residence Tent on King St., age 8 months, born in "City," time of residence "Life," P.G. DeSaussure, buried in Public cemetery. *News and Courier*, September 9, 1886: "V.D. Bright, colored infant, eight days old, died in a tent on King street from exposure."

34. Brown, Daniel Moses. Colored male, died September 3, 1886, of "injuries received - earthquake," residence City Hospital, age 40 years, born in "City," time of residence "Life," R.B. Hanahan, buried in Emanuel cemetery. *News and Courier*, September 5: "Ex-Policeman Moses D. Brown, colored, died at his residence at the Old United States Courthouse in Meeting Street yesterday...." [obituary follows]. "Moses Brown, colored, janitor of the government building, leg amputated." *Atlanta Constitution*, September 4, 1886. City death certificate: occupation: janitor; place of death: City Hospital; cause of death: Bright's disease; primary cause: Earthquake injury & rheumatism;

secondary: shock from amputation. Buried September 4, 1886, Emanuel Cemetery.

35. Brown, Hattie. Colored female, died September 10, 1886, of "trismus nascentum [infant tetanus], exposure, and injuries received - earthquake," residence 91 Church St., age 6 days, born in "City," time of residence "Life," P.G. DeSaussure, buried in Public cemetery.

36. Brown, James. Colored male, died September 1, 1886, of "injuries received - earthquake," residence 68 Vanderhorst St., age 64, born in "City," time of residence "Life," C.H. Rivers, Deputy Coroner, buried in Rikersville cemetery.

37. *Buchanan, Mrs. Mary. "Mrs. Mary Buchanan, white, aged 50 years, died at No. 26 New street. Mrs. Buchanan had been an invalid for years and was suffering from rheumatism. She was taken out of her building after the shock on Tuesday night, and died from exposure and fright." *News and Courier*, September 7, 1886; also *Atlanta Constitution*, September 7, 1886.

38. *Bull, James. "James Bull, colored, two years old, George street, scrofula and exposure. *News and Courier*, September 12, 1886, citing city burial permit.

39. Burrows, Mrs. Tenie. White female, died September 7, 1886, of "shock, heart disease, and injuries received - earthquake," residence 30 Laurens St., age 45, born in Virginia, time of residence 23 years, H.M. Cleckley, buried in Magnolia cemetery.

40. Cade, Bertha. White female, died September 3, 1886, of "injuries received - earthquake" and "exposure," residence 55 Spring St., age 9 years, born in "City," time of residence "Life," John L. Dawson, C.H. Rivers, Deputy Coroner, buried in St. Lawrence cemetery. *News and Courier*, September 5: "Little Bertha Cade, a daughter of Mr. H. L. Cade, was buried yesterday...." [obituary follows]. "Was fatally crushed by the crowd fleeing the South Carolina Railway station on Tuesday night." *News and Courier*, September 11, 1886.

41. Carmins, Archey. Colored male, died October 5, 1886 of "cold and exposure - earthquake," residence 168 St. Phillip St., age 12 years, born in "City," time of residence not stated, F.O. Alleman, buried in Public cemetery.

42. *Collins, Martha. City death certificate: Colored female, age 2 years, 11 months, died September 5 of "diarrhea and asthma," [but Dr. P. Gourdin DeSaussure noted that "exposure was the real cause of death"], place of death 71 Charlotte St., buried Field of Rest. Not listed as an earthquake death in "Return of Deaths."

43. Cook, John. Colored male, died August 31, 1886, of "injuries received - earthquake," residence 104 Tradd St., age 23, born in Alabama, time of residence not stated, C.H. Rivers, Deputy Coroner, buried in Rikersville Baptist cemetery. *News and Courier*, September 3, 1886: "John Cook, colored, while passing the corner of Friend and Tradd Streets, was crushed by a falling wall. He resided at 110 Tradd street, and was buried yesterday." *News and Courier*, September 3, 1886: "John Cook, colored, 23 years, laborer, of Alabama, 104 Tradd street." *News and Courier*, September 3, 1886: "John Cook, colored, fisherman, residence Tradd street, near Friend."

44. David, Angeline, Mrs. Colored female, died August 31, 1886, of "injuries received - earthquake," residence 68 Nassau St., age 38, born in "City," time of residence "Life," A.C. McClennan, Deputy Coroner, buried in Unity & Friendship Cemetery. *News and Courier*, September 1, 1886: "A colored woman named Angeline David and residing at No. 68 Nassau street died of fright at 11 o'clock on Tuesday night. She was in the house when the shock came but remained unhurt. The fright, however, was so great that after an hour of violent struggling she died." *News and Courier*, September 3, 1886: "Mrs. David, 68 Nassau street, died three-quarters of an hour from a nervous shock."

45. Deas, William. Colored male, died August 31, 1886, of "accident — earthquake," residence East Bay, age 15, born "City," time of residence "Life," C.H. Rivers, Deputy Coroner, buried in Public Cemetery.

46. Defrates, Joseph King. White Male, died September 14, 1886, of "Gastro Enteritis, exposure and injuries received - earthquake," residence St. Xavier Infirmary, age 50 years, born in New York, time of residence 8 years, J.J. Edwards, buried in

St. Lawrence cemetery. Additional story: *News and Courier*, September 16, 1886.

47. Edgar, Robert. Colored male, died September 24, 1886, of "pneumonia, exposure and earthquake," residence 120 Columbus St., age 1 year, 2 months, 5 days, born in "City," time of residence "Life," B.M. Lebby, buried in St. Lawrence cemetery.

48. *Ellis, Thomas. black male, Summerville. (*See also* Allen, John).

49. Esnard, Mrs. Ann. White female, died September 18, 1886, of "constipation, exposure - earthquake," residence 84 Line St., age 42, born in Savannah, Ga., time of residence 28 years, C. L. Meyer, buried in St. Lawrence cemetery.

50. Fleming, Grace. Colored female, died August 31, 1886, of "injuries received - earthquake," residence 79 King St., age 18, born in S.C., time of residence 8 years, C.H. Rivers, Deputy Coroner, buried in John's Island, S.C. *News and Courier*, September 3, 1886: "It was not until yesterday that the death of three persons, all of whom were killed on the same spot, and at the same time, were made known to a Reporter. These were Aleck Miller, of 97 King Street, Grace Fleming, of 99 King street, and Joe Rodolf, who lived in King street, opposite Weim's Court. All were colored persons." *News and Courier*, September 3, 1886: "Grace Fleming, colored, 18 years, nurse, of South Carolina, 79 King Street."

51. *Foster, Lucy. "Lucy Foster, colored, a widow, about 50 years old, was frightened to death. She was a stout, healthy woman. Her sister on returning from St. Philip's Church found her lying across the bed dead." *News and Courier*, September 5, 1886.

52. *Fraser, Lizzie. "On Mary street, near Meeting, a house was shaken down and all the inmates escaped, except a little colored child named Lizzie Fraser, who was crushed to death." *News and Courier*, September 1, 1886.

53. Glover, Anna. Colored female, died August 31, 1886, of "injuries received - earthquake," residence Friend St., age 36, born in "City," time of residence "Life," C.H. Rivers, Deputy

Coroner, buried in Rikersville Baptist Cemetery. *See also* Glover, Annie in list of injured, below.

54. *Glover, Sarah. Listed in "Additional Deaths," *Atlanta Constitution*, September 4, 1886.

55. Gordon, William C. White male, died September 11, 1886, of "malarial fever, exposure, and earthquake," residence A. Street, near Harris', age 48 years, 10 months, 12 days, born in "City," time of residence "Life," W.D. Crum, buried in Magnolia cemetery.

56. Gourdin, Jane. Colored female, died September 1, 1886, of "injuries received - earthquake," residence City Hospital, age 10 years, born in "City," time of residence "Life," R. B. Hanahan, buried in Mount Holly, S.C.

57. Grant, Isabel. Colored female, died September 7, 1886, of "exposure and injuries received - earthquake," residence 3 Trapman St., age 1 year 3 months, born in "City," time of residence "Life," J. L. Danson, Jr., buried in James Island, S.C. *News and Courier*, September 9, 1886: "Isabel Grant, colored infant, one month old, 3 Trapman street, congestion and exposure."

58. Grant, Lizzie. Colored female, died September 6, 1886, of "exposure and injuries received - earthquake," residence Washington Square, age 22 years, born in S.C., time of residence 2 years, P.G. DeSaussure, buried in Public cemetery. Atlanta *Constitution*, September 7, 1886: "Today died Lizzie Grant, of heart disease brought on by fright and exposure." "On the 6th (yesterday) a death occurred on Washington Square, the victim being Lizzie Grant, colored, aged 22 years. She was living on the [Sullivan's] Island and came to the city on Wednesday. The immediate cause of death was heart disease, brought on by fright and exposure." *News and Courier*, September 7, 1886.

59. Grant, William. Colored male, died August 31, 1886, of "accident - earthquake," residence 290 East Bay St., age 13, born "City," time of residence "Life," C.H. Rivers, Deputy Coroner, buried in Cain's Ground.

60. Haig, Lucinda. Colored female, died September 10, 1886, of "dentition, exposure, and injuries received - earthquake,"

residence Oak St., age 1 year, 6 months, 9 days, born in "City," time of residence "Life," Edward Mazyck, buried in Brandt's Farm.

61. Harleston, Eliza. Colored female, died September 10, 1886, of "congestion of lungs, exposure, and injuries received - earthquake," residence 45 Radcliffe St., age 14 days, born in "City," time of residence "Life," Edward Mazyck, buried in Rikersville Baptist cemetery.

62. Harris, Anna/Hannah. Colored female, died August 31, 1886, of "injuries received - earthquake," residence Wentworth St., age 60, born in S.C., time of residence 25 years, C.H. Rivers, Deputy Coroner, buried in Rikersville Baptist Cemetery. *News and Courier*, September 1, 1886: "Anna Harris, a colored woman, about 63 years old, occupying a small house in Mr. Webb's yard at [illegible] Wentworth street, was killed while lying in bed by the falling of the roof Tuesday night." *News and Courier*, September 3, 1886: "Hannah Harris, 60 years, of Beaufort, Wentworth street."

63. Hogan, Ella. White female, died September 11, 1886, of "epilepsy, exposure, and injuries received - earthquake," residence Coming St., age 66 years, 6 months, 11 days, born in "City," time of residence "Life," R.B. Rhett, buried in St. Lawrence Cemetery.

64. Irvine, Julius. Colored male, died September 6, 1886, of "exposure and injuries received - earthquake," residence George St., age 5 months, 14 days, born in "City," time of residence "Life," J.J. Edwards, buried in Public cemetery. *News and Courier*, September 9, 1886: "Julius Irvin, colored infant, five months old, Bernard's lot, bronchitis and exposure."

65. *Ivy, Henry. "Two trains were wrecked [near Langley, S.C.], the engineers of which, Messrs. Brisenden and Reynolds, were severely injured, and the colored firemen, Jack Simmons and Henry Ivy, killed." *News and Courier*, September 3, 1886.

66. Jackson, Rosa. Colored female, died September 2, 1886, of "injuries received - earthquake," residence City Hospital, age 50 years, born in "City," time of residence "Life," R.B. Hanahan, buried in Rikersville cemetery.

67. *Jacobs, Lavinia. "Lavinia Jacobs, colored, Chalmers street, killed." *News and Courier*, September 1, 1886.

68. Jenkins, Alexander. Colored male, died September 5, 1886, of "exposure" and "injuries received - earthquake," residence 2 Union St., age 5 months, born in "City," time of residence "Life," C.H. Rivers, Deputy Coroner, buried in Emanuel cemetery. "The only deaths resulting from the earthquake reported yesterday [September 5] were those of Alexander Jenkins, colored, five months old, and Julia Riley, two years old, both from exposure." *News and Courier*, September 6, 1886.

69. Jenkins, Eddie. Colored male, died October 8, 1886, of "malarial fever & exposure - earthquake," residence 36 Reid St., age [blank], born in "not stated," time of residence not stated, H.W. Cleckley, buried in Strawberry [a hamlet near the] N.E.R.R.

70. Jenkins, Mattie. Colored female, died August 31, 1886, of "accident - earthquake," residence 5 Tradd St., age 11 years, born in S.C., time of residence 8 years, C.H. Rivers, Deputy Coroner, buried in Public Cemetery. *News and Courier*, September 3, 1886: "Mattie Jenkins, 11 years, of Summerville, S.C., 5 Tradd St."

71. Jones, Samuel. Colored male, died September 18, 1886 of "whooping cough, exposure and earthquake," residence Lee St., age 2 years, born in "City," time of residence "Life," B.M. Lebby, buried in Wesley Cemetery.

72. Knocks, Ella. Colored female, died September 6, 1886, of "measles, exposure, and injuries received - earthquake," residence 85 Calhoun St., age 4 years, born in Georgetown, S.C., time of residence 6 months, P.G. DeSaussure, buried in Field of Rest. "Ellen Knox, colored, four years old, 85 Calhoun street, died yesterday from the effects of exposure." *News and Courier*, September 7, 1886.

73. Knocks, Ida. Colored female, died September 9, 1886, of "marasmus [emaciation], exposure, and injuries received - earthquake," residence 85 Calhoun St., age 3 months, born in "City," time of residence "Life," P.G. DeSaussure, buried in Field of Rest. "Ida Knox, colored infant, three months old, at No. 85 Calhoun St." *News and Courier*, September 11, 1886.

74. Lawrence, Lavinia. Colored female, died September 10, 1886 of "debility, exposure, and injuries received - earthquake," residence 29 Henrietta St., age 1 month, 7 days, born in "City," time of residence "Life," P.G. Desaussure, buried in Remley's Point, S.C.

75. *Lee, Mary. Listed in "Additional Deaths," *Atlanta Constitution*, September 4, 1886.

76. *Lynch, Maurice J. *News and Courier*, September 1: "Mr. M.J. Lynch was desperately hurt in front of his son's store in Meeting street. A stone of great weight feel upon him and broke one of his legs, if not both. He was taken to a place of safety in what was feared to be a dying condition. Father Duffy was with him." *News and Courier,* September 3: "Crushed in his Door Step. Mr. M. J. Lynch, who was killed on Tuesday night and was reported in The *News and Courier* of Wednesday as fatally injured died before midnight [longer story follows.]" "Lynch, Maurice J., Saloon, 119 Meeting," *Southern Business Guide, 1885-1886,* 722.

77. Lynch, Patrick. White male, died August 31, 1886 of "accident - earthquake," residence 15 Wentworth St., age 70, born in Ireland, time of residence 40 years, J.L. Ancrum, buried St. Lawrence Cemetery. St. Lawrence Cemetery interment records: "Burial 1 September 1886, Patrick Lynch, born in Ireland, aged 70, residence of 18 Wentworth St., Accident as cause of death, Fr. Dougherty was officiating pastor, [buried in] section 4, range 4, number 8."

78. McMannon, Ann. White female, died September 30, 1886, of "asthma & exposure - earthquake," residence 57 Radcliffe St., age 61, born in Ireland, time of residence 37 years, Edmund Ravenel, buried in St. Lawrence cemetery.

79. Meynardie, B. P. White male, died August 31, 1886, of "accident — earthquake," residence 83 America St., age 86, born in S.C., residence "Life," C.H. Rivers, Deputy Coroner, buried Magnolia Cemetery.

80. Middleton, Sarah (Mrs. Jacob). Colored female, died August 31, 1886, of "accident — earthquake," Court House Square, age 25, born in S.C., time of residence 4 years, C.H.

Rivers, Deputy Coroner, buried Beaufort, S.C. *News and Courier*, September 1, 1886: "At 1 o'clock this morning the body of a colored woman was taken from the débris at the corner of Meeting and Broad streets, near the lamppost. It proved to be the body of Mrs. Jacob Middleton. The unfortunate woman must have been passing just as the east wall of the building [The Guard House] fell." *News and Courier*, September 3: "Sarah Middleton, 25, washer, of Beaufort, Courthouse Square.

81. *Middleton, Susan. *News and Courier*, September 1: "Among those who were killed were Susan Middleton, colored, and another colored woman, both of whom were crushed under the fallen portico of the Main Stationhouse." [*See* Sarah Middleton, above].

82. Miller, Alexander. Colored male, died August 31, 1886, of "injuries received - earthquake," residence 97 King St., age 19, born in "City," time of residence "Life," C.H. Rivers, Deputy Coroner, buried in Emanuel cemetery. *News and Courier*, September 3, 1886: "It was not until yesterday that the death of three persons, all of whom were killed on the same spot, and at the same time, were made known to a Reporter. These were Aleck Miller, of 97 King Street, Grace Fleming, of 99 King street, and Joe Rodolf, who lived in King street, opposite Weim's Court. All were colored persons." *News and Courier*, September 3, 1886: "Alex. Miller, colored, 19 years, laborer, city, 97 King street."

83. Mitchell, Alice. Colored female, died September 30, 1886, of "cold from exposure - earthquake," residence 4 Sheppard St., age 10 years, born in "City," time of residence "Life," J.P. DeVeaux, buried in Col[ored] Lutheran cemetery.

84. Mitchell, Mary F. Colored female, died September 13, 1886, of "cholera infantum, exposure and earthquake," residence 37 Marsh St., age 16 months, born in "City," time of residence "Life," Edmund Ravenel, buried in Bathsheba cemetery.

85. *Murphy, C.H. Augusta, Georgia. "C.H. Murphy, of N.W. Murphy & Co., wholesale and retail shoe dealers, was terribly frightened by the Tuesday shock, and has since been prostrated. Today he sat on the floor of his room, coolly loaded

a thirty-eight calibre pistol, and deliberately blew out his brains. His mother was the only person in the house at the time, and this shock, together with the earthquake scare, will end her life in all probability." *Atlanta Constitution*, September 4, 1886.

86. Murray, Rosa Lee. Colored female, died September 1, 1886, of "injuries received - earthquake," residence 89 St. Philip St., age 26, born in S.C., time of residence [blank], C.H. Rivers, Deputy Coroner, buried in Sumter, S.C.

87. Nathans, Nancy. Colored female, died September 25, 1886, of "mental aberration, exhaustion - earthquake," residence Meeting St., age 18, born in S.C., time of residence 9 years, buried in Horlbeck's Farm.

88. Nickelby, Olive C. Colored female, died August 31, 1886, of "accident - earthquake," residence 47 Church St., age 11, born in S.C., time of residence 9 years, C.H. Rivers, Deputy Coroner, buried Rikersville Baptist Church Cemetery. *News and Courier*, September 1, 1886: "Olive Nickelby, mulatto girl, 11 years old killed at Judge Bryan's yard, Church street." *News and Courier*, September 3, 1886: "Olive C. Nickelby, colored, 11 years, of South Carolina, 47 Church street."

89. Nowell, Miss M.C. White female, died September 8, 1886, of "shock and exposure and injuries received - earthquake," residence 15 Thomas St., age 50, born in "City," time of residence "Life," W.H. Huger, buried in St. Paul's church cemetery. *News and Courier*, September 10, 1886: "Miss M. C. Nowell, white, fifty years old, 15 Thomas street, earthquake shock."

90. Pawley, Leonard F. Colored male, died September 23, 1886, of "congestion lungs, exposure and earthquake," residence Radcliffe St., age 7 years, 1 month, 23 days, born in "City," time of residence "Life," M.G. Champlin, buried in Rikersville Baptist cemetery.

91. Pennington, Clarence S. Colored male, died October 6, 1886, of "enteritis & exposure - earthquake," residence 16 Cannon St., age 7 years, born in "City," time of residence not stated, C.L. Meyer, buried in Centennary cemetery.

92. Pinckney, Marie. Colored female, died September 1, 1886, of fever-shock and "injuries received — earthquake,"

residence Vanderhorst St., age 16, born in "City," time of residence "Life," Edmund Mazyck, buried in Coosawhatchie, S.C.

93. *Powell, Mr. "One of the most pitiful cases in the city is that of Mrs. Powell, whose husband was killed at the corner of Church and Elliott streets on Tuesday night." *News and Courier*, September 10, 1886.

94. Powers, Peter. White male, died August 31, 1886, of "accident - earthquake," 104 Church St., age 44, born [blank], time of residence 20 years, C.H. Rivers, Deputy Coroner, buried Bethany cemetery. *News and Courier*, September 3, 1886: lists age as 44 years and occupation as ship carpenter.

95. Pugh, Mary A. Colored female, died October 4, 1886, of "liver inflammation & exposure - earthquake," residence 18 America St., age 40 years, born in "City," time of residence "Life," A.C. McClennan, buried in Field of Rest.

96. Ramos, Jose E. White male, died September 14, 1886, of "malarial fever and shock - earthquake," residence Meeting St., age 74, born in Havanna, Cuba, time of residence 40 years, C.H. Rivers, Deputy Coroner, buried in St. Lawrence cemetery.

97. *Rector, Florence. Black female. City *Yearbook*, 1886, p. 90: "Only two victims [at the City Hospital on Queen Street] were caught, a colored man, Robert Ridoff, and a colored woman, Florence Rector, who perished under the falling bricks of the buildings known as the colored wards." Also noted in *Atlanta Constitution*, September 4, 1886.

98. Redoff, Robert. Colored male, died August 31 of "accident — earthquake," residence City Hospital, age 43, born in S.C., time of residence not stated, H.B. Hanahan, buried in Public Cemetery. *News and Courier*, September 1, 1886: "At the Hospital was also the body of a colored man, who was killed by a falling piazza." *News and Courier*, September 3: "Robert Redoff, colored, 43 years, laborer, of South Carolina, City Hospital." City *Yearbook*, 1886, p. 90: "Only two victims [at the City Hospital on Queen Street] were caught, a colored man, Robert Ridoff, and a colored woman, Florence Rector, who perished under the falling bricks of the buildings known as the colored wards."

99. Redoff/Rodolf, Joseph. Colored male, died August 31, 1886, of "accident - earthquake," residence King St., age 23 years, 4 months, born in "City," time of residence "Life," C.H. Rivers, Deputy Coroner, buried in Field of Rest. *News and Courier*, September 3: "Joseph Redoff, colored, 23 years, laborer, city, King street;" also *News and Courier*, September 3, 1886: "It was not until yesterday that the death of three persons, all of whom were killed on the same spot, and at the same time, were made known to a Reporter. These were Aleck Miller, of 97 King Street, Grace Fleming, of 99 King street, and Joe Rodolf, who lived in King street, opposite Weim's Court. All were colored persons."

100. Reston, Flora. Colored female, died September 2, 1886, of "injuries received - earthquake," residence City Hospital, age 16 years, born in "City," time of residence "Life," R.B. Hanahan, buried in Public cemetery. *See also* Rector, Flora. This may be the same woman.

101. Richardson, Jefferson C. White male, died September 14, 1886, of "injuries received - earthquake," residence Wharf St., age 60 years, 1 month, 11 days, born in "City," time of residence "Life," H.W. DeSaussure, buried in Magnolia Cemetery. "Died at the residence of Mr. S. J. Pregnall on Wharf St. yesterday [two-paragraph obituary]", *News and Courier*, September 15, 1886.

102. Riley, Julia. Colored female, died September 5, 1886, of "injuries received - earthquake," and "shock," residence 2 Weims Court, age 2 years, 9 months, 6 days, born in "City," time of residence "Life," J.L. Dawson, Jr., buried in Mt. Pleasant, S.C. "The only deaths resulting from the earthquake reported yesterday [September 5] were those of Alexander Jenkins, colored, five months old, and Julia Riley, two years old, both from exposure." *News and Courier*, September 6, 1886. "On the 5th inst. Julia Riley, colored, died of marasmus and exposure." Atlanta *Constitution*, September 7, 1886; same, *News and Courier*, September 7, 1886.

103. Riley, Sarah. Colored female, died September 8, 1886, of "scrofula, exposure, and injuries received - earthquake,"

residence 49 George St., age 2 years, born in "City," time of residence "Life," J.J. Edwards, buried in Rikersville Baptist cemetery. *News and Courier*, September 10, 1886: "Sarah Riley, colored, two years old, 49 Green street, scrofula and exposure."

104. Rivers, Sarah. Colored female, died September 10, 1886, of "injuries received - earthquake," residence City Hospital, age 55 years, born in "S.C.," time of residence 8 years, B.A. Pyatt, buried in James Island, S.C. *News and Courier*, September 12, 1886: "fracture of the skull."

105. Roberts, Eugenia. Colored female, died August 31, 1886, of "injuries received - earthquake," residence 6 St. Michael's Alley, age 1 year, 3 months, born in "City," time of residence "Life," J.L. Dawson, Jr., buried in Emanuel Cemetery.

106. Robson, Ainsley H. White male, died August 31, 1886, of accident - earthquake," residence Coming St., age 28 years, 8 months, 14 days, born in Charleston, time of residence "life," Magnolia Cemetery. *News and Courier*, September 1: "Mr. Ainsley Robson was killed by the falling of a piazza"; also September 11, 1886. *See also* George H. Moffett Family Papers, South Carolina Historical Society.

107. Sampson, Joseph. White male, died September 24, 1886, of "shock and exposure - earthquake," residence King St., age 73, born in England, time of residence 50 years, C.L. Meyer, buried in K.K. Beth Elohim cemetery.

108. *Sawyer, Zera B. Black female. "Zera B., daughter of Isaac Sawyer, colored, barber, 132 Calhoun street." *News and Courier*, September 3, 1886. "132 Calhoun St. Daughter of Isaac Sawyer, barber, killed and two children injured in the ruins." *News and Courier*, September 5, 1886.

109. Sawyer, Zerubbabel [son of Isaac Sawyer]. Colored male, died August 31, 1886, of "accident - earthquake," residence 132 Calhoun St., age 1 year, 10 months, born in Charleston, time of residence "life," C.H. Rivers, Deputy Coroner, buried Centennary Cemetery. *News and Courier*, September 3: "A child of Isaac Sawyer, colored, living in Calhoun street one door east of Meeting Street was killed on Tuesday night."

110. Shier, Mary Alice. White female, died September 17, 1886, of "congestion lungs & exposure - earthquake," residence 56 Bull St., age 1 year, 14 days, born in "City," time of residence "Life," R.L. Brodie, buried in St. Lawrence cemetery.

111. *Simmons, Jack. *See* Ivy, Henry.

112. Simons, Clarissa. Colored female, died August 31, 1886, of "injuries received - earthquake," residence 58 Tradd St., age 70, born in South Carolina, time of residence 1 year, 6 months, C.H. Rivers, Deputy Coroner, buried in Beaufort, S.C. *News and Courier*, September 3, 1886: "Clarissa Simons, [illegible], Beaufort, S.C., 58 Tradd street."

113. Skillings, Charlotte. Colored female, died September 14, 1886, of "fever and exposure and earthquake," residence 13 Ann St., age 6 years, born in S.C., time of residence not stated, C.H. Rivers, Deputy Coroner, buried in Christ Church Parish.

114. Smalls, Hannah. Colored female, died August 31, 1886, of "injuries received - earthquake," residence 61 Judith St., age 40, born in "City," time of residence "Life," C.H. Rivers, Deputy Coroner, buried in Col[ored] Bethel [Methodist] cemetery. *News and Courier*, September 3, 1886: "Hannah Smalls, colored, 40 years, washer, city, 61 Judith street."

115. *Smalls, Julia. *News and Courier*, September 1, 1886: "Julia Smalls, colored, infant killed."

116. Smalls, Mary Jane. Colored female, died September 6, 1886, of "exposure and injuries received - earthquake," residence King St., age 8 years, 5 months, 13 days, time of residence "Life," J.L. Danson, Jr., buried in Rikersville cemetery.

117. Smith, Amelia, infant of. Colored female, died October 3, 1886, of "cold & exposure - earthquake," residence 201 Coming St., age 1 year, 5 months, born in "City," time of residence "Life," J.P. DeVeaux, buried in Emanuel cemetery.

118. *Thompson, W. F. "W. F. Thompson, colored, 35 years old, hemorrhage and exposure." *News and Courier*, September 9, 1886.

119. Torck, Miss Annie. White female, died September 1, 1886, of "injuries received - earthquake," residence 13 Vendue Range, age 20, born in S.C., time of residence "life," T.B.

McDow, buried Bethany Cemetery. *News and Courier*, September 1: "The house of Mrs. Annie Torck, on Secession Street, fell in and wounded her, it is thought, fatally."

120. Ward, Anna Rebecca. Colored female, died August 31, 1886, of "injuries received - earthquake," residence 62 Beaufain St., age 32, born in Savannah, Ga., time of residence 12 years, C.H. Rivers, Deputy Coroner, buried in Field of Rest. [This may be the woman noted in the *News and Courier* on September 1 as "A colored woman in Beaufain street killed."]

121. White, Caesar. Colored male, died September 7, 1886, of "measles, exposure, and injuries received - earthquake," residence 8 Cromwell's Alley, age 1 year, 6 months, born in S.C., time of residence not stated, C.H. Rivers, Deputy Coroner, buried in James Island, S.C. *News and Courier*, September 9, 1886: "Caesar White, a colored infant, two years old, 8 Cromwell's alley, measles and exposure."

122. Williams, Josephine. Colored female, died September 9, 1886, of "gastritis, exposure, and injuries received - earthquake," residence 146 Coming St., age 1 year, 8 months, 10 days, born in "City," time of residence "Life," W.D. Crum, buried in Centennary cemetery.

123. Wilson, Thomas. Colored male, died August 31, 1886, of "accident - earthquake," residence 124 Coming St., age 70, born "City," time of residence "Life," C.H. Rivers, Deputy Coroner, buried in Public Cemetery.

124. Young, Rebecca. Colored female, died September 25, 1886, of "shock & exposure - earthquake," residence 89 Tradd St., age 3 years, 11 months, 3 days, born in "City," time of residence "Life," J.F.M. Geddings, buried in Field of Rest.

Part II

Other Earthquake Casualties
August 31 - November 30, 1886

The only known statement about the number of persons injured by the 1886 earthquake was found in the Trenton (New

Jersey) *Times* of September 2, 1886, which reprinted the material from an unnamed newspaper. It cited "the loss of from thirty to forty lives [and] the maiming of over one hundred persons." At that date, all such numbers were guesswork, but even during the 120 years since the earthquake, no compilation of non-fatal casualties has ever been attempted. The list below was drawn chiefly from newspapers and personal correspondence. Each entry represents at least one casualty, and frequently more. Again, unless otherwise noted, all injuries took place in Charleston. Like the list of fatalities above, this record will be forever incomplete.

1. Unidentified children. During a shock at 12:56 p.m., November 5, 1886: "All the schools were in session at the time of the shock. One pupil in the colored school was crushed in a panic of pupils, and several pupils in the Shaw school [Shaw Memorial Colored School], colored, were injured by falling plastering. No one was seriously injured." *Atlanta Constitution,* November 6, 1886.

2. Unidentified mental patients in South Carolina and Georgia. "The state lunatic asylum [South Carolina Lunatic Asylum in Columbia] now contains nearly seven hundred inmates, more than it has ever sheltered before. Not a few of the more recent arrivals are thought to be victims of the earthquake excitement." *Atlanta Constitution,* November 23, 1886. "At least two deaths are reported from fright, and numbers of people have become insane." *Atlanta Constitution,* September 3, 1886.

3. Unidentified females. Augusta, Georgia. "Two ladies lie at the point of death from fright." Savannah (Georgia) *Morning News,* September 2, 1886, citing Augusta, Georgia newspaper of September 1, 1886.

4. Unidentified female. "A colored girl of thirteen with a fractured limb was brought in yesterday." *News and Courier,* September 6, 1886.

5. Unidentified female. "A colored woman living in the kitchen of Capt. G.D. Bryan's house on Church street...was badly hurt about 11 o'clock by the wall of the house on the

corner of Water and Atlantic streets falling on her." *News and Courier,* September 3, 1886.

6. Unidentified females in church #1. Savannah, Georgia. "[In Savannah] Revivals were in progress in two negro churches, and in the panic several women fainted and fell and were severely injured by the crowds who rushed over them." Dutton, 326-327, citing *New York Tribune* reprint of Savannah newspaper notice.

7. Unidentified females in church #2. Savannah, Georgia. "[In Savannah] Revivals were in progress in two negro churches, and in the panic several women fainted and fell and were severely injured by the crowds who rushed over them." Dutton, 326-327, citing *New York Tribune* reprint of a Savannah newspaper notice.

8. Unidentified female. "A colored woman, in an unconscious condition was found in front of Market in Meeting street." *News and Courier,* September 3, 1886.

9. Unidentified female and male. "About that time two gentlemen who were near the Pavillion Hotel heard piercing cries for help. They went down Hasel street in the direction of the cries, and found a white man and woman half-buried in the ruins of the Lazarus building. These were extricated and sent to the Hospital in a wagon furnished by Mr. Pickett." *News and Courier,* September 1, 1886. "Dr. Chazal informed a reporter that he had been called away to attend to two persons who had been injured at the house of Mrs. Lazarus, at 64 Hasel street. No particulars could be had." *News and Courier,* September 1, 1886.

10. Unidentified female. "An old white lady with nerves completely shattered by fright...[was] brought in yesterday." *News and Courier,* September 6, 1886.

11. Unidentified female and at least eleven other people. "Dr. Buist informed a Reporter that there were in the City Park at the corner of Wentworth and Meeting streets no less than 12 wounded persons who had already received his professional attention. Among them was a young girl whose leg was broken." *News and Courier,* September 1 & 3, 1886.

12. Unidentified female. "In one place lay an old lady very ill from typhoid fever, whose condition had been seriously aggravated by the terrors of the night." *News and Courier,* September 1, 1886.

13. Unidentified female. "Mr. J.C.E. Richardson, living at 12 Friend street, was seriously injured in the head by his house falling in upon him. His condition is very critical. His colored servant was likewise dangerously hurt, several of her limbs being broken." *News and Courier,* September 1, 1886.

14. Unidentified male. Langley, S.C. "The engineer [of a train] stuck to his post, and though the engine was completely submerged, managed to escape with two broken legs." *Atlanta Constitution,* September 3, 1886.

15. Unidentified male. "[A man] living on King street, opposite Dr. Collins' drug store was struck on the head with falling bricks. He is suffering from compression of the brain and fractured ribs and is expected to die." *News and Courier,* September 3, 1886.

16. Unidentified male. "A colored man, whose name could not be learned, was crushed by a falling wall in Coming opposite Montague street yesterday. He was engaged at work on the premises, and was caught under the wall as it fell. He was taken out and received prompt medical attention. His injuries are thought to be fatal." *Pickens* [South Carolina] *Sentinel,* September 18, 1886.

17. Unidentified male. "Just at the first shock a wagon bearing two men was passing the corner of Reid and King streets, when the gable end of a store on that corner fell and covered the unfortunates with the debris, one of whom was killed. It was very late when the body was extricated, and it was impossible to ascertain the person's name." *News and Courier,* September 1, 1886.

18. Unidentified patients. "Ten patients severely and some mortally wounded are in the Hospital, six white and four colored. They present a terrible spectacle." *News and Courier,* September 1, 1886.

19. Unidentified female. "A servant in the kitchen was

buried by the falling roof and timbers [at the home of George W. Williams, Jr., No. 2 Atlantic St.]; it required an hour to extricate her." *News and Courier*, October 2, 1886.

20. Unidentified females. Savannah, Georgia. "Two white women in different parts of the city leaped from second story windows, and suffered broken bones, but were not fatally hurt." Dutton, 326-327, citing *New York Tribune* reprint of a Savannah newspaper notice.

21. Aarons, Harry. "Harry Aarons and Lillie Aarons, white children, cut and bruised." *Atlanta Constitution*, September 4, 1886. "Doing well" in City Hospital, *News and Courier*, September 5, 1886.

22. Aarons, Lillie. *See* Aarons, Harry.

23. Arnold, Mr. near Summerville, South Carolina. "The engineer Burns, and his fireman Arnold, colored, had been badly injured by the tremendous leap which the train took in the dark under the unseen influence of the shock that dismantled Charleston." *News and Courier*, September 3, 1886.

24. Baynard, Mrs. "Mrs. Baynard, St. Phillip street, near the public school, fell down the steps and dislocated her hip bone." *News and Courier*, September 1, 1886.

25. Behre, Mr. C. The E.R.C. considered the case of "Mr. C. Behre who had been paralyzed the night of the earthquake and whose house had been seriously damaged." E.R.C. minutes, September 17, 1886.

26. Bing, Amelia. Savannah, Georgia. "Among those who were injured at the First African Church were Amelia Bing, Hagar Robinson, Anna Matthews, two ribs broken. Benj. Creamer was also badly injured." *News and Courier*, September 5, 1886.

27. Bissell, Mr. C.S. "Mr. C.S. Bissell was in the office of Capt. H.A. DeSaussure when the first shock occurred. They both ran out, but after Capt. DeSaussure had gained the street, the door slammed, cutting off Mr. Bissell from this mode of retreat. The ceilings and bricks fell thickly over him, striking him on the shoulder and in the left side after he fell. He ultimately succeeded in opening the door and got into the street

but not until his hands had been much cut." *News and Courier,* September 3, 1886.

28. Blainey, Adeline. "Adelaine Blainey, who resides in Lee street, is at the point of death. She is a colored woman, aged about 60 years, and broke her leg while trying to escape from her building on Tuesday night [August 31, 1886]. She is not expected to live through the night." *News and Courier*, September 7, 1886; *Atlanta Constitution*, September 7, 1886.

29. Blake, Mr. "Mr. Blake, struck in the back." *Berkeley Gazette*, September 4, 1886.

30. Bradley, Lizzie. "Lizzie Bradley, a colored girl about 18 years of age, living at 119 King street, was badly hurt by a falling piazza during the lower King street fire." *News and Courier,* September 3, 1886.

31. Brisenden, Mr. Aiken County. "Two trains were wrecked [near Langley, S.C.], the engineers of which, Messrs. Brisenden and Reynolds, were severely injured, and the colored firemen, Jack Simmons and Henry Ivy, killed." *News and Courier*, September 3, 1886.

32. Brown, Moses. "Moses Brown, colored, was badly wounded on King street by falling bricks." *News and Courier,* September 1, 1886. "Moses Brown, colored, janitor of Government building in Meeting Street, leg amputated [in City Hospital]. Condition critical." *News and Courier,* September 5, 1886.

33. Bulwinkley, B. B. "B. B. Bulwinkley, head bruised and hip injured. *Atlanta Constitution*, September 4, 1886. "Condition improved" in City Hospital, *News and Courier,* September 5, 1886.

34. Burns, Mr. *See* Arnold, Mr.

35. Calhoun, James. "James Calhoun, seriously injured in the face and jaw." *Atlanta Constitution*, September 4, 1886.

36. Campbell, Josephine. "Josephine Campbell, colored, scalp wound." *Atlanta Constitution,* September 4, 1886. "Improving [in City Hospital]," *News and Courier,* September 5, 1886.

37. Columbus, Mr. "Mr. Columbus, struck on the arm." *Berkeley Gazette*, September 4, 1886.

38. Creamer, Benjamin. Savannah, Georgia. *See* Bing, Amelia.

39. Davis, James. "James Davis, arm broken and shoulder-blade out of place." *Berkeley Gazette*, September 4, 1886.

40. Days, Susan. "Susan Days was seriously injured at 6 Philadelphia street." *News and Courier*, September 1, 1886.

41. Delany, Mary. "Mary Delany, colored, amputated leg." *Atlanta Constitution*, September 4, 1886. "Condition favorable [in City Hospital]," *News and Courier*, September 5, 1886.

42. DeLyon, Ms. Savannah, Georgia. "On Price street, near McDonough street, a colored woman named DeLyon jumped out of a second story window and broke her leg." *News and Courier*, September 5, 1886.

43. DeSaussure, Capt. H.A. "Capt H.A. DeSaussure was painfully injured in head while at his office." *News and Courier*, September 1, 1886.

44. Dorn, James. "James Dorn, who lives over Oldenbuttel's restaurant, had his arm broken and his shoulder-blade knocked out of place." *News and Courier*, September 1, 1886.

45. Drayton, Mary Ann. "Mary Ann Drayton, colored, 1 Philadelphia street, sustained painful injuries on body and lower limbs." *News and Courier*, September 1, 1886.

46. Dunn children. Four children of Sgt. J. P. Dunn, a police officer, were injured in their bedroom by a falling roof at their three-story home at 17 Tradd Street." *News and Courier*, September 12, 1886.

47. Eads, William. "William Eads, ankle fractured." *Atlanta Constitution*, September 4, 1886. "William Eads, knee dislocated." *Atlanta Constitution*, September 4, 1886.

48. Elliott, Lucy. "Miss Lucy Elliot, ankle fractured." *Atlanta Constitution*, September 4, 1886.

49. Flowers, Eliza. "Eliza Flowers, colored, compound fracture of leg." *Atlanta Constitution*, September 4, 1886. "Condition favorable [in City Hospital]," *News and Courier*, September 5, 1886.

50. Flynn, James. "James Flynn, injured on head, shoulder and arm." *Berkeley Gazette*, September 4, 1886. "Mr. M.J.

Flynn, a compositor on The News and Courier, jumped from the window of the composing room into the side alley when the first shock was felt. He sustained serious injuries about the shoulder, and on the arm and head." *News and Courier,* September 1, 1886.

51. Forrest, John. "John Forrest, white, was taken to the City Hospital yesterday suffering from a contusion of the leg, received at the American Hotel during the earthquake on Tuesday night." *News and Courier*, September 5, 1886.

52. Fowler, Isaiah. "Isaiah Fowler, colored man living at 122 King street, was seriously injured by falling timbers during the lower King street fire. He sustained a severe gash in the head, besides internal and external injuries, resulting in paralysis of the right arm." *News and Courier,* September 3, 1886.

53. Frazer, Sarah. "Sarah Frazer, colored, living on the corner of Washington and Vernon streets, received a superficial scalp wound, necessitating the use of three or four stitches to draw it together, and was also injured in the left ankle." *News and Courier,* September 3, 1886.

54. Gadsden, James. "James Gadsden, colored, fractured skull. Condition favorable [at City Hospital]." *News and Courier,* September 5, 1886.

55. Gadsden, Lucy Jones. "Lucy Jones Gadsden, colored, fracture of skull." *Atlanta Constitution,* September 4, 1886.

56. Galliot, Mrs. E. and daughter. "Mrs. E. Galliot, colored, was struck by a brick opposite the Pavillion Hotel, and her head badly injured. Her daughter was also badly hurt in the same way." *News and Courier,* September 1, 1886.

57. George, Ann. "Ann George, a colored girl fourteen years old, has been brought up to the Lunatic Asylum for treatment. Her home is near Smoak's Crossroads, Colleton County. She had previously shown a tendency toward insanity, but the earthquake shock of Tuesday last week completely shattered her mind, and she is now raving with a strong homicidal tendency." *News and Courier*, September 11, 1886.

58. Glover, Annie. "Annie Glover, who was hurt by the falling of Mr. Richardson's house in Friend street was injured."

News and Courier, September 3, 1886.

59. Goutvenier, Mr. J.O. "Mr. J.O Goutvenier, foreman of engine No. 2, was taken with a paralytic stroke, brought on by the strain and excitement while superintending his company at the fire in Legaré street. He is not expected to survive." *News and Courier,* September 3, 1886. "Messrs. Goutvenier and Mackin, foreman and callman respectively, of Engine No. 2, were each presented yesterday [December 24, 1886] by their comrades with handsome testimonials. Both of these gentlemen were injured while doing their duty on the night of the earthquake." *News and Courier,* December 25, 1886.

60. Grant, Rebecca. Savannah, Georgia. "At Coakley's boarding house, 201 South Broad street, a lamp was shaken down from a table and exploded, starting quite a blaze in a bed-room. Rebecca Grant, a colored woman in Mr. Coakley's employ, ran out of the kitchen, when a brick from a falling chimney struck her on the head, knocking her senseless. She was severely injured." *News and Courier,* September 5, 1886.

61. Haight children. Two children of police Pvt. A. B. Haight were injured when their house collapsed on their family August 31, 1886." *News and Courier,* September 16, 1886.

62. Hamilton, Kate. "Kate Hamilton, colored, 6 Philadelphia street, sustained painful injuries in the head from falling bricks." *News and Courier,* September 1, 1886.

63. Hammond, Mr. Isaac/Samuel. "Mr. Isaac Hammond, is thought to be fatally wounded, both his hips and legs being broken, and also his left arm. He said to Mr. Poulnot that he did not know whether he jumped from the three-story window on Broad street or was thrown out. He crawled from the sidewalk to the middle of the road, and on being removed uttered the most heartrending shrieks." *News and Courier,* September 1, 1886. "Mr. Samuel Hammond, who was badly injured the night of the 31st, while jumping from the window of his residence in Broad street, is recovering slowly. Besides being badly bruised about the head and face, his thigh was broken, which will necessitate his confinement to bed for some time. He is at the Mansion House in Broad street." *News and Courier,* September

16, 1886; also another story, same issue.

64. Harden, James. "James Harden, an old colored laborer, was seriously, if not fatally, crushed by a falling wall in upper King street yesterday morning.... His injuries are serious, and will probably prove fatal." *News and Courier*, September 17, 1886.

65. Heidt, Mr. "Mr. Heidt, Chalmers street, seriously injured." *News and Courier*, September 1, 1886.

66. Hicks, Mr. S. D. "Mr. S. D. Hicks, from J. B. Pace Tobacco Company, of Richmond, Va., who was injured by a brick from a fallen wall, has recovered from it and has gone again on his trip." *News and Courier*, September 7, 1886.

67. Holman, Edward. "[Holman] was aboard the Summerville train on his way to Lincolnville when the shock came and wrecked the train in the neighborhood of Ten-Mile Hill. He was struck in several places by the upsetting of the train, but manfully assisted the passengers and helped to extricate the men from the engine." *News and Courier*, September 11, 1886.

68. Holmes, William. "Wm. Holmes, who was crushed by a falling cornice on Tuesday last, was attended by Dr. W.P. Ravenel, and not by Dr. P. G. DeSaussure, as was stated at the time." *News and Courier*, September 10, 1886.

69. Howard, Isabella. "Isabella Howard, Ravenel court, seriously injured." *News and Courier*, September 1, 1886.

70. Jenkins, Fortuna. "Crushed by a Staircase." Also mentions injuries to a "colored child" named Julia Richardson. *News and Courier*, September 14, 1886 [microfilm very difficult to read]. Ms. Jenkins was probably black.

71. Jenkins, Isaac. "Isaac Jenkins, colored, wall fell on him in Cow alley. Seriously damaged all over body." *News and Courier*, September 1, 1886.

72. Jessen (female). "A young girl named Jessen, was also injured at the same place [Secession Street]. Upon being taken home she commenced bleeding internally. It is thought she will die." *News and Courier*, September 1, 1886.

73. Jones, Priscilla. "No. 25 Middle Street, chimney from

adjoining house fell down, splitting head of Priscilla Jones, colored, 12 years old; girl attended by Dr. T. L. Ogier." *News and Courier,* September 5, 1886.

74. Lively, Edmund (and unknown men). "Mr. Edmund Lively, of Richmond, Va., who boards at 205 5th street in that city, was walking in front of the City Hospital. The side of a house fell on him and badly injured his back and head. He crawled from under the debris and saw some men at the store on the corner of Mazyck and Queen streets, whom he supposed to have been killed, as he left them lying on the sidewalk. He staggered on as far as the City Hall Park, and there fell completely overcome." *News and Courier,* September 1, 1886. "Everett Lively, of Richmond, Va., reported wounded, is now going about." *Atlanta Constitution,* September 4, 1886.

75. Mackin, Mr. "Messrs. Goutvenier and Mackin, foreman and callman respectively, of Engine No. 2, were each presented yesterday [December 24, 1886] by their comrades with handsome testimonials. Both of these gentlemen were injured while doing their duty on the night of the earthquake." *News and Courier,* December 25, 1886.

76. Magwood, Helen. "Mr. Henry Magwood, who resides with his two sisters in Franklin street, near Queen, was injured in the leg by a falling piazza while escaping from his house on the night of the earthquake. His sister, Miss Helen Magwood, was also severely injured, several of her ribs having been broken and her head badly cut." *News and Courier,* September 7, 1886.

77. Magwood, Henry. *See* Magwood, Helen.

78. Martin, Mrs. Robert. "Mrs. Robert Martin, wife of the shoe merchant in Market Street, near King, was badly hurt." *News and Courier,* September 1, 1886.

79. Matthews, Anna. *See* Bing, Amelia.

80. Mitchell, Ellen. "Ellen Mitchell, mulatto, serious internal injuries, residence Judge Bryan's yard." *News and Courier,* September 1, 1886.

81. Molen, W.M. Mt. Pleasant, S.C. (?) "Mr. W.M. Molen, foot and arms." *Berkeley Gazette,* September 4, 1886.

82. O'Connor, Miss. "Not a single person [on Sullivan's Island] was injured, with the exception of Miss O'Connor, whose ankle was badly sprained in getting out of the [Moultrie] house." *News and Courier,* September 3, 1886.

83. Oakman, J.G. Augusta, Georgia. "J.G. Oakman, collector of this city is crazed and roams the streets in a wild way, speaking of the crookedness of the earth, and the terrible warnings sent by the Almighty. His friends are caring for him." *Atlanta Constitution*, September 4, 1886.

84. Oldenbuttel, Mr. "The nephew of J.W. Oldenbuttel, East Bay, is seriously injured." *News and Courier,* September 1, 1886.

85. Oliver, Charles. "Charles Oliver and James Strobel, two colored men, were employed on Thursday [September 9] to pull down a dangerous two-story brick out-building belonging to Mr. C. M. Seignious, adjoining Mr. Bremer's lot, No. 88 Hasel street. They were in the second story of the house, and had succeeded in pulling down most of the wall, when at an unlucky moment a huge mass of brick tumbled on them, carrying down the wooden joists and everything with it. The men were extricated from the ruins insensible. Oliver was severely cut on the scalp and lips and Strobel badly bruised on the hips and chest.... Both men are doing well...." *News and Courier,* September 11, 1886.

86. Owens child. "A young boy 15 years of age, a son of Mr. John R. Owens was walking near the corner of Morris and St. Philip street when the first shock came on Tuesday night. The little fellow was picked up from the earth by the force of the undulation and hurled to the other side of the street." *News and Courier,* September 3, 1886.

87. Oxlade, Anne Amelia. "No. 42 Middle street, two-and-an-half story brick residence of Oswald McMillan, dangerously cracked throughout, dangerous; Miss Anne Amelia Oxlade, severely injured on face and body." *News and Courier,* September 5, 1886.

88. Palmer, Dave. "Dave Palmer, colored, had his right leg fractured during the fire at 123 King street and was sent to the City Hospital." *News and Courier,* September 3, 1886.

89. Palmer, Mamie. "Miss Mamie Palmer, residence 17 John street received numerous internal injuries by the falling piazza." *News and Courier,* September 1, 1886.

90. Pope, Joseph. Savannah, Georgia. "Mr. Joseph Pope, at the corner of Taylor and West Broad streets, was struck on the head as he ran out of the house into the yard by a brick which fell from the chimney." *News and Courier*, September 5, 1886.

91. Powell, Mrs. "One of the most pitiful cases in the city is that of Mrs. Powell, whose husband was killed at the corner of Church and Elliott streets on Tuesday night, and who was herself badly wounded. She was provided with a home at the Park House, on Meeting Street, yesterday and was furnished with rations by the relief committee." *News and Courier*, September 10, 1886.

92. Pratt, W. "W. Pratt jumped out of a window 155 Calhoun, leg broken." *News and Courier,* September 1, 1886.

93. Reynolds, Mr. *See* Brisenden, Mr.

94. Richardson, J.C.E. "Mr. J.C.E. Richardson, living at 12 Friend street, was seriously injured in the head by his house falling in upon him. His condition is very critical. His colored servant was likewise dangerously hurt, several of her limbs being broken." *News and Courier,* September 1, 1886.

95. Richardson, Julia. *See* Jenkins, Fortuna.

96. Rivers, Charles. "A colored man named Charles Rivers, living at No. 3 King street, was struck by a falling wall near the corner of King street and the Battery, and his left leg was broken below the knee. His wife Sarah Rivers, was also badly wounded at the same time." *News and Courier,* September 1, 1886. "Charles Rivers, colored, amputated leg." *Atlanta Constitution*, September 4, 1886. "Condition favorable [in City Hospital]." *News and Courier,* September 5, 1886.

97. Rivers, Sarah. *See* Rivers, Charles.

98. Robinson, Hagar. *See* Bing, Amelia.

99. Robson, Mary. "The Robsons' piazzas tore off clean-Ainsley & his two sisters Mary & Sallie buried in the ruins-they . . . [were] but only bruised; he [was] struck upon the

head. . . . He was trapped in the wreckage for three-quarters of an hour before he died." Elizabeth Moffett, the Robsons' neighbor, to her daughter, Anna.

100. Robson, Sallie. *See* Robson, Mary.

101. Sanders, Maggie. "Maggie Sanders, white, amputated foot." *Atlanta Constitution*, September 4, 1886. "Doing very well [in City Hospital]": *News and Courier,* September 5, 1886.

102. Sawyer children. "132 Calhoun St. Daughter of Isaac Sawyer, barber, killed and two children injured in the ruins." *News and Courier,* September 5, 1886.

103. Schillee, Willis. "Willis Schillee, who clerks in J. W. Oldenbuttel's restaurant, was seriously injured by a cornice of the house falling on his hip; he was taken down on Boyce's wharf and attended by a doctor, who said that no bones were broken, but his spine is seriously injured." *News and Courier,* September 1, 1886.

104. Scharlock, Mrs. "No. 41 [Calhoun St.], corner Calhoun and East Bay, brick [house], gable down, back piazza fallen, numerous cracks, dangerous. Dr. Scharlock's wife fell with the piazza." *News and Courier,* September 5, 1886.

105. Schur, B. "One of the most painful accidents which occurred the night of the 31st of August happened to Mr. B. Schur, who kept a cigar store on Broad Street. He was asleep in the back room of his store, when the roof fell in and inflicted painful injuries all over his body. He is at present being tended by friends in the neighborhood of his store." *News and Courier,* September 11, 1886.

106. Shackelford, Anna. "Mrs. Shackelford, Miss Anna Shackelford, Lewis Shackelford and Neese Shackelford were more or less painfully injured, the last more than the others, but none dangerously." *News and Courier,* September 3, 1886.

107. Shackelford, Lewis. *See* Shackelford, Anna.

108. Shackelford, Mrs. *See* Shackelford, Anna.

109. Shackelford, Neese. *See* Shackelford, Anna.

110. Simpson, Fred. "Fred Simpson, colored, of New York, double fracture of the left leg." *Atlanta Constitution,* September 4, 1886.

111. Smalls, Silvey (and two unidentified children). "Silvey Smalls and two children were asleep in the kitchen on the premises of Mrs. Yates on Church street. The roof of the kitchen fell in, burying the three. Silvey and one of the children were painfully but not seriously bruised." *News and Courier,* September 3, 1886.

112. Smith, Mrs. T.J. Savannah, Georgia. "Mrs. T.J. Smith, at the corner of Broughton and West Broad, seized her 15 months old child, got out of the window on the Broughton street side on the awning and jumped to the street with the girl in her arms. She was badly injured." *News and Courier,* September 5, 1886.

113. Steel, Maggie. Augusta, Georgia. "Maggie Steel, a factory operative's daughter, has lost her mind, and is hopelessly insane, roaming about the house and crying aloud for help. She seems to experience shock after shock." *Atlanta Constitution,* September 4, 1886.

114. Strobel, James. *See* Oliver, Charles.

115. Tierney, Martin. "Mr. Martin Tierney, No. 21 Meeting street, sustained very painful injuries of the head, ankle and hips, from the falling bricks of his kitchen." *News and Courier,* September 3, 1886.

116. Torrent, Mrs. J. A. Beaufort County. "Mrs. J.A. Torrent was much prostrated by the earthquake shocks last week, and, we are sorry to say, she is still very sick." *Palmetto Post,* Port Royal, S.C., September 9, 1886.

117. Wainger, Frank M. "Frank M. Wainger, white, Hanover street opposite Hampton court, was struck by falling bricks yesterday and badly injured." *News and Courier,* September 5, 1886.

118. Walker, Sam. "Sam Walker, colored, arm and leg fractured." *Atlanta Constitution,* September 4, 1886.

119. Watkins, Sam. "Sam Watkins, colored, arm and leg fractured. Condition favorable [in City Hospital]." *News and Courier,* September 5, 1886.

120. Webb, George. "George Webb, skull fractured." *Atlanta Constitution,* September 4, 1886.

121. Wilkins, Annie. "Annie Wilkins, colored, living at 100 King street, was painfully injured on the head and right shoulder from falling brick." *News and Courier,* September 3, 1886.

122. Williams, Mary. "Mary Williams, colored, face and head badly bruised." *Atlanta Constitution,* September 4, 1886. "Improving [in City Hospital]." *News and Courier,* September 5, 1886.

123. Williams, Mr. "The Knights of Labor…were holding a meeting…light shock was felt while they were in session. The stampede was so violent that a knight named Williams was trampled on and had his arm and leg broken." *Atlanta Constitution,* September 3, 1886.

124. Williams, Mrs. "Mrs. Williams jumped from the second story of her residence at the corner of Wentworth and Meetings streets, and injured her spine badly. She is the daughter of Mrs. M. J. Williams." *News and Courier,* September 1, 1886.

125. Wilson, James M. "No. 89 [Meeting Street], the brick residence of Mr. James M. Wilson, will have to be pulled down. There is a huge gap on the second story exposing the room of Mr. Wilson, from which he barely escaped with his life and very much bruised." *News and Courier,* September 6, 1886.

ACKNOWLEDGMENTS

Although I first learned about the Great Charleston Earthquake in 1982, while working as the newest staff member of the South Carolina Historical Society, the impetus to start researching and writing this book came twenty years later from Robert Fleming, then manager of Charleston's Barnes & Noble bookstore on Sam Rittenberg Boulevard. Because of his deep interest in "de quake," and his incessant prodding, I decided to see if sufficient source material was available to do the subject justice. I was not disappointed. Indeed, initially, I was overwhelmed.

A grant from the Post and Courier Foundation, administered by the Historic Charleston Foundation, enabled me to research and explore the complex and highly effective local government response to the earthquake, which I present here as The Charleston Disaster Recovery Model. The grant also provided funds to make copies of the model available to all South Carolina county disaster management supervisors and to the main county libraries and enabled me to conduct earthquake awareness talks throughout Charleston County. At the Post and Courier Foundation, Pierre Manigault's insightful questions about the disaster recovery process helped me shape my inquiries.

Because of the immensity and complexity of the data upon which this book is based, it would never have been possible to write it without the assistance at Corinthian Books of Diane Anderson, senior editor, with assistance from Elizabeth Burnett, Ph.D., historical editor; researchers Monica Biddix, Heidi Bradley, Kristine Dudley, Jessica Lancia, Elizabeth E. Mullaney, and Nadia Shamsedin; and typesetter Steve McCardell; graphic artist Rob Johnson; and publicist Rebecca Imholz.

Special thanks are due to Dr. Pradeep Talwani, professor of geophysics at the University of South Carolina, director of the South Carolina Seismic Network, and the nation's foremost authority on the 1886 Charleston earthquake, for generously sharing his time and knowledge. At the College of Charleston's Department of Geology and Environmental Geosciences, Dr. Norman S. Levine, Dr. Steven C. Jaumé, and Dr. Briget Doyle shared with me their extensive knowledge and databases of Charleston earthquake information. Also at the college, Dr. Robert Stockton advised and guided me through the intricacies of Charleston's rich architectural history. Finally, Dr. Susan E. Hough, USGS geophysicist and editor-in-chief of *Seismological Research Letters*, explained to me her findings on remotely triggered earthquakes. I also owe a debt of gratitude to USGS senior geographer Dr. K. Eric Anderson for introducing me to the earthquake- and geoscience-oriented members of this outstanding technical advice team, and for his lucid comments on the text of this book.

As always, numerous scholars, professional specialists, and gracious, knowledgeable people from Charleston and "from off" shared unstintingly their support, enthusiasm, knowledge, documents, and images.[1] Heading the list of indispensable allies in writing this book are the information professionals of the Charleston County Library system. Particular thanks are due to Nicholas Butler, Ph.D., special collections manager; Marianne Cawley, South Carolina Room manager, and her marvelous staff; Linda Stewart and her interlibrary loan colleagues, and at Mt. Pleasant, Reference Librarian Susan Frohnsdorff and her peerless and cheerful colleagues.

Elsewhere in the Palmetto State, I am indebted to John L. All, historian, Mt. Pleasant Presbyterian Church; Ann Taylor Andrus, historian, Bethel United Methodist Church; Elizabeth B. Asnip, Pawley's Island; Robert Barrett, archivist, St. Helena's Episcopal Church, Beaufort; Raejean and Frank Beattie, Hopsewee Plantation, Georgetown; Cecil Bennett; Sister Anne Francis Campbell, Sisters of Charity of Our Lady of Mercy; Ellen Chamberlain, library director, University of South Carolina-Beaufort; Glen Clayton, Special Collections, Furman University Library; Clare Cochran; Grace Cordial, South Carolina Room librarian, Beaufort County Library; Barbara Doyle, historian, Middleton Place Foundation;

Patricia C. Doyle, Georgetown; Tammie L. Dreher, earthquake coordinator, South Carolina Emergency Preparedness Division, Columbia; Kathy DuLaney, president, McCormick County Historical Society; Bobby F. Edmonds, historian, McCormick County; Cynthia Ellis, chief financial officer, Historic Charleston Foundation; Eric Emerson, Ph.D., and his staff at my *alma mater*, The South Carolina Historical Society; Karen Emmons, archivist/librarian, Historic Charleston Foundation; Martha Erwin; Brian Fahey, archivist, Catholic Diocese of Charleston; Marie E. Ferrara, head, Special Collections, College of Charleston Library; Jim Fitch, director, The Rice Museum, Georgetown; Mary Giles, archivist *emeritus*, Catholic Diocese of Charleston; Marge Grove, senior vice president, Charleston Chamber of Commerce; Anne W. Hagood, chairman of the board, Barnwell County Museum; Kenneth Harrell, director, Dorchester County Emergency Services Department; Diane Henegar, owner, General Francis Marion Bamberg House, Bamberg; A. V. Huff, Ph.D., professor of history, Furman University; Patty Jack, Education and Research Department, Drayton Hall; Monty Jones; Glenn Keyes, architect; Richard F. King; Jennifer Lankford, chamber docent, City of Charleston Council Chamber; Katrina Lawrimore, director, Kaminski House Museum, Georgetown; Maxine F. Lutz, Historic Beaufort Foundation; Susan McMillan, historian, Conway; Janet E. Meleney, archivist, Clarendon County Archives; Cathy Michael; Alma Montague, Confederate Home and College, Charleston; Ethel Trenholm Nepveaux, historian; Rose Marie Oakes; Jason Patno, director, Berkeley County Emergency Preparedness; Arnold Penuel; Valerie Perry, associate director of museums, Historic Charleston Foundation; Alberta Quattlebaum, historian, Pawley's Island; Jeff G. Reid, Jr.; Hal Rigby, vice president, Summerville Preservation Society; Jeanne S. Robinson; Katharine S. Robinson, executive director, Historic Charleston Foundation; Harold Robling, historian, St. Paul's Episcopal Church, Summerville; Mary Julia Royall, Mt. Pleasant historian; Catherine Sadler, librarian, and the staff of the Charleston Library Society; The Reverend Dow Sanderson, rector, Holy Communion Episcopal Church, Charleston; Taylor Shelby, director, The Powder Magazine Museum; Anne Sheriff; Carl H. Simmons, Director of Building Inspections, County of Charleston; Dr. Jack and Anneliese Simmons; Lacy Sproul; Mary Edna Sullivan, curator, Middleton Place; Evan R. Thompson, executive director, Historic Beaufort Foundation; Jane Thornhill; Bo Turner, vestryman, Old St. Andrews's Episcopal Church, Charleston; Libby Wallace, research director, The *Post and Courier*; Margaret T. Wannamaker; Melanie Whittington, Department of Earth and Atmospheric Sciences, Saint Louis University Earthquake Center; George W. Williams, historian, St. Michael's Episcopal Church, Charleston; Terry Williams, records manager, Charleston Water System; Matthew Wise, records specialist, City of Charleston Department of Archives and Records; Elizabeth Jenkins Young, preservationist, folklorist, and tour guide, Charleston; and Pauline Zidlick, curator, Barnwell County Museum.

I would also like to thank those people from all over the United States who contributed their help and goodwill. They include Mary Domingos, Gold Bar, Washington; Tom Fetters, railroad historian, Lombard, Illinois; Richard D. Kaplan, Metuchen, New Jersey; Virginia Mescher, Ragged Soldier Sutlery and Vintage Volumes; Christopher Murphy, Ph.D., Department of History and Anthropology, Augusta State University, Augusta, Georgia; Blanchard Smith, Alexandria, Virginia; and Ned Wells, Hendersonville, North Carolina. And last but not by any means least, I owe thanks to a small but fascinating group of people who shared with me their knowledge and images of the R.M.S. *Etruria*, including Allen J. Baden; Arthur and Claire Blundell, Bishops Stortford, Hertfordshire, England; Dave Dermon, III; Mike Dovey, Solihull, England; Keith Droste, Anaheim, California; Richard C. Faber, Jr., New York; Paul K. Gjenvik, director, Gjenvick-Gjønvik Archives (www.steamships.org), Atlanta, Georgia; and David Shipman.

1 According to Mary Julia Royall of Mt. Pleasant (and patron saint of Lowcountry historians), "from off" is a Gullah term denoting anyone not from the Lowcountry before "de wah;" "from away" is the Geechee (white English speech, flavored by Gullah) equivalent term with the same meaning.

SOURCE NOTES

PREFACE

1 *San Francisco Chronicle*, May 6, 1906.

2 The U.S. Federal Population Census of June 30, 1880, found the population of Charleston to be 49,784. According to a private census made for the 1885-1886 city directory, Charleston was home to 27,605 (46%) whites and 32,540 (54%) blacks, for a total population of 60,145 residents. 40,000 people sleeping outdoors: Charleston, SC: *News and Courier*, September 7, 1886. Hereafter cited as *News and Courier*.

3 Clarence Edward Dutton, *The Charleston Earthquake of August 31, 1886. By Capt. Clarence Edward Dutton, U.S. Ordnance Corps.* U.S. Geological Survey Ninth Annual Report, 1887-1888, 219. Hereafter cited as Dutton. Omaha and Bermuda: New York *Times*, September 1, 1886; all others: Dutton, *passim*.

4 Dutton, 219.

5 Dutton, 219. The conversion factor (multiplier) used here to convert 1886 dollars to their equivalent buying power in 2006 dollars is 20.1. Where present-day dollar equivalents are useful for clarity, the 1886 amount will be followed by the 2006 equivalent in parentheses. Multiplier source: "Inflation Conversion Factors for Years 1665 to estimated 2015, online at http://oregonstate.edu/Dept/pol_sci/fac/sahr/infcf16652005.pdf October 8, 2005.

6 Leonardo Seeber and John G. Armbruster, "The 1886-1889 Aftershocks of the Charleston, South Carolina Earthquake: A Widespread Burst of Seismicity," *Journal of Geophysical Research*, 92 (1987): 2663-2696. Hereafter cited as Seeber and Armbruster.

7 Carl W. Stover and Jerry L. Coffman, *Seismicity of the United States, 1568-1989* (Washington, DC: U.S. Geological Survey Professional Paper 1527, 1993), 353-354. Hereafter cited as Stover and Coffman.

8 This small (magnitude 2.6) earthquake occurred at 1:02 p.m. on November 19, 2005. Its source was located two miles west and three miles beneath Summerville. National Earthquake Information Center, USGS.

CHAPTER 1

As When a Cat Trots Across the Floor

1 R.B. Whorton and R. Clary, unpublished data, 1972, in G. A. Bollinger and T. R. Visvanathan, "The Seismicity of South Carolina Prior to 1886," in Douglas W. Rankin, ed., *Studies Related to the Charleston, South Carolina Earthquake of 1886 – A Preliminary Report*. Geological Survey Professional Paper 1028. Washington, DC: Government Printing Office, 1977, 40. Hereafter cited as Bollinger and Visvanathan.

2 Dutton, 230.

3 Dutton, 230.

4 *City of Charleston Yearbook*, 1886 (Charleston, SC: Walker, Evans & Cogswell Co., 1886), 375. Hereafter cited as City *Yearbook*, 1886.

5 Federal Population Census, 1880, 152 Calhoun Street (north side), Charleston.

6 Kenneth E. Peters and Robert B. Herrmann, eds., *First-hand Observations of the Charleston Earthquake of August 31, 1886, and Other Earthquake Materials. Reports of W. J. McGee, Earle Sloan, Gabriel E. Manigault, Simon Newcomb, and Others* (Columbia, SC: Bulletin 41, South Carolina Geological Survey, 1986), 106. Hereafter cited as Peters and Herrmann.

7 City *Yearbook*, 1886, 375.

8 Peters and Herrmann, 96; Dutton, 231.

9 City *Yearbook*, 1886, 372; *News and Courier*, August 28, 1886.

10 City *Yearbook*, 1886, 372; *News and Courier*, August 28, 1886.

11 Peters and Herrmann, 112.

12 City *Yearbook*, 1886, 373.

13 *Atlanta Constitution*, August 28, 1886.

14 *Atlanta Constitution*, August 28, 1886.

15 *Atlanta Constitution*, August 29, 1886.

16 City *Yearbook*, 1886, 373n; Stephen Taber, "Seismic Activity in the Atlantic Coastal Plain near Charleston, South Carolina," *Bulletin of the Seismological Society of America*, 4 (1914): 150, hereafter cited as Taber; Dutton, 231.

17 Peters and Herrmann, 112.
18 Stover and Coffman, 348.
19 Peters and Herrmann, 112.
20 *Atlanta Constitution*, August 29, 1886.
21 *Atlanta Constitution*, August 29, 1886.
22 City *Yearbook*, 1886, 374.
23 Peters and Herrmann, 112.
24 Peters and Herrmann, 106.
25 City *Yearbook*, 1886, 374; *News and Courier*, September 3, 1886.
26 Peters and Herrmann, 112.
27 Frank R. Whilden, *News and Courier*, September 2, 1930.
28 Robert G. Chisolm, Charleston, to the Rev. E. T. Horn, Pennsylvania, September 1, 1886. "The Charleston Earthquake as told in letters from Robert G. Chisolm." Typed transcript, Collection #43/190, South Carolina Historical Society, Charleston, S.C. Hereafter referred to as S.C.H.S.
29 Edward Laight Wells, personal narrative of the Charleston earthquake (typescript), 1. Edward Laight Wells Papers, 1858-1924. Collection #1136.00, S.C.H.S.
30 Undated *News and Courier* clipping, 1937, S.C.H.S. vertical file #30-4.
31 *The Berkeley Gazette* (Mt. Pleasant, S.C.), September 4, 1886.
32 In South Carolina, the term "hop" was in use as early as the Civil War period to describe a group of people assembled for informal dancing.
33 James Sprunt, "The Earthquake of 1886," *Chronicles of Cape Fear River, 1660-1916* (Raleigh, NC: Edwards & Broughton Publishing, 1916), 538. Hereafter cited as Sprunt. R.M.S. stands for Royal Mail Ship, a vessel licensed to carry mail abroad under contract to the British Royal Mail. The *Etruria* and its 285 crew members accommodated 550 first-class and 800 third-class (steerage) passengers.
34 The damage estimates for the 1885 hurricane have been estimated variously between $1.5 and $2 million ($30,150,000 - $40,200,000 in present (2006) dollars).
35 City *Yearbook*, 1886, 229.
36 City *Yearbook*, 1886, 229.
37 City *Yearbook*, 1886, 232-233. Courtenay's European points of call have not been identified, but probably included his ancestral homeland, Ireland.
38 Anthony Toomer Porter, *Led On! Step by Step! Scenes from Clerical, Military, Educational, and Plantation Life in the South, 1828-1898. An Autobiography* (New York: G. P. Putnam Sons, 1898), 392. Hereafter cited as Porter.
39 Porter, 392.
40 Porter, 392.

CHAPTER 2
Terror on Broad Street

1 Charleston City Council minutes, August 10, 1886. Hereafter cited as City Council minutes.
2 City *Yearbook*, 1886, 432-433.
3 Charleston's street numbering system was changed in 1886. The present-day addresses are used here to the extent it has been possible to determine them.
4 George Armstrong Wauchope, *The Writers of South Carolina* (Columbia, SC: The State Co., 1910), 282-290.
5 Except as noted, this chapter is based chiefly on Carl McKinley's account of the earthquake in the *News and Courier*, from a modified account reprinted in Dutton, 212-225 and the City *Yearbook*, 1886; augmented with details from Jonathan Poston's *The Buildings of Charleston* (Columbia, SC: The University of South Carolina Press, 1997), hereafter cited as Poston; and from Arthur Mazÿck and Gene Waddell, Charleston in 1883 (Easley, SC: Southern Historical Press, 1983), hereafter cited as Mazÿck and Waddell.
6 Dutton, 213.
7 Dutton, 213.
8 There are conflicting reports (9:51 and 9:55 p.m.) as to when the hands of St. Michael's clock stopped. Given that there was no recorded shock at 9:55, the earlier time is the most likely.
9 Coping installed 1838: *News and Courier*, September 11, 1886.
10 Mazÿck and Waddell, figure 48.
11 *News and Courier*, September 1, 1886; Mazÿck and Waddell, figure 12.
12 Poston, 162-163; *News and Courier*, September 1 & 3, 1886.
13 Peters and Herrmann, 28.
14 Mazÿck and Waddell, 12, and text accompanying illustration 14; Poston, 186; and Robert P. Stockton, *The Great Shock: The Effects of the 1886 Earthquake*

on the Built Environment of Charleston, South Carolina (Easley, SC: Southern Historical Press, 1986), 37, hereafter cited as Stockton.

15 Hibernian Society Hall: Peters and Herrmann, 18; Mazÿck and Waddell, illustration 10.

16 Dutton, 263.

17 9 Glebe St.: Poston, 524-525; family anecdote and photographs from Jeanne S. Robinson, Mt. Pleasant, SC, letter to the author, June 10, 2005.

18 Dutton, 263.

19 "A very interesting feature to the stranger are the buzzards [turkey vultures] which congregate here [in Market Hall] in large numbers, ready to scramble over the scraps thrown to them, and sometimes not waiting for this but will steal the meat from the stands whenever the butcher's back is turned. They are very tame, and it is a fine to injure them." *The Charleston Guide and Business Directory for 1885-6*, 42.

20 Mazÿck and Waddell, 14-15.

21 Poston, 502-503.

22 Harriet Kinloch Smith, Charleston, to Miss Louisa ("Luly") Porcher, September 1886, in Edward M. Gilbreath, "Letter Provides Dramatic Account of 1886 Quake," *News and Courier*, August 12, 2004.

23 Elizabeth S. Moffett, Charleston, to her daughter, Anna, September 1, 1886. George H. Moffett Family Papers, Collection #11/306/16, S.C.H.S. See *In Memoriam* for Ainsley Robson's death record; also *News and Courier*, September 11, 1886; and an undated newspaper clipping, Charleston County Library vertical files, for the details of his funeral.

24 Elizabeth S. Moffett, Charleston, to her daughter, Anna, September 2, 1886. George H. Moffett Family Papers, Collection #11/306/16, S.C.H.S. In 1886, many large Charleston homes had stables and carriage houses on their downtown properties.

25 Elizabeth S. Moffett, Charleston, to her daughter, Anna, September 2, 1886. George H. Moffett Family Papers, Collection #11/306/16, S.C.H.S. During the siege of Charleston from 1861 to 1865, civilians evacuated the peninsula and moved their families and slaves to places like Fairfield, SC, 145 miles inland, just north of Columbia. Nevertheless, Union troops plundered these upcountry refuges at the end of the war.

26 Susan Pringle, 27 King Street, to My Dear Cousins, September 8, 1886. Pringle family papers, 1745-1897. Collection #1083.00, S.C.H.S.

27 *News and Courier*, September 3, 1886.

28 Now known as the Bull-Huger House, it is now 34 Meeting Street.

29 *News and Courier*, September 3, 1886. In *The Dwelling Houses of Charleston, South Carolina*, written by Alice Ravenel Huger Smith (Philadelphia. J.B. Lippincott Co. 1917), 81, the author recounts a death very much like Alexander's, and it may be him she had in mind when she said, "The fatal accident on the Huger steps was an incident of the earthquake of 1886. A parapet, which had replaced the bull's-eye, was thrown off, and a portion of it fell upon and crushed an unfortunate young Englishman. He was visiting at the house and attempted to run out while the shock was most violent."

30 The Main Guard House was damaged beyond repair and was torn down. In 1896-97, the present Post Office was constructed on the site. *News and Courier*, September 1, 1886; Return of Deaths, August 31, 1886; Poston, 168.

31 *City Directory*, 1889.

32 Fleming, Miller, and Rodolf: Return of Deaths, August 31, 1886; *News and Courier*, September 1, 1886. 100 King St. as address: *News and Courier*, October 1, 1886.

33 L. E. "Ned" Cantwell, Charleston, to Cousin Marie, September 10, 1886. "Earthquake Echoes," 1886-ca. 1950. Collection #43/0189, S.C.H.S.

34 *News and Courier*, September 1 and 3, 1886; undated *News and Courier* clipping, Charleston County Public Library vertical file; and *Southern Business Guide, 1885-1886*, 722. Patrick Lynch, age 70, born Ireland, buried in St. Lawrence Cemetery, is also listed in the Return of Deaths, August 31, 1886. His relation to Maurice J. Lynch has not been established.

35 *News and Courier*, September 1, 1886. No whites with residences listed on Hasell Street or "City Hospital" appear

in the Return of Deaths for the earthquake period.

36 *The Berkeley Gazette*, September 4, 1886.

37 Return of Deaths, September 3, 1886; *News and Courier*, September 5 & 11, 1886.

38 *The Berkeley Gazette*, September 4, 1886.

39 *The Berkeley Gazette*, September 4, 1886.

40 *The Berkeley Gazette*, September 4, 1886.

41 City *Yearbook*, 1886, 347.

CHAPTER 3

Panic in the Pines

1 Barbara Lynch Hill, Summerville. *A Sesquicentennial Edition of the History of The Flower Town in the Pines* (Summerville, SC: privately published, 1997), 86. Hereafter cited as Hill.

2 Clarice and Lang Foster, eds., *Beth's Pineland Village* (Summerville, SC: Summerville Preservation Society, 1988), 79. Hereafter cited as *Beth's Pineland Village*.

3 Peters and Herrmann, 106.

4 Peters and Herrmann, 106.

5 Dutton 272.

6 Both women's stories: *Beth's Pineland Village*, 81.

7 Peters and Herrmann, 112.

8 Peters and Herrmann, 112.

9 Peters and Herrmann, 112. John B. Gadsden wrote his recollections for the USGS researchers on October 11, 1886, while they were still fresh and clear in his mind.

10 Peters and Herrmann, 112.

11 Peters and Herrmann, 112.

12 Peters and Herrmann, 112.

13 City *Yearbook*, 1886, 376.

14 Sarah Boylston, "Trainman Tells of Earthquake," *News and Courier*, August 30, 1936. Hereafter cited as Boylston.

15 Robert G. Chisolm, Charleston, to the Rev. E. T. Horn, Pennsylvania, September 2, 1886. "The Charleston Earthquake as told in letters from Robert G. Chisolm, ca. 1890." Collection #43/190, S.C.H.S.

16 Boylston.

17 Boylston.

18 Boylston.

19 Boylston.

20 Dutton, 273.

21 City *Yearbook*, 1886, 377.

22 Tighe: 1880 Federal Population Census, Summerville; 1920 Federal Population Census, Washington, D.C.

23 City *Yearbook*, 1886, 377.

24 *Beth's Pineland Village*, 82.

CHAPTER 4

Chaos in the Palmetto State

1 Omaha and Bermuda: New York *Times*, September 1, 1886.

2 *Harper's Weekly*, September 11, 1886.

3 Dutton, 497-509.

4 Mt. Pleasant's 1886 population: Petrona Royall McIver, *History of Mount Pleasant, South Carolina* (Charleston, SC: Ashley Printing and Publishing Co., 1960), 93.

5 *Berkeley Gazette*, September 4, 1886.

6 *Berkeley Gazette*, September 4, 1886.

7 *Berkeley Gazette*, September 4, 1886.

8 *Berkeley Gazette*, September 4, 1886.

9 *Berkeley Gazette*, September 4, 1886.

10 *Berkeley Gazette*, September 4, 1886.

11 *Berkeley Gazette*, September 4, 1886.

12 *Berkeley Gazette*, September 4, 1886.

13 *Berkeley Gazette*, September 4, 1886.

14 *Berkeley Gazette*, September 4, 1886.

15 *Berkeley Gazette*, September 4, 1886.

16 *Berkeley Gazette*, September 4, 1886.

17 *Berkeley Gazette*, September 11, 1886.

18 Dutton, 508.

19 Dutton, 322.

20 *Berkeley Gazette*, September 11, 1886.

21 *Berkeley Gazette*, September 7, 1886.

22 *Berkeley Gazette*, September 11, 1886.

23 Dutton, 322-323.

24 Walter Hazard, editor, *The Georgetown Enquirer*, September 8, 1886.

25 Dutton, 502.

26 Dutton, 323, 502.

27 Dutton, 500.

28 "Many Legends Born in 1886 Earthquake," *News and Courier*, September 2, 1957.

29 Cooper's story was first published in 1951 in the Mullins *Enterprise*. This telling was adapted from Betty Molnar, "Earthquake Brings Revival," *Lowcountry Companion*, July-October 2005 and Ruby Miles, "Horry County Residents' Memories 'Jarred' Over Earthquake of 1886," *The State* (Columbia, SC), August 20, 1967.

30 Dutton, 507.

31 *Harper's New Monthly Magazine*, December 1875, 9.

32 Pinckney, 159.

33 Peters and Herrmann, 1986.

34 John Drayton Hastie, *The Story of Historic Magnolia Plantation and Its Gardens* (Charleston, SC: privately published, 1948), 4.

35 Patty Jack, Education Department, Drayton Hall, to the author, March 27, 2006.

36 *News and Courier*, September 9, 1886; Glen Keyes, supervising architect for the 2005 restoration, to the author, March 28, 2006; and Bo Turner, vestryman of St. Andrews's, to the author, March 28, 2006.

37 *News and Courier*, September 5, 1886; Dutton, 323, 499.

38 Dutton, 304-306.

39 Dutton, 497; City *Yearbook*, 1886, 379; *News and Courier*, September 5, 1886.

40 Robert Barrett, Archivist of St. Helena's Church, stated that the church's records and scrapbooks make no mention of the earthquake, and that during a major restoration project at the church in 1998-2000, no evidence of earthquake damage was noted, nor were any iron reinforcing rods found. Robert Barrett, personal communication with the author, March 10, 2006.

41 Courthouse cracks: *The Palmetto Post*, Port Royal, SC, September 9, 1886. Hereafter cited as *The Palmetto Post;* Liollio Architects, "The Old Beaufort College Building: A Historical Structure Report. Architectural Development and Historical Alterations. Prepared by Liollio Architects, 1998." University of South Carolina Library - Beaufort, Beaufort, SC.

42 Maxine F. Lutz, Historic Beaufort Foundation, to the author, April 20, 2006.

43 E. B. Rodgers, "Reminiscences of Beaufort Storms. Read before the Beaufort County Historical Society, June 1950," Beaufort County Library vertical files. Hereafter cited as E.B. Rodgers. Rodgers, born in 1884, was two years old during the earthquake. The stories he records were told to him late in life by his elders. *The Palmetto Post*, Port Royal, SC, September 9, 1886. Waddell and his former master, Tucker: "Letter from Rev. H. H. Tucker, D.D., to the Negro Baptists of Georgia Holding a Centennial Celebration in Savannah, June 6-18, 1888," online March 10, 2006 at www.reformedreader.org/history/love/chapter18.htm.

44 *The Palmetto Post*, September 9, 1886.

45 E.B. Rodgers.

46 *The Palmetto Post*, September 9, 1886.

47 *The Palmetto Post*, September 9, 1886.

48 Dutton, 506; *The Palmetto Post*, September 2, 1886.

49 Dutton, 503.

50 W.P.A. interview with "Mom" Agnes James, Claussen, SC, in G. Wayne King, *Rise Up So Early: A History of Florence County, South Carolina* (Spartanburg, SC: The Reprint Company, 1981), 122-123.

51 Dutton, 501; *News and Courier*, September 10, 1886.

52 *News and Courier*, November 17, 1957.

53 Sarah Ciples (Niles) Goodwyn, Columbia, September 13, 1886, to My dear child [Margaret Anna (Goodwyn) Legaré]. Collection of Jeff G. Reid, Jr. Maggie Dunlap could not be identified.

54 Dutton, 497.

55 Accommodations: *News and Courier*, September 10, 1886.

56 Kenneth E. Peters, "Aiken's 'Rock of Ages,' in William S. Brockington, Jr. and Judith T. VanSteenburg, eds., *Historical Sketches on Aiken* (Aiken, SC: Aiken Sesquicentennial Committee, 1985), 51-52.

57 Bobby F. Edmonds, *The Making of McCormick County* (McCormick, SC: Cedar Hill Unlimited, 1999), 371-372.

58 Langley Mill Pond train wreck: *News and Courier*, September 3, 1886; *Atlanta Constitution*, September 2 & 3, 1886; Peters and Herrmann, 98. The names of the South Carolina Railway firemen in the Langley and Horse Creek wrecks vary from report to report, and have not been positively identified. The two black firemen who drowned are identified as Toy and Simmons in the *News and Courier*, September 3, 1886. On September 2, the *Constitution* stated, "the fireman and engineer were both killed." On September 3, the *Constitution* stated that in this wreck, "both the engineer [Reynolds] and the fireman [Ivy] were killed."

59 *Atlanta Constitution*, September 2, 1886.

60 Major J.M. Reilly, Augusta, GA, to Chief of Ordnance, Washington, DC, September 2, 1886. Augusta State University Library, Augusta, GA.

61 *The Pickens Sentinel*, September 9, 1886.

62 *The Pickens Sentinel*, September 9, 1886.

63 Webb Garrison, *A Treasury of Carolina Tales* (Nashville, TN: Rutledge Hill Press, 1996), 62-64.

64 Interview by R.V. Williams with Sam Lewis, Spartanburg, SC, February 28, 1939. South Carolina Writers' Project, W.P.A.

65 Diary of Mary Davis Brown, 1822-1903, York County, SC. Collection of Cathy Michael.

CHAPTER 5

Between Hell and Daylight

1 "The Electric Light, which had disappeared two years ago [1881] from Charleston, has again made its appearance, a number of our citizens having put up a new plant of the Thompson Houston System, which is working very satisfactorily." City *Year Book*, 1886, 123. In 1884, an electric generator was installed to light the State House in Columbia, although the city itself "did not convert from gas to electric street lights until 1887." Walter Edgar, *South Carolina: A History* (Columbia, SC: University of South Carolina Press, 1998), 467. Hereafter cited as Edgar.

2 City *Yearbook*, 1886, 123.

3 Peters and Herrmann, 107.

4 City Directory, 1885-86, 76-77.

5 City *Yearbook*, 1886, 115.

6 City Directory, 1885-86, 77.

7 City *Year Book*, 1886, 109.

8 *Charleston As It Is After The Earthquake Shock of August 31, 1886* (Charleston, SC: no publisher, 1886), 3. Hereafter cited as *Charleston As It Is*.

9 The *News and Courier*, September 9 & 11, 1886, described the runaway fire horses; also *Charleston As It Is*, 3.

10 "*Charleston As It Is*, 52; 1880 Federal Population Census, Charleston, South Carolina, Fourth Ward.

11 Julian L. Selby, *Memorabilia and Anecdotal Reminiscences of Columbia, S.C., and Incidents Connected Therewith* (Columbia, SC: The R. L. Bryan Co., 1905), 111. Hereafter cited as Selby.

12 City *Yearbook*, 1886, 113-114; *News and Courier*, September 1, 1886. The sources used here for determining who owned and occupied the various burned buildings included the 1886 City *Yearbook* and both available city directories. These sources frequently disagreed with each other as to ownership. Those seeking definitive information on these addresses will have to consult the land records at the Register of Mesne Conveyances office.

13 City *Yearbook*, 1886, 113; *News and Courier*, September 1, 1886.

14 *News and Courier*, September 3, 1886, 3; Elizabeth S. Moffett, Charleston, to her daughter, Anna, September 2, 1886. George H. Moffett Family Papers, S.C.H.S.

15 City Directory, 1890. Charleston, SC: Southern Directory and Publishing Co., 1890. The Legaré Street house fire: *News and Courier*, September 1, 1886.

16 *Southern Business Guide*, 1885-86 (Cincinnati, OH: United States Central Publishing Company, 1886), 732. Hereafter cited as *Southern Business Guide*.

17 Morris: 1885-6 City Directory, 26, 192, 198, 201, 213, 220-221; *Southern Business Guide*, 1885-86, 724; *Charleston As It Is*, 58, 60.

18 Andrews: *Southern Business Guide*, 1885-86, 717; City *Yearbook*, 1886, 114.

19 City *Yearbook*, 1886, 114.

20 Karish: 1885-6 City Directory, 179; "Carish": *Southern Business Guide*, 1885-86, 719.

21 Castion: 1885-6 City Directory, 176-177, 241; *Southern Business Guide*, 719.

22 McDevitt: 1885-6 City Directory, 172; *Southern Business Guide*, 1885-86, 724. Dohen: *Southern Business Guide*, 1885-86, 720.

23 The 1885-6 City Directory, 179, shows A. Swalsky, a clothing retailer, at 102 and 130 King; the City *Yearbook* for 1886, 114, shows that Morris owned and occupied it.

24 Sewalskey: *Southern Business Guide*, 1885-86, 728.

25 The addresses listed in the newspaper reports sometimes conflict with city directory listings, as the latter are, in some cases, a year or so out of date. The *News and Courier* names the two tenements as 120 and 129 King Street.

26 City Directory, 1889. Insurance: *News and Courier*, September 3, 1886;

Joiner: the 1885-6 City Directory, 172, shows him at 196 Meeting St., but he evidently moved to King Street just prior to the fire.

27 City Directory, 1889. L.P. Goutvenier: 1885-6 City Directory, 186; *Southern Business Guide*, 1885-86, 721.

28 1888 Sanborn Fire Map.

29 *Southern Business Guide*, 1885-86, 719; H. C. Stockdell, "To the [Insurance] Companies Transacting Business in Charleston, S.C.: Herewith we hand you final report of the Inspectors, together with a 'List of Buildings That Should Come Down.'" Atlanta, GA: no publisher, December 11, 1886. Hereafter cited as Stockdell Report.

30 *Charleston As It Is*, 3; City *Yearbook*, 114.

31 *Charleston As It Is*, 3.

32 McComb: *Southern Business Guide*, 1885-86, 724; 1885-6 City Directory, 187; *City Yearbook*, 1886, 114..

33 Stockdell Report, December 11, 1886.

34 City *Yearbook*, 1886, 113-114. Burnam: *Southern Business Guide*, 1885-86, 719.

35 *Charleston City Directory, 1889.* Charleston, SC: Southern Directory and Publishing Co., 1889. Hereafter cited as City Directory, 1889. I.L. Mintz/Mentz: *News and Courier*, September 1, 1886; 1885-6 City Directory; Bull: 1885-6 City Directory, 184. L.L. Mintz: *News and Courier*, December 10, 1886.

36 *News and Courier*, September 1, 1886. The 1885-6 City Directory lists T. Schadaressi & Bros. fruit store at 125 Meeting, with two other family members and their fruit stores at 51 Church St. and 195 Meeting Street. No King Street addresses are listed for this family.

37 *The Pickens Sentinel*, September 9, 1886, reprinting from the *Columbia Daily Record*.

38 *Charleston As It Is*, 3.

39 Selby, 110.

40 City *Yearbook*, 1886, 107.

41 "A Fearful Night. One Lady's Experience—A Graphic Account of the Earthquake," Bangor, Maine *Daily Whig and Courier*, September 27, 1886.

42 *News and Courier*, September 1, 1886; *Charleston As It Is*, 16; Poston, 420.

43 Undated *News and Courier* clipping, Charleston County Public Library vertical files.

CHAPTER 6
The Survivors

1 *Charleston As It Is*, 3.

2 Walter J. Fraser, *Charleston! Charleston! The History of a Southern City* (Columbia, SC: University of South Carolina Press, 1991), 16. Hereafter cited as Fraser.

3 Fraser, 232.

4 Edgar, 233.

5 *The Charleston Patriot*, June 6, 1835.

6 Poston, 412.

7 Richard N. Côté, *Mary's World: Love, War, and Family Ties in Nineteenth-century Charleston* (Mt. Pleasant, SC: Corinthian Books, 1999), 194, hereafter cited as Côté, *Mary's World*; City *Yearbook*, 1880, 306-307.

8 Frances Butler Simkins and Robert Hilliard Woody, *South Carolina During Reconstruction* (Chapel Hill: University of North Carolina Press, 1932), 10-11.

9 W. Chris Phelps, *The Bombardment of Charleston, 1863-1865* (Gretna, LA: Pelican Publishing Co., 2002), 27. Hereafter cited as Phelps.

10 Phelps, 11.

11 E. Milby Burton, *The Siege of Charleston, 1861-1865* (Columbia, SC: University of South Carolina Press, 1970), 321.

12 Poston, 587 and 601.

13 Mrs. St. Julien Ravenel, *Charleston. The Place and the People* (New York: The Macmillan Co., 1927), 505.

14 Edgar, 58.

15 Edgar, 10, 161.

16 *News and Courier*, August 27, 1985; Edgar, *passim*.

17 City *Yearbook*, 1885, 378.

18 City *Yearbook*, 1885, 388.

19 *Atlanta Constitution*, September 3, 1886.

20 Because of conflicting accounts, there is no way to make an accurate list of the hundreds of aftershocks which were felt in Charleston and the rest of the state in the months after the first (main) shock at 9:51 p.m. on August 31, 1886, nor is there any way to measure their relative magnitude. The reason is simple: a shock felt as "strong" in one part of the city (or Summerville) might be felt weakly or not at all two miles away (as in Mt. Pleasant).

In addition, many of the shocks were not recorded until days or months after they happened, which decreases the precision of the record. In this book, I have generally used the data from the Dutton Report and its supporting sources, supplemented by contemporary personal records. Ultimately, what is important to know is that between 1698 and 1980, South Carolina has recorded at least 2,000 earthquake shocks. After August 31, 1886, the aftershocks were frequent (as many as one or more per day in the first month), continued to cause or aggravate property damage, and kept the entire population on edge and under severe stress well into 1887. They continue to this day, although at reduced frequency.

21 *Charleston As It Is*, 12.

22 Selby, 111-112. He was probably referring to the libation center owned by J.D.E. Meyer, a block down Chalmers Street at its intersection with Church Street. 1885-86 City Directory, 216.

23 Selby, 113.

24 *News and Courier*, September 2, 1957.

25 *Atlanta Constitution*, September 6, 1886.

26 Undated *News and Courier* clipping, 1937, S.C.H.S. vertical file #30-4.

27 *Atlanta Constitution*, September 3, 1886.

28 Undated *News and Courier* clipping, S.C.H.S. vertical file #30-4.

29 Charles R. Rowe, *Pages of History: 200 Years of the Post and Courier* (Charleston, SC: Evening Post Publishing Co., 2003), 71. Hereafter cited as Rowe.

30 Rowe, 73.

31 The "sticks" were composing sticks: narrow, foot-long V-shaped sticks of wood that held the pieces of metal type that made up the lines of text.

32 *Hartford Times* in the *News and Courier*, September 10, 1886.

33 City *Yearbook*, 1886, 356.

34 1885-6 City Directory, 82.

35 Mazÿck and Waddell, 32.

36 City *Yearbook*, 1886, 88-89; 1885-6 City Directory, 187.

37 Stockton, 45.

38 1885-6 City Directory, 82.

39 City *Yearbook*, 1886, 96-97.

40 Mazÿck and Waddell, 9.

41 Stockton, 44.

42 Mazÿck and Waddell, text accompanying illustrations 19 and 20; Stockton, 44.

43 *Charleston As It Is*, 38.

44 Mazÿck and Waddell, 8; Stockton, 45.

45 City *Yearbook*, 1886, 99-101.

46 Potter's field earthquake victim burials: City of Charleston, "Return of Deaths Within the City of Charleston," August-October, 1886. Hereafter cited as Return of Deaths.

47 The former female academy was completed in 1808 by Nathaniel Russell, a prominent New England merchant who came to Charleston in 1765 and "quickly amassed a large fortune." In 1857 the Neoclassical mansion was purchased by Gov. R.F.W. Allston. In 1870 his heirs sold it to the Sisters of Charity of Our Lady of Mercy, who owned it for the next thirty-eight years. The Nathaniel Russell House was purchased in 1955 by the Historic Charleston Foundation, meticulously restored, and is now open to the public as a house museum. Poston, 261-262.

48 Mazÿck and Waddell, text accompanying illustration 53; Stockton, 38.

49 *By-laws of the Orphan House of Charleston, South Carolina, Revised and Adopted by the Board of Commissioners 4 April 1861, Approved by the City Council of Charleston 23 April 1861* (Charleston: Steam-Power Presses of Evans & Cogswell, 1861), *passim*.

50 Father's death: *News and Courier*, September 10, 1886; Wilhelmina's admission: Susan L. King, *History and Records of the Charleston Orphan House*, (Columbia, SC: *South Carolina Magazine of Ancestral Research*, 1994), II:93.

51 City *Yearbook*, 79-80; *News and Courier*, September 1, 1886.

52 Porter, 398.

53 Porter, 398-399. In E. B. Rodgers "Reminiscences of Beaufort Storms," a paper read before the Beaufort County Historical Society, June 1950 (Beaufort County Library - vertical files), the man with the beaver hat is identified as being Mr. James Crofut.

54 *News and Courier*, September 12, 1947.

55 Andrew Robinson and Pradeep Talwani, "Building damage at Charleston, South Carolina, associated with the

1886 earthquake," Bulletin of the Seismological Society of America, 73 (April 1983): 643. Hereafter cited as Robinson and Talwani.

56 City *Yearbook*, 1886, 358.

57 City *Yearbook*, 1886, 358-359.

58 Robert G. Chisolm, Charleston, to the Rev. E. T. Horn, Pennsylvania, September 2, 1886. "The Charleston Earthquake as told in letters from Robert G. Chisolm," Collection #43/190, S.C.H.S.

59 Trenton, New Jersey *Times*, September 2, 1886.

60 *News and Courier*, September 5 & 9, 1886. Reports that Dawson dispatched a mounted courier to follow the roadbed of the South Carolina Railway north until he found an operating telegraph office are probably also true, but the telegraphic report that first reached the newspaper's Columbia bureau was sent over a Southern Telegraph wire directly from Charleston at 9:15 p.m.

61 *The Christian Herald and Signs of Our Times*, September 16, 1886.

CHAPTER 7

The First Rays of Hope

1 Trenton, New Jersey *Times*, September 2, 1886.

2 The Charleston Gas-Light Company's successor, the South Carolina Electric and Gas Company, maintains an office there now. Poston, 402.

3 William B. Guerard: *Southern Business Guide*, 1885-86, 721.

4 Rough Minutes & Correspondence of the Executive Relief Committee and Bound Minutes of the Executive Relief Committee, 1886-1887 (Microfilm #HIS-2004-001, City of Charleston Division of Records Management), September 13, 1886. Hereafter cited as E.R.C. Minutes.

5 E.R.C. minutes, September 14, 1886.

6 E.R.C. minutes, September 6, 1886.

7 E.R.C. minutes, September 6, 1886.

8 E.R.C. minutes, September 8, 1886.

9 E.R.C. minutes, September 8, 1886.

10 E.R.C. minutes, September 8, 1886.

11 *Atlanta Constitution*, September 7, 1886.

12 Augustine Thomas Smythe, Charleston, to My Wife, September 8, 1886. Augustine Thomas Smythe Papers, Collection # 1209.03.02.04, S.C.H.S.

CHAPTER 8

Tent City

1 *Atlanta Constitution*, September 7, 1886.

2 *The Confederate Baptist*, May 4, 1864. Plantation files, Darlington County Historical Commission.

3 Caroline W. Bristow, "Former Summervillian Recalls Quake of 1886," *News & Courier*, August 30, 1959.

4 *Atlanta Constitution*, September 4, 1886.

5 Pickens *Sentinel*, September 18, 1886.

6 *Berkeley Gazette*, September 11, 1886.

7 *Atlanta Constitution*, September 7, 1886.

8 *Atlanta Constitution*, September 7, 1886.

9 Selby, 112.

10 E.R.C. minutes, September 6, 1886.

11 *Atlanta Constitution*, September 7, 1886.

12 E.R.C. minutes, September 6, 1886.

13 E.R.C. minutes, September 6, 1886.

14 *Atlanta Constitution*, September 4, 1886.

15 E.R.C. minutes, September 8, 1886.

16 Arkansas City, Arkansas *Republican*, November 6, 1886.

17 *The Christian Herald and Signs of Our Times*, September 16, 1886, 589.

18 Edward T. Horn, Catasauqua, Pennsylvania, to Robert G. Chisholm, Charleston, September 6, 1886. The Charleston Earthquake, as told in letters from Robert Chisholm, ca. 1890. Collection #43/0190, S.C.H.S.

19 *Atlanta Constitution*, September 7, 1886.

20 *Atlanta Constitution*, September 8, 1886.

21 *Atlanta Constitution*, September 11, 1886.

22 E.R.C. minutes, September 9, 1886.

23 *Atlanta Constitution*, September 6, 1886.

24 Return of Deaths, September 1, 1886.

25 Heslin: *Southern Business Guide*, 1885-86, 721; *News and Courier*, September 8, 1886.

26 Tamsberg: *Southern Business Guide*, 1885-86, 730; *News and Courier*, September 8, 1886.

27 U.S.S. *Wistaria*: online at http://www.uscg.mil/hq/g-cp/history/WEBCUTTERS/Wistaria_1882.html April 22, 2006.

28 Selby, 113.

29 Selby, 114.

30 E.R.C. minutes September 14, 1886.

31 E.R.C. minutes, September 14 to November 27, 1886.

32 Diary of an unknown Charleston resident (fragment), August 31-September 16, 1886. Ficken-Ball miscellany, 1782-1933. Collection #0225.00, S.C.H.S.

CHAPTER 9

Trouble in the Streets

1 William James, *The Varieties of Religious Experience* (New York: New American Library, 1958), 24.

2 From 1882 to 1895, Mt. Pleasant was the county seat of Berkeley County.

3 *Berkeley Gazette*, September 11, 1886.

4 Porter, 395-397.

5 Porter, 395-397.

6 Porter, 395-397.

7 Porter, 395-397.

8 Porter, 395-397.

9 Stockton, 45.

10 Robbery: *News and Courier*, September 6, 1886; Ortmann's saloon: *City Directory*, 1885-1886, 216; *Southern Business Guide*, 1885-1886, 727.

CHAPTER 10

The March of Science

1 Mary C. Rabbitt for the United States Geological Survey, *Minerals, Lands, and Geology for the Common Defence and General Welfare, II (1879-1904)*, United States Government Printing Office, 1980: 100. Hereafter cited as Rabitt.

2 Rabbitt, 98.

3 Online at http://www.powellmuseum.org/MajorPowell.html February 20, 2006.

4 Online at http://3dparks.wr.U.S.G.S..gov/3Dcanyons/html/hillers.htm on February 20, 2006.

5 Rabbitt, 78.

6 Rabbitt, 33.

7 Rabbitt, 52.

8 Peters and Herrmann, 13.

9 Peters and Herrmann, 14.

10 *Atlanta Constitution*, September 6, 1886.

11 Don D. Fowler, *The Western Photographs of John K. Hillers: Myself in the Water* (Washington, D.C.: Smithsonian Institution, 1989), 15. Hereafter cited as Fowler.

12 A number of professional photographers, plus newspaper and magazine sketch artists from the U.S. and abroad, created over 300 images of the earthquake damage in the Charleston-Summerville area in 1886. McGee's complaint about photographic equipment shortages is consistent with the use of photos in the Dutton Report made by other photographers. Fowler, 160.

13 Charles E. Dutton and Edward Everett Hayden, "Abstract of the Results of the Investigation of the Charleston Earthquake," *Science* [Supplement], May 20, 1887. Hereafter cited as Dutton and Hayden.

14 *A Plan of Charles Towne From a Survey of Edwd. Crisp, in 1704*. Map Division, Library of Congress.

15 *The Ichnography of Charles-Town at High Water*. London, 1739. Map Division, Library of Congress.

16 Peters and Herrmann, 1-3.

17 "Earthquake Phenomena," *The Manufacturer and Builder*, 18 (January 1886): 13.

18 *New York Star*, reprinted in the *News and Courier*, September 15, 1886. Mother Carey's Chickens is the sailors' nickname for the storm petrel seabird. "Mother Carey" is a corruption of *Mater Cara*, one of the epithets for Mary, the mother of Christ, used by the Spanish and Portuguese sailors....

19 *Atlanta Constitution*, August 29, 1886.

20 *News and Courier*, September 14, 1886, reprinting a report from the *New York World*.

21 *Berkeley Gazette*, September 11, 1886.

22 *The Illustrated London News*, September 11, 1886.

23 *Atlanta Constitution*, September 28, 1886.

24 *Berkeley Gazette*, September 11, 1886.

25 *News and Courier*, September 20, 1886; Matthew 24:7 (KJV).

26 *The Christian Herald and Signs of Our Times*, September 16, 1886.

27 Wiggins: *Atlanta Constitution*, September 22, 23, and 25, 1886; *News and Courier*, September 30, 1886; and *Arkansas City Republican*, Saturday, October 2, 1886.

28 *Atlanta Constitution*, September 4, 1886.

29 *Atlanta Constitution*, September 4, 1886.

30 *Atlanta Constitution*, September 4, 1886.

31 McGee entered South Carolina on the Northeastern Railroad, which

connected with the Seaboard Coast Line at Florence, S. C. After his arrival in Charleston, he made his preliminary notes on the geological effects along the route of the South Carolina Railway tracks between Charleston and Summerville on two rail trips to that place: on September 3-4, when he spent the night in Summerville, and again on September 5, when he traveled there and back the same day. After the departure of McGee and Mendenhall, Charlestonian Earl Sloan retraced their routes, expanded the search areas, examined the area west of Charleston along the Charleston and Savannah Railroad, made more precise measurements, and recorded more detailed observations. The surviving field notes of McGee and Sloan are published in Peters and Herrmann; the summary, with professionally drawn illustrations, is published in Dutton. The account presented here is a composite of what the team found, with sources for significant details individually cited.

32 The Dutton report sometimes contradicts itself or gives conflicting versions of the same event. In the case of the South Carolina Railway warehouses, p. 257 says they moved ten feet, and on p. 269 says *a* warehouse (probably referring to one of the former) was moved eight feet nine inches. This is not the result of sloppy investigative work but rather reflects the difference between the initial damage survey notes and the detailed examinations that followed.

33 Peters and Herrmann, 18.

34 Dutton, 283.

35 Peters and Herrmann, 18.

36 Dutton, 283.

37 Dutton, 283.

38 Dutton, 283.

39 Peters and Herrmann, 24.

40 Peters and Herrmann, 23.

41 Dutton, 284.

42 Peters and Herrmann, 22.

43 Peters and Herrmann, 18.

44 Dutton, 285.

45 Dutton, 286, 294.

46 Dutton, 286.

47 Peters and Herrmann, 19.

CHAPTER 11
Ground Zero

1 Peters and Herrmann, 19.

2 Dutton, 274.

3 Dutton, 274.

4 Peters and Herrmann, 19.

5 Unless otherwise noted, all of McGee's data is from the transcription of his field notes in Peters and Herrmann, 6-38.

6 Dutton, 277.

7 The *Columbia Daily Record*, September 3, 1886.

8 *Beth's Pineland Village*, 32; Elizabeth Ann Poyas, *Days of Yore, or, Shadows of the Past* (Charleston, SC: W. G. Mazÿck, 1870), 6.

9 Dutton, 274.

10 Based on McGee's field notes, the *News and Courier's* September 5 edition was in error when it stated that McGee "was the guest of Col. Gregg" for the night.

11 Peters and Herrmann, 20.

12 Peters and Herrmann, 18-20.

13 *Beth's Pineland Village*, 32.

14 Dutton, 274-275.

15 Peters and Herrmann, 19.

16 The tower of the parish church of St. George, Dorchester, was partially restored in the 1960s and extensive archaeological investigations were carried out in the village in the 1990s. Dorchester village is now open to the public as Colonial Dorchester State Historical Site.

17 Pradeep Talwani, *Macroseismic Effects of the 1886 Charleston Earthquake* (Columbia, SC: Department of Geological Sciences, University of South Carolina, 2004), 3-4.

18 Dutton, 278.

19 Fractures at 84 King Street: Peters and Herrmann, 11.

20 Ebaugh is identified as the informant in the *News and Courier*, September 5, 1886.

21 *Atlanta Constitution*, September 5, 1886.

22 *Atlanta Constitution*, September 5, 1886.

23 Ebaugh's implication that Summerville was home to 3,000 residents was based on a guess, as just under 2,000 would probably have been the maximum population at the time.

24 *Atlanta Constitution*, September 6, 1886.

25 *Atlanta Constitution*, September 6, 1886.

CHAPTER 12

In Search of the Holy Grail

1 Peters and Herrmann, 40.

2 Hemphill, *Men of Mark in South Carolina*, III:389. Hereafter cited as *Men of Mark.*

3 *Men of Mark*, III: 389-390.

4 Sloan's biographical information: Peters and Herrmann, 40-43; *Men of Mark*, III:388-390; Hardy Crawford Geddings, ed., *Who's Who in South Carolina* (Columbia, SC: McCaw, 1921), 175; Yates Snowden, *History of South Carolina* (New York: Lewis Publishing Co., 1920), III:72; and T. W. Vaughan, "Memorial of Earle Sloan," *Bulletin of the Geological Society of America*, 40:57-61.

5 *Men of Mark*, III:390.

6 Peters and Herrmann, 47.

7 Peters and Herrmann, 50.

8 Dutton, 256-257.

9 Earle Sloan, Charleston, to William J. McGee, Washington, October 19, 1886, in Peters and Herrmann, 46.

10 Peters and Herrmann, 11.

11 Peters and Herrmann, 49.

12 Peters and Herrmann, 49.

13 Peters and Herrmann, 49.

14 *Atlanta Constitution*, September 14, 1886.

15 Dutton, 304-305.

16 Peters and Herrmann, 53.

17 Dutton, 210.

CHAPTER 13

The Measure of the Man

1 Julia Courtenay Campbell, *The Courtenay Family: Some Branches in America* (Charlottesville, Virginia: privately printed, 1964), ii. Hereafter cited as *The Courtenay Family*.

2 "The Courtenays of Ireland," in The Newry (S.C.) *Reporter*, December 3, 1907. His mother: 1880 Federal Population Census.

3 Courtenay's obituary, *News and Courier*, March 18, 1909.

4 *The Courtenay Family*, 21.

5 *The Courtenay Family*, 34.

6 1860 Federal Population Census. Slave Schedules.

7 Courtenay's military service information: *Combined Military Service Records of Confederate Soldiers Who Served in Organizations from South Carolina*, and from Randolph W. Kirkland, Jr., *Dark Hours: South Carolina Soldiers, Sailors, and Citizens who were held in Federal Prisons during the War for Southern Independence, 1861-1865* (Charleston, SC: South Carolina Historical Society, 2002).

8 Courtenay's obituary, *News and Courier*, March 18, 1909.

9 The details of this feud and aborted duel—many of which conflict with Courtenay's military service records—may be found in "Correspondence," a fifteen-page compilation of letters privately printed for Courtenay's friends shortly after September 8, 1865. *Confederate Imprints, 1861-1865*, Reel 83, no. 2564, page 4. New Haven, CT: Research Publications, 1974. Hereafter cited as Courtenay, "Correspondence."

10 Richard B. McCaslin, *Portraits of Conflict: A Photographic History of South Carolina in the Civil War* (Fayetteville, AR: The University of Arkansas Press, 1994), 223-224, 308. Denied access to reading material: courtesy of Virginia Mescher, "Pick Up a Good Book: Literacy and Popular Literature of the Civil War." Archived online in *Virginia's Veranda*, February 2005 at www.raggedsoldier.com, 2006.

11 The *Charleston Courier*, March 30, 1865.

12 Courtenay, "Correspondence," 14.

13 *The Courtenay Family*, 34.

14 *The Courtenay Family*, 31.

15 Courtenay's obituary, *News and Courier*, March 18, 1909.

16 95 Ashley Avenue: Mazÿck and Waddell, text accompanying illustration #71; Poston, 487-488.

17 Chamber of Commerce service dates courtesy of Marge Grove, Senior Vice President, Charleston Chamber of Commerce, to the author, June 29, 2005.

18 Courtenay's obituary, *News and Courier*, March 18, 1909.

19 1880 Federal Population Census.

20 William Way, *The History of Grace Church, Charleston, South Carolina* (Charleston, SC: Grace Church, 1948), 165.

21 City Council minutes, December 19, 1883.

22 "In Memoriam," City *Yearbook*, 1908, 365-367.

23 Mazÿck and Waddell, opp. plate 30.

24 "In Memoriam," City *Yearbook*, 1908, 365-367.

25 Theo. G., Wm. P. Clyde & Co., New York, to William A. Courtenay, September 4, 1886. William A. Courtenay Collection, Charleston Library Society.

26 *The Christian Herald and Signs of Our Times*, New York, September 16, 1886, 578.

27 Telegram, Francis W. Dawson to William A. Courtenay, c/o W.P. Clyde, Broadway, September 3 (overwritten as September 4), 1886. William A. Courtenay Collection, Charleston Library Society.

28 City Council Minutes, September 14, 1886.

29 Sprunt, 538.

30 Julian M. Bacot diary, September 7, 1886. Bacot-Huger Collection, #11/57/4, S.C.H.S.

CHAPTER 14
Courtenay Takes Command

1 Terry & Co. were listed at 9 Market Street in *The Charleston Guide and Business Directory For 1885-6*, 342. Their change in address may have taken place before or as a result of the earthquake.

2 City *Yearbook*, 1886, 111, 114. A false alarm was received on September 11, 1886, showing that the fire alarm telegraph system was again operational.

3 Peters and Herrmann, 107.

4 *News and Courier*, September 6, 1886; *Beesley's Illustrated Guide*, 6.

5 Beasley and Lockwood: *News and Courier*, September 14, 1886. The *News and Courier*, September 9, 1886, stated that the clock stopped at 9:54 p.m., but the author believes this to be in error. The height of St. Michael's steeple was 186' 3" before the earthquake. The tower settled eight inches after the quake, reducing the height above street level 185'7": *Beesley's Illustrated Guide*, 5.

6 *Atlanta Constitution*, September 4, 1886.

7 *Atlanta Constitution*, September 3, 1886.

8 Mazÿck and Waddell, xii.

9 Edward T. Horn, Charleston, to Dr. B. M. Schmucker, Pottstown, Pa., September 7, 1886. Chisolm, Robert George, 1831-1907. The Charleston earthquake: as told in letters from Robert Chisolm, ca. 1890. Collection #43/190, S.C.H.S.

10 *Atlanta Constitution*, September 5, 1886.

11 *Frank Leslie's Illustrated Newspaper*, September 18, 1886. The name of the recipient of this letter is not stated, but a telegraph to Andrew Simonds with essentially the same content was published in New York on September 9 and reprinted in the *News and Courier* on September 10.

CHAPTER 15
Finding the Money

1 The biographical information about Dawson has been drawn from *Reminiscences of Confederate Service, 1861-1865, by Francis W. Dawson* (Charleston, SC, News and Courier Book Press, 1882; reprinted with an introduction by Bell I. Wiley, Baton Rouge, LA: Louisiana State University Press, 1980); The Augusta *Chronicle, The Charleston Earthquake. 1886. F. W. Dawson* (articles originally reproduced in the Augusta *Chronicle* from various other newspapers, 1886); P. McNeely, "Francis W. Dawson: Dueling with the Code of Honor," a paper presented at the Symposium on the 19th Century Press, the Civil War, and Free Expression, University of Tennessee at Chattanooga, 1999; and E. Culpepper Clark, *Francis Warrington Dawson and the Politics of Restoration: South Carolina, 1874-1889* (Tuscaloosa, AL: University of Alabama Press, 1980); the records of the C.S.S. *Beaufort*, online at http://cssvirginia.org/vacsn3/crew/beaufort, April 29, 2005; and Rowe, *Pages of History*.

2 *Atlanta Constitution*, July 7, 1886.

3 *Atlanta Constitution*, November 18, 1885.

4 Atlanta *Constitution*, July 6, 1886.

5 Kenneth E. Peters, "Disaster Relief Efforts Connected with the 1886 Charleston Earthquake. A paper presented for the Fourth Citadel Conference for the South, April 13, 1985," 5. Hereafter cited as Peters, Disaster Relief.

6 Springfield, Massachusetts *Republican* in the *News and Courier*, September 10, 1886.

7 *News and Courier*, September 4, 1886; Columbia *Daily Register*, September 4, 1886.

8 Peters, Disaster Relief, 6.

9 Peters, Disaster Relief, 6.

10 City council minutes, September 14, 1886.

11 City *Yearbook*, 1886, 90-91.

12 Refugee exodus estimates: *The Illustrated London News*, September 11, 1886, 276.

13 *Atlanta Constitution*, September 6, 1886.

14 City council minutes, September 14, 1886.

15 *Charleston As It Is*, 4-5.

16 Dutton, 414-415.

17 E.R.C. minutes, November 24, 1886.

18 Dutton, 470.

19 1886 Earthquake. Miscellaneous correspondence, Charleston County Public Library.

20 City of Charleston. Executive Relief Committee. *The Earthquake, 1886. Exhibits showing the receipts and disbursements, and the applications for relief, with the awards and refusals of the Executive Relief Committee in over 2,000 cases of house owners and 317 cases of application for losses in personal property*. Charleston, SC: Lucas, Richardson & Co., 1887. Hereafter as E.R.C. Final Report.

21 Baltimore *City Paper*, July 19, 2000.

22 Dutton, 453.

23 Dutton, 453.

24 E.R.C. Final Report.

25 Dutton, 332n1.

26 E.R.C. Final Report.

27 Pedro Pablo Rodríguez, "José Marti and the Charleston Earthquake," in CubaNow.net, online May 10, 2006 at http://www.cubanow.net/global/loader.php?secc=5&cont=stories/num00/terr_Charleston.htm; also José Marti, "El Terremoto de Charleston," in Enrique Anderson Imbert and Eugenio Florit, eds., *Literatura Hispanoamericana* (New York: Holt, Rinehart and Winston, 1970), II:46-52.

28 T. T. Curnick, Uxbridge, Mass., to William A Courtenay, September 20, 1886. 1886 Earthquake. Miscellaneous correspondence. Charleston County Public Library.

29 Eva Bates, Danbury, CT, to the Mayor of Charleston, September 27, 1886. 1886 Earthquake. Miscellaneous correspondence. Charleston County Public Library.

30 Joseph H. Oppenheim, Savannah, Ga., to William A. Courtenay, September 17, 1886. 1886 Earthquake. Miscellaneous correspondence. Charleston County Public Library.

31 Edward E. Bacon, Saccarappa, Me., to Mayor Courtenay, October 4, 1886. 1886 Earthquake. Miscellaneous correspondence. Charleston County Public Library.

32 George F. Trostel, Buffalo, New York, to William A. Courtenay, September 26, 1886. Joseph H. Oppenheim, Savannah, Ga., to William A. Courtenay, September 17, 1886. 1886 Earthquake. Miscellaneous correspondence. Charleston County Public Library.

33 C.E. Goodrich, Nyack, New York, to William A. Courtenay, September 3, 1886. 1886 Earthquake. Miscellaneous correspondence. Charleston County Public Library.

34 *News and Courier*, September 15, 1886; *Atlanta Constitution*, September 14, 1886.

35 *Atlanta Constitution*, September 4, 1886.

36 *News and Courier*, September 15 & 22, 1886; Richard C. Madden, *Catholics in South Carolina: a History* (Lanham, MD: University Press of America, 1985), 142.

37 Porter, 395.

38 The Rev. Phil Bryant, pastor, French Huguenot Church, in Joyce Bagwell, *Low Country Quake Tales* (Easley, SC: Southern Historical Press, 1986), 47; Peters, *Disaster Relief Efforts*, 15.

39 Sol Breibart, "Charleston Jews and the earthquake of 1886," *Charleston Jewish Journal*, March 1994, 18. Hereafter cited as Breibart.

40 Peters, *Disaster Relief Efforts*, 14.

41 O.G. Marjenhoff, Charleston, to William A. Courtenay, September 12, 1886. 1886 Earthquake. Miscellaneous correspondence. Charleston County Public Library.

42 Peters, *Disaster Relief Efforts*, 14.

43 Mitchell (South Dakota) *Daily Republican*, October 19, 1886.

44 Report of the Special Committee appointed by the Chamber of Commerce of the State of New York to obtain relief for the sufferers by the earthquake at Charleston, SC. (New York: Press of the Chamber of Commerce, 1886), 11.

45 E.R.C. minutes, October 1, 1886.

46 Except as noted, the biographical information for the Rev. William Henry Harrison Heard is drawn chiefly from his autobiography, *William H. Heard, From Slavery to the Bishopric in the A. M. E.*

Church (Philadelphia, PA: A. M. E. Book Concern, 1928). The name of Heard's mother is spelled both as Pathenia and Parthenia in the book.

47 *Atlanta Constitution*, September 8, 1886.

48 Cleveland *Gazette*, September 18, 1886.

49 Cleveland *Gazette*, September 25, 1886.

50 Cleveland *Gazette*, September 25, 1886.

51 Peters, 15.

52 City Council minutes, October 1, 1886.

CHAPTER 16

Bricks, Mortar, and Earthquake Bolts

1 Peters, 7.

2 Andrew Robinson and Pradeep Talwani, "Building Damage at Charleston, South Carolina, Associated with the 1886 Earthquake," *Bulletin of the Seismological Society of America*, 73 (April 1983): 646. Hereafter cited as Robinson and Talwani.

3 Robinson and Talwani, 633.

4 Robinson and Talwani, 638.

5 Robinson and Talwani, 634.

6 H. C. Stockdell, Record of Earthquake Damages (Atlanta, GA, Winham & Lester Publishers, 1886), structure #2722. The author used the Excel spreadsheet version developed by the College of Charleston's Department of Geology and Environmental Geosciences; also Côté, *Mary's World*, 354-355.

7 Dutton, 230.

8 Peters and Herrmann, 88.

9 Peters and Herrmann, 88.

10 Peters and Herrmann, 88.

11 Robinson and Talwani, 647.

12 Brooklyn *Daily Eagle*, April 1, 1885; also Jacob A. Riis, *The Battle with the Slum* (New York: The Macmillan Co., 1902), 14-15. The name was also spelled Buddensieck.

13 Peters and Herrmann, 88.

14 Robinson and Talwani, 646.

15 Stockton, 15.

16 E.R.C. minutes, September 13, 1886.

17 City Council minutes, September 14, 1886.

18 Peters and Herrmann, 113.

19 The Summerville Committee. *Report of the Committee for the Relief of the Sufferers by the Earthquakes in Summerville and Vicinity* (Charleston, SC: The News and Courier Book Presses, 1886), *passim.*

20 *Berkeley Gazette*, September 11, 1886.

21 E.R.C. minutes, September 18, 1886.

22 Online May 22, 2006 at http://www.chezjag.net/badmunster/badmunstNavbar.htm

23 National Register of Historic Places Inventory-Nomination Form for Mulberry Plantation, Berkeley County, South Carolina. The property was listed in the National Register October 15, 1966.

24 Poston, 90; also Taylor Shelby, director, The Powder Magazine Museum, to the author, June 3, 2006.

25 Online February 2, 2006 at http://www.preservationdirectory.com/historicproperties_wheatlands.html.

26 H. Kittridge, Cor. Barron and Washington Streets, New York, to William A. Courtenay, September 6, 1886, *News and Courier*, September 9, 1886.

27 Minutes of the Seventh Annual Stockholders Meeting of the City of Charleston Water Works Company, February 24, 1887.

28 John L. All, historian, Mt. Pleasant Presbyterian Church, to the author, May 20, 2006.

29 Jack Simmons, M.D., Charleston, to the author, May 21, 2006.

30 Diary of Julian M. Bacot, September 24, 1886. Bacot-Huger Collection, S.C.H.S.

31 Augustine T. Smythe, Charleston, to my dear wife, September 6, 1886. Augustine T. Smythe papers, S.C.H.S.

32 Augustine T. Smythe, Charleston, to my dear Wenna (daughter Hannah), September 11, 1886. Augustine T. Smythe papers, S.C.H.S.

33 City Council minutes, September 14, 1886.

CHAPTER 17

Success and Celebration

1 City *Yearbook*, 1886, 430-433.

2 Stover and Coffman, 350.

3 Stover and Coffman, 350.

4 City Council minutes, December 14, 1886.

5 Minutes of the Seventh Annual Stockholders Meeting of the City of Charleston Water Works Company, February 24, 1887. Charleston Commissioners of Public Works.

6 Thomas Ravenel Porcher papers, S.C.H.S.

7 New York *Times*, November 1, 1887.

8 City Council minutes, December 12, 1887.

CHAPTER 18

How the Grail Was Found

1 Dutton and Hayden, 489-501.

2 Dutton and Hayden, 490

3 City *Yearbook* , 1886, 408.

4 *Palmetto Post*, September 9, 1886.

5 Dutton and Hayden, 490.

6 Dutton and Hayden, 497.

7 Dutton and Hayden, 497.

8 Dutton and Hayden, 501.

9 Dutton, 390.

10 Dutton, 319-320.

11 Dutton, 211.

12 George D. Louderback, "The Personal Record of Ada M. Trotter of Certain Aftershocks of the Charleston Earthquake of 1886," *Bulletin of the Seismological Society of America*, 34 (1944): 206. Hereafter cited as Louderback.

13 Online at http://www.arlingtoncemetery.net/anitanew.htm June 8, 2006.

14 Rowe, 72-73.

15 Gail Moore Morrison, "I Shall Not Pass This Way Again: The Contributions of William Ashmead Courtenay." Online at www.webcom.com/scourt/thisway.htm on January 30, 2006. Hereafter cited as Morrison.

16 Morrison.

17 Morrison; *News and Courier*, March 18, 1909; and "In Memoriam," Charleston *Yearbook*, 1908, 365-367.

18 Taber, 11.

19 For non-geologists, *The Handy Geology Book*, by Patricia Barnes-Svarney and Thomas E. Svarney, provides a reliable guide to the basic concepts of geology and seismology.

20 Douglas W. Rankin, ed., *Studies Related to the Charleston, South Carolina Earthquake of 1886 – A Preliminary Report*. Geological Survey Professional Paper 1028.(Washington, DC: G.P.O., 1977), 5. Hereafter cited as Rankin.

21 The 1912-1983 seismicity reports: Stover and Coffman, 351-354.

22 Rankin, 1.

23 The MPSSZ was named by geologists Arthur C. Tarr, Pradeep Talwani, Susan Rhea, and David Amick.

24 Pradeep Talwani, interview with the author, February 6, 2006.

25 Rankin, 1.

26 Rankin, 1.

27 Rankin, 13.

28 Rankin, 1.

29 Pradeep Talwani, "Internally consistent patterns of seismicity near Charleston, South Carolina," *Geology* 10:654.

30 *Proceedings of Conference XX: A Workshop on The 1886 Charleston, South Carolina Earthquake and Its Implications for the Future* (Reston, Virginia: USGS Open Report 83-843, 1983), 200. Hereafter cited as USGS Open Report 83-843.

31 USGS Open Report 83-843, 25.

32 USGS Open Report 83-843, 26.

33 USGS Open Report 83-843, 142; *Atlanta Constitution*, September 26, 1886.

34 USGS Open Report 83-843, 239.

35 USGS Open Report 83-843, 49.

36 USGS Open Report 83-843, 37.

37 Charles Lindbergh, ed., *Earthquake Hazards, Risk, and Mitigation in South Carolina and the Southeastern United States* (Charleston, S.C.: The South Carolina Seismic Safety Consortium, 1986).

38 Pradeep Talwani to the author, June 15, 2006.

39 Pradeep Talwani and William T. Schaeffer, "Recurrence rates of large earthquakes in the South Carolina Coastal Plain based on paleoliquefaction data" *Journal of Geophysical Research*, 106:6621.

40 Interviews with Pradeep Talwani by the author, February 6, and June 15, 2006.

CHAPTER 20

The Next Big One

1 Robert Mills, *Statistics of South Carolina* (Charleston, SC: Hurlbut and Lloyd, 1826), 64.

2 Nuttli, 1.

3 Otto Nuttli, cited in Rankin, 1.

4 Carl H. Simmons, Director, Department of Building Inspection, County of Charleston, to the author, June 14, 2006. Hereafter cited as Simmons interview.

5 Simmons interview.

6 Pradeep Talwani, ed., South Carolina Seismic Network Bulletin(s), 2001-2004. Online April 19, 2006 at http://scsn.seis.sc.edu/bulletin.

CHAPTER 21
In Memoriam

1 The "Record of Deaths Within the City of Charleston" is preserved in the South Carolina Room of the Charleston County Library, Charleston. Deaths from the earthquake in August, September, October, and November were clearly marked by the city officials and kept separate from those that resulted from natural causes.

2 "Earthquake Hazard Reduction in the Central and Eastern United States: A Time for Examination and Action," *Monograph 5. Socioeconomic Impacts. 1993 National Earthquake Conference, May 3-5, 1993.* Memphis, Tennessee: Central United States Earthquake Consortium, 1993, 20-21. The section of the monograph that named U. S. earthquakes from the colonial period to present with high death tolls listed San Francisco, California, 1906 (700 deaths, now known to be over 4,000); Unimake Island, Alaska, 1946 (173 deaths); Prince William Sound, Alaska, 1964 (131 deaths); and Mona Passage, Puerto Rico, 1918 (116 deaths). It ignored altogether the Charleston, South Carolina earthquake of 1886 (83 deaths known as of 1993; 125 known as of this publication); then moved on to note 64 deaths in the 1971 San Fernando, California, earthquake and 67 in the 1989 Loma Prieta, California, earthquake.

Illustration Sources

Over the last decade, I have assembled a large collection of images of Charleston and the South Carolina Lowcountry. Images made by me, taken from my own collection, or donated to me as scans are cited as RNC. If a shortened version of the title of a publication is shown, the full title is provided in the bibliography. In some cases, it was necessary to request formal permission to publish images. This permission was graciously granted, and citations appear as requested. Abbreviations used are shown below.

CAII: *Charleston As It Is After The Earthquake Shock of August 31, 1886*, courtesy Mary Julia Royall; CATE: *Charleston After The Earthquake, August 31st, 1886;* CCPL: Charleston County Public Library Special Collections; CLS: The Charleston Library Society, Charleston, SC; CofC: Special Collections, College of Charleston, SC; Dutton: Clarence E. Dutton, ed., *The Charleston Earthquake of August 31, 1886* (USGS); FL: *Frank Leslie's Illustrated Newspaper;* GMA/CAA: Gibbes Museum of Art / Carolina Art Association, Charleston, SC; HCF: Historic Charleston Foundation, Charleston, SC; HW: *Harper's Weekly;* ILN: *Illustrated London News;* LC: The Library of Congress; LG: *The London Graphic;* MPF: Middleton Place Foundation, Inc., Charleston, SC; PCA: The Post and Courier archives, Charleston, SC; SLU: Saint Louis University Department of Earth and Atmospheric Sciences, Earthquake Center, St. Louis, MO; SM: *Scribner's Monthly;* USGS: U. S. Geological Survey Photo Library, Denver, CO.

Front cover: Meeting Street: HW, September 11, 1886; William A. Courtenay, RNC; front flap: Hibernian Hall, RNC; back cover: a dah and her charges, RNC; St. Michael's Church, RNC; William M. Bird & Co., USGS; back flap author photograph: Ron Anton Rocz, Charleston.

Preface, xiv: A Glimpse of Charleston, ILN, September 11, 1886. Text page 6: Houses on South Battery: SM, June 1874; 10: The Simmons-Edwards House: SM, June 1874; 11: William A. Courtenay, RNC; 12: R.M.S. *Etruria* at Liverpool, Arthur and Claire Blundell, Bishops Stortford, Hertfordshire, England; 17: Carlyle McKinley, PCA; 19: The *News and Courier* building, HW, September 11, 1886; 21: W. W. Smith's, HW, September 11, 1886; 23: St. Michael's Church, Dutton, plate X; 25: Hibernian Society Hall, RNC; 26: The Charleston Hotel, HW, September 11, 1886; 27: John and Mary Robinson, collection of Jeanne S. Robinson; 28: Charleston eagles, HW, September 11, 1886; 29: Steitz's fruit store, HW, September 11, 1886; 31: Falling chimneys, *L'Illustrazione Italiana*, Rome, September 18, 1886; 32: Susan Pringle, Miles Brewton House Collection; 34: Main Guard House, USGS; 50: northbound derailment, USGS; 52: rails bent, Dutton, fig. 19, p. 285; 53: repairing the track, HW, September 11, 1886; 59: isoseismal map: Stover and Coffman, *Seismicity of the United States, 1568-1989* (Revised), USGS; 70: fissure near the Ashley River: Dutton, plate XXIII; 72: burned ruins, MPF; 74: displaced tracks, Dutton, plate XXV; 76: Beaufort Arsenal, RNC; 80: Sarah Goodwyn, collection of Jeff G. Reid, Jr.; 82: train wreck at Langley Mill Pond, SLU; 84: concealment shoe: courtesy Dr. Christopher Murphy, Augusta State University, Augusta, Georgia; 87: Mary Davis Brown, collection of Cathy Michael; 91: volunteer firemen, LC; 91: fire insurance plaque, RNC; 94: Francis L. O'Neill, CAII; 96: the burnt district, FL, September 18, 1886; 108: the dead and the wounded, *L'Illustrazione Italiana*, Rome, September 18, 1886; 112: terrified survivors, HW, September 11, 1886; 116: Meeting Street, LC; 117: John Rutledge House, LC; 119: aftermath of the Great Cyclone, RNC; 125: The Front Wall Gone, HW, September 11, 1886; 128: Charleston City Hospital, USGS; 132: Roper Hospital, Dutton, plate XII; 133: The Medical College, CATE; 135: Charleston Orphan House, LC; 148: Camping out on King Street, LG, September 25, 1886; 154: waiting for food, HW, September 18, 1886; 164: distributing the tents, HW, September 18, 1886; 165: ruins of the Cathedral: LC; 166: tents at Wragg Square, RNC; 167: a dah and her charges, RNC; 168: camping in misery, HW, September 11, 1886; 169: tents on Rutledge Avenue, RNC; 174:

wooden shelters in Marion Square, FL, September 18, 1886; 175: sheds and tents in Marion Square, CAII; 177: camp at Washington Park, USGS; 179, camp at Artesian Park, CAII; 186: a black policeman, SM, June 1874; 190: a prayer meeting, HW, September 18, 1886; 191: The Rev. Anthony Toomer Porter, Anthony Toomer Porter, *Led On! Step by Step*; 196: an outdoor worship service: HW, September 18, 1886; 198: the Charleston City Jail, USGS; 210: William John McGee, *The National Cyclopedia of American Biography*; 214: Thomas Corwin Mendenhall, National Academy of Sciences; 214: guests flee the Charleston Hotel, FL, September 18, 1886; 217: John Karl "Jack" Hillers, USGS; 220: Plan of Charles Towne in 1704: LC; 221: Creeks and marshes in Charleston, GMA/CAA; 224: The Last Stand, *Puck, 1899*, RNC; 236: a large craterlet, USGS; 237: sinkhole under a house, USGS; 243: Mr. Brown's hotel, RNC; 244: Gen. John C. Minott's house, USGS; 247: Vose's inn, RNC; 247: St. Paul's Episcopal Church, RNC; 248: Dr. Smith's tombstone, RNC; 249: drawing of Smith's tombstone, USGS, in Peters and Herrmann, *First-Hand Observations of The Charleston Earthquake*; 249: tower of St. George's, Dorchester, Dutton, plate XXII; 250: Fort Dorchester: RNC; 251: house at Lincolnville: USGS; 252: McGee at First (Scots) Presbyterian Church, USGS; 259: Earle Sloan, *Men of Mark in South Carolina*; 263: fallen chimneys, Dutton, fig. 26, p. 299; 265: a craterlet, USGS; 268: Church of St. James, Goose Creek, USGS; 270: Sloan's field map, City of Charleston *Yearbook*, 1886; 271: Sloan's 1886 isoseismal map, Dutton, plate XXVI; 275: *Etruria's* passenger list, Paul K. Gjenvick, director, Gjenvick-Gjønvik Archives; 281: William A. Courtenay, RNC; 284: pilot boat greets the Etruria, RNC; 295: stabilizing St. Michael's Church, RNC; 301: the constant aftershocks, FL, September 18, 1886; 313: The U.S. Post Office, RNC; 318: Francis W. Dawson, PCA; 328: charity cyclone, FL, September 11, 1886; 329: Clara Barton, LC; 332: earthquake tourist booklet, RNC; 334: Earthquake Beer advertisement, CAII; 335: "For the Charleston Sufferers, CCPL; 346: The Rev. William F. H. Heard, *From Slavery to the Bishopric in the A.M.E. Church. An Autobiography;* 355: William Bird & Co., USGS; 356: the earthquake spared some, RNC; 361: inside an earthquake-shattered house, CLS; 362: Gabriel Manigault, M.D.: CofC; 363: bricks laid in Flemish bond, RNC; 367: aftermath of "Buddensiek mortar," RNC; 369: original vs. made land: GMA/CAA; 376: Army tents shelter Summerville victims: SLU; 380: Unitarian Church, CATE; 384: earthquake bolt gib plates, RNC; 385: earthquake bolt at Middleburg plantation, RNC; 385: reinforcing rods at Mulberry Plantation, HCF; 391: William M. Bird engraving, RNC; 398: Mayor Courtenay's letter, RNC; 404: Gala Week, HW, November 12, 1886; 411: Isoseismal map, Dutton, plate XXIX; 416: Clarence E. Dutton, USGS; 421: bust of William A. Courtenay, Collection of City Hall, Charleston, SC; 430: Woodstock Fault, courtesy of Dr. Pradeep Talwani.

BIBLIOGRAPHY

PRIMARY SOURCES

The author would like to thank the following organizations for permission to publish from their collections.

Augusta State University, Augusta, GA.

- Augusta Arsenal correspondence.

Beaufort County Library, Beaufort, SC.

- Rogers, E.B. "Reminiscences of Beaufort Storms."

Charleston County Public Library, South Carolina Room, Charleston, SC.

- City of Charleston. Earthquake Records. Roll # HIS 2004-0001.
- City of Charleston. Records of the Executive Relief Committee for the Earthquake of 1886.
- City of Charleston. "Return of Deaths Within the City of Charleston, S.C.," 1886.

Charleston Library Society, Charleston, SC.

- William A. Courtenay Collection. Scrapbook, "The Cyclone of 1885 and Earthquake 1886."

City of Charleston, Department of Archives, Charleston, SC.

- Bixby, W. H. *U.S. Government Commission Report on Examination of Buildings in Charleston, S.C., Injured by the Recent Earthquake of August 1886.*
- City Council Proceedings, 1885-1888.
- Executive Relief Committee. Minutes and other Records, 1886-1887.

City of Charleston, Commissioners of Public Works, Charleston, SC.

- *Minutes of the Seventh Annual Stockholders Meeting of the City of Charleston Water Works Company, February 24, 1887.*

College of Charleston, Special Collections, Charleston, SC.

- Bacot family papers, 1860-1938.
- Frank R. Fisher. Notes, 1882-1902.
- Frank R. Fisher Papers. "The Earthquake of August 31st 1886."
- Josephine Rhett Bacot Papers.

Darlington County Historical Commission, Darlington, SC.

- Plantation files.

Diocese of Charleston (Roman Catholic Church), Diocesan Archives, Charleston, SC.

- Episcopal correspondence, 1886-1887.
- Laity vertical file.
- Priests' vertical file.
- St. Lawrence Cemetery records.

Gibbes Museum of Art / Carolina Art Association, Charleston, SC.

- Map, "Creeks and Marshes, Charleston Peninsula."

Library of Congress. Manuscript Division.

- Combined Military Service Records of Confederate Soldiers Who Served in Organizations from South Carolina.
- Federal Population Census records, 1850-1930.
- "Shocking Earthquakes." Broadside, February 7, 1812. Boston, MA: printed and sold at the Printing-Office, Corner of Theatre-Alley, February 7, 1812. Printed Ephemera Collection, portfolio 49, folder 16.
- Slave Narratives from the Federal Writers' Project, 1936-1938. South Carolina Narratives.

Library of Congress. Map Division.

- A Map of Charleston . . . Prepared to Accompany Mayor Courtenay's Centennial Address, August 13, 1883.
- A Plan of Charles Town from a Survey of Edw. Crisp, Esq. in 1704.

Middleton Place Foundation, Charleston, SC.

- Photograph Collection.

Private collections.

- Diary of Mary Davis Brown, 1822-1903, York County, SC.

Rockefeller Family Archives, Rockefeller Archives Center, Rockefeller University, North Tarrytown, NY.

- John D. Rockefeller's Charity Index Cards, c. 1863-1903.

Saint Louis University, St. Louis, MO. Earthquake Center.

- Charleston Earthquake Image Collection.

South Carolina Historical Society, Charleston, SC.

- Agricultural Society of South Carolina. Minutes, 1880-1935, and Minutes of the Executive Committee, 1879-1887. Collection #0251.00.
- Alston-Pringle-Frost papers, 1693-1990. Collection #1285.00.
- Bacot-Huger Collection. Collection #30-29-1.
- Bacot-Huger Collection. Julius M. Bacot Diary. Collection #11/57/4.
- Benseman family record, 1851-1874. Collection #43/2062.
- Cantwell, L. E. Earthquake Echoes, 1886-ca. 1950. Collection # 43/198.
- Celia. Letter to "Aunt May," September 5, 1886. Collection 43/609.
- Chisholm, Robert George, 1831-1907. The Charleston Earthquake as told in the Letters from Robert G. Chisolm. Collection #43/0190.
- Courtenay, William Ashmead, 1831-1908. Collection 43/353.
- Cumming, Thomas John. Papers. Collection #43-0365.
- deSaussure, Henry Alexander, 1851-1903. Collection # 43/2144.
- Diary of an unknown Charleston resident (fragment), August 31-September 16, 1886.
- Ficken-Ball miscellany, 1782-1933. Collection #0225.00.
- Heriot, Benjamin George. Commonplace book, 1859-1903. Collection #34/0110.
- Middleton, Harriott, 1828-1905. Family Papers, 1861-1874. Collection # 1168.02.08.
- Moffett, George H. Family Papers, 1830-1904. Collection #11/306/16.
- Moubray, William. Letters, 1812-1819. Collection #43/2076.
- Murden, Eliza Crawley. Collection #43/569.
- Ravenel, Thomas Porcher. Papers, 1810-1904. Collection # 1171.02.02.
- Riecke, Anthony W. Scrapbooks, 1869-1886. Collection # 34/388-390.
- Smythe, Augustine Thomas. Papers, 1853-1938. Collection # 1209.03.02.04.
- Wells, Edward Laight. Papers, 1858-1924. Collection #1136.00.

University of South Carolina, South Caroliniana Library, Columbia, SC.

- Sanborn Map Company, *Insurance Maps of Charleston.* Online.
- William Ashmead Courtenay Collection.

University of South Carolina Library - Beaufort, Beaufort, SC.

- "The Old Beaufort College Building: A Historical Structure Report. Architectural Development and Historical Alterations. Prepared by Liollio Architects, 1998."

PERIODICALS

The American Missionary, Boston, MA.
The *Arkansas City Republican,* Arkansas City, AR.
The *Atlanta Constitution*, Atlanta, GA
The *Baltimore Sun*, Baltimore, MD.
The *Bangor Daily Whig and Courier,* Bangor, ME.

The *Berkeley Gazette*, Mt. Pleasant, SC.
The *Brooklyn Daily Eagle*, Brooklyn, NY.
The *Charleston Courier*, Charleston, SC.
The Christian Herald and Signs of Our Times, New York, NY.
The Christian Illustrated Weekly, New York, NY.
Citypaperonline, Baltimore, MD.
The *Cleveland Gazette*, Cleveland, OH.
The *Columbia Daily Record*, Columbia, SC.
The *Columbia Daily Register*, Columbia, SC.
The *Evening Post*, Charleston, SC.
Frank Leslie's Illustrated Newspaper, New York, NY.
The Georgetown Enquirer, Georgetown, SC.
L'Illustrazione Italiana, Rome, Italy.
The *Mitchell Daily Republican*, Mitchell, SD.
The *News and Courier*, Charleston, SC.
The *New York Commercial Advertiser*, New York, NY.
The *New York Times*, New York, NY.
The Palmetto Post, Port Royal, SC.
The Pickens Sentinel, Pickens, SC.
Scribner's Monthly, New York, NY.
South Carolina Seismic Network Bulletin. Online.
The State, Columbia, SC.
The *Times*, Trenton, NJ.
The *Times and Democrat*, Orangeburg, SC.

SECONDARY SOURCES

"A Fearful Night. One Lady's Experience—A Graphic Account of the Earthquake." The *Bangor* (ME) *Daily Whig and Courier*, September 27, 1886.

Annual Trade Review, 1886-1887. Charleston, SC: Walker, Evans, & Cogswell Co., 1877.

Bagwell, Joyce. *Low Country Quake Tales*. Easley, SC: Southern Historical Press, 1986.

Bakun, W.H., and M.G. Hopper. "Magnitudes and Locations of the 1811-1812 New Madrid, Missouri, and the 1886 Charleston, South Carolina, Earthquakes." *Bulletin of the Seismological Society of America*, 94:64-75.

Bakun, W.H., A. C. Johnston, and M. G. Hopper. *Modified Mercalli Intensities (MMI) for Large Earthquakes near New Madrid, Missouri, in 1811-1812, and near Charleston, South Carolina, in 1886*. Washington, DC: USGS Report 01-184, 2002.

Beesley, Charles Norbury. *Beesley's Illustrated Guide to St. Michael's Church, Charleston, S.C.* 2nd ed. Charleston, SC: Southern Printing & Publishing Co., 1908.

Bollinger, Gilbert A. "The Earthquake at Charleston in 1886." Presented at the B.S.S.C. Meeting, Charleston, SC, February 13, 1985.

Bollinger, Gilbert A., and T. R. Visvanathan. "The Seismicity of South Carolina Prior to 1886." In *Studies Related to the Charleston, South Carolina Earthquake of 1886—A Preliminary Report*, edited by Douglas W. Rankin. USGS Professional Paper 1028. Washington, DC: U.S. Government Printing Office, 1977.

Breibart, Sol. "Charleston Jews and the Earthquake of 1886." *The Charleston Jewish Journal*, March 1994, 18.

Bristow, Caroline W. "Former Summervillian Recalls Quake of 1886." The (Charleston, SC) *News and Courier*, August 30, 1959.

By-laws of the Orphan House of Charleston, South Carolina, Revised and Adopted by the Board of Commissioners 4 April 1861, Approved by the City Council of Charleston 23 April 1861. Charleston, SC: Steam-Power Presses of Evans & Cogswell, 1861.

Carlyle McKinley. Three hundred copies, printed from Jenson Old-Style Roman Type, on Van Gilder Hand-made paper, for Hon. Wm. A. Courtenay, LL. D., for the private use of family and friends. November, 1904. Privately published, 1904.

Charleston As It Is After The Earthquake Shock of August 31, 1886. Charleston, SC: publisher unknown, 1886.

The Charleston Earthquake. 1886. F.W. Dawson. Augusta, GA: *The Augusta Chronicle*, September 1886.

Chowns, T. M., and C. T. Williams. "Pre-Cretaceous rocks beneath the Georgia Coastal Plain: regional implications." In *Studies Related to the Charleston, South Carolina Earthquake of 1886: tectonics and seismicity*, edited by G.S. Gohn. USGS Professional Paper 1313. Washington, DC: U.S. Government Printing Office, 1983.

City of Charleston. *Year Book of the City of Charleston*, 1885, 1886, 1887, and 1888.

City of Charleston Executive Relief Committee. *The Earthquake, 1886. Exhibits showing the receipts and disbursements, and the applications for relief, with the awards and refusals of the Executive Relief Committee in over 2,000 cases of house owners and 317 cases of application for losses in personal property*. Charleston, SC: Lucas, Richardson & Co., 1887.

Clark, E. Culpepper. *Francis Warrington Dawson and the Politics of Restoration: South Carolina, 1874-1889*. Tuscaloosa, AL: University of Alabama Press, 1980.

Clayton, Richard. "Pages from the Past: Earthquake Rattles Lighthouse." *Lighthouse Digest Magazine*, March 2000.

Comprehensive Seismic Risk and Vulnerability Study for the State of South Carolina. Final Report. West Columbia, SC: South Carolina Emergency Preparedness Division, 2001.

Confederate Imprints, 1861-1865. Reel 83, no. 2564. New Haven, CT: Research Publications, 1974.

Côté, Richard N. *Mary's World: Love, War, and Family Ties in Nineteenth-century Charleston.* Mt. Pleasant, SC: Corinthian Books, 2000.

Courtenay, William Ashmead, and Julia Courtenay Campbell. *The Courtenay Family: Some Branches in America.* Charlottesville, VA: privately printed, 1964.

Dawson, Francis W. *Reminiscences of Confederate Service, 1861 – 1865*, edited by Bell I. Wiley. Baton Rouge, LA: Louisiana State University Press, 1980.

Derrick, Samuel M. *Centennial History of [the] South Carolina Railroad.* Columbia, SC: The State Co., 1930.

Dutton, Clarence Edward, and Edward Everett Hayden. "Abstract of the Results of the Investigation of the Charleston Earthquake." *Science* [Supplement], May 20, 1887, 489-501.

Dutton, Clarence Edward. *The Charleston Earthquake of August 31, 1886.* USGS Ninth Annual Report, 1887-1888. Washington, DC: U.S. Government Printing Office, 1888.

"Earthquake Phenomena." *The Manufacturer and Builder*, 18:13.

Edgar, Walter. *South Carolina: A History*. Columbia, SC: University of South Carolina Press, 1998.

Edmonds, Bobby F. *The Making of McCormick County.* McCormick, SC: Cedar Hill Unlimited, 1999.

Elton, David J., M. Eeri, and James R. Martin II. "Dynamic Site Periods in Charleston, SC." *Earthquake Spectra*, 5:703-734.

"Foundation Makes Exciting Find!" *Middleton Place Foundation Notebook*, 27:1.

Fowler, Don D. *The Western Photographs of John K. Hillers: Myself in the Water.* Washington, DC: Smithsonian Institution, 1989.

Fraser, Walter J. *Charleston! Charleston! The History of a Southern City.* Columbia, SC: University of South Carolina Press, 1991.

Freeman, John Ripley. *Earthquake Damage and Earthquake Insurance: Studies of a rational basis for earthquake insurance, also studies of engineering data for earthquake resisting construction. By John Ripley Freeman, engineer.* New York: McGraw Hill, 1932.

From the Past to the Present into the Future: City of Charleston Fire Department 1882-1984. Charleston, SC: Charleston Fire Department, 1984.

Garrison, Webb. *A Treasury of Carolina Tales.* Nashville, TN: Rutledge Hill Press, 1994.
Geddings, Hardy Crawford, ed. "Earle Sloan." *Who's Who in South Carolina.* Columbia, SC: McCaw, 1921.
Hastie, John Drayton. *The Story of Historic Magnolia Plantation and Its Gardens.* Charleston, SC: John Drayton Hastie, 1984.
Heard, William Henry Harrison. *From Slavery to the Bishopric in the A.M.E. Church. An Autobiography.* Philadelphia, PA: A.M.E. Book Concern, 1928.
Heisser, Manning Reynolds. *I Remember.* Charleston, SC: privately printed, 1989.
Hemphill, J. C., ed. *Men of Mark in South Carolina.* Washington, DC: Men of Mark Publishing Co., 1907.
Hill, Barbara Lynch. *Summerville. A Sesquicentennial Edition of the History of The Flower Town in the Pines.* Summerville, SC: privately published, 1997.
Historical Sketch of the Confederate Home and College, Charleston, South Carolina, 1867-1921. Charleston, SC: Walker, Evans, and Cogswell Co., 1921.
Historic Society of Bamberg County. *History of Bamberg County, South Carolina. Commemorating One Hundred Years, 1897-1997.* Spartanburg, SC: published for The Historic Society of Bamberg by The Reprint Company, 2003.
Hough, Susan E. "Understanding Earthquakes Makes for Messy Science." *Los Angeles Times*, May 13, 1994.
Hough, Susan E., Leonardo Seeber, and John G. Armbruster. "Intraplate Triggered Earthquakes: Observations and Interpretation." *Bulletin of the Seismological Society of America*, 93:2212-2221.
Housel, Louis V. "An Earthquake Experience." *Scribner's Magazine*, March 1878, 662-672.
Hu, Ke, Sarah L. Glassman, and Pradeep Talwani. "Magnitudes of Prehistoric Earthquakes in the South Carolina Coastal Plain from Geotechnical Data." Paper No. 2. Grant Report 00HQGR0032, USGS, Washington, DC, 2000.
James, William. *The Varieties of Religious Experience.* New York: New American Library, 1958.
Johnson, John. *After the Earthquake. A New-Year Sermon, Preached Sunday, January 9th, 1887, to the Congregation of St. Philip's Church, Charleston, S.C.* Charleston, SC: The News and Courier Book Presses, 1887.
Jones, N.P., E. K. Noji, G. S. Smith, and R.M. Wagner. Monograph 5: Socioeconomic Impacts, Chapter 5, 19-53. 1993 National Earthquake Conference. Central United States Earthquake Consortium, Memphis, TN, May 2-5, 1993.
Katcher, Philip. *The Army of Robert E. Lee.* London: Arms and Armour Press, 1996.
King, G. Wayne. *Rise Up So Early: A History of Florence County, South Carolina.* Spartanburg, SC: The Reprint Company, 1981.
King, Susan L. *History and Records of the Charleston Orphan House.* 2 vols. Easley, SC: Southern Historical Press, 1984.
Kirkland, Randolph W., Jr. *Dark Hours: South Carolina Soldiers, Sailors, and Citizens who were held in Federal Prisons during the War for Southern Independence, 1861-1865.* Charleston, SC: South Carolina Historical Society, 2002.
Klinck, Isabella Strybling. *Personal Experience of the Great Charleston Earthquake.* North Liberty, IA: Ice Cube Press, 2003.
Lindbergh, Charles, ed. *Earthquake Hazards, Risk, and Mitigation in South Carolina and the Southeastern United States.* Prepared by The South Carolina Seismic Safety Consortium in Participation With The Southeastern United States Seismic Safety Consortium. Charleston, SC: The Citadel, 1986.
Louderback, George D. "The personal record of Ada M. Trotter of certain after-shocks of the Charleston (S.C.) earthquake of 1886." *Bulletin of the Seismological Society of America*, 34:199-206.
Madden, Richard C. *Catholics in South Carolina: a History.* Lanham, MD: University Press of America, 1985.

Martí, José. "El Terremoto de Charleston." In *Literatura Hispanoamericana* (rev. ed.), edited by Enrique Anderson Imbert and Eugenio Florit. New York: Holt, Rinehart and Winston, 1970.

Mazÿck, Arthur, and Gene Waddell. *Charleston in 1883*. Easley, SC: Southern Historical Press, 1983.

McCaslin, Richard B. *Portraits of Conflict: A Photographic History of South Carolina in the Civil War.* Fayetteville, AR: The University of Arkansas Press, 1994.

McIver, Petrona Royall. *History of Mount Pleasant, South Carolina.* Charleston, SC: Ashley Printing and Publishing Co., 1960.

McKinley, Carlyle. "A Descriptive Narrative of the Earthquake of August 31, 1886." City of Charleston *Yearbook*, 1886, 344-441.

McKinley, Carlyle. *An Appeal to Pharaoh: The Negro Problem and its Radical Solution.* New York: Fords, Howard & Hulbert, 1889.

McKinley, Carlyle. "The August Cyclone—1885." City of Charleston *Yearbook*, 1885, 371-388.

Mendenhall, Thomas C. "Seismoscopes and Seismological Investigations." *The American Journal of Science*, February 1888.

Mescher, Virginia. "Pick Up a Good Book: Literacy and Popular Literature of the Civil War." Online in *Virginia's Veranda* at www.raggedsoldier.com, January 25, 2006.

Mills, Robert. *Statistics of South Carolina, Including a View of its Natural Civil & Military History, General & Particular.* Charleston, SC: Hurlbut and Lloyd, 1826.

Morrison, Gail Moore. "I Shall Not Pass This Way Again: The Contributions of William Ashmead Courtenay." Online at www.webcom.com/scourt/thisway.htm on January 30, 2006.

New York State Chamber of Commerce. *Report of the Special Committee appointed by the Chamber of Commerce of the State of New York to obtain relief for the sufferers by the earthquake at Charleston, S.C.* New York: Press of the Chamber of Commerce, 1886.

Nishenko, N.P., and Gilbert A. Bollinger. "Forecasting Damaging Earthquakes in the Central and Eastern United States." *Science*, 249:1412-1416.

Nuttli, Otto W., Gilbert A. Bollinger, and Robert B. Herrmann. *The 1886 Charleston, South Carolina, Earthquake— A 1986 Perspective.* USGS Circular 1985. Washington, DC: U.S. Government Printing Office, 1986.

Obermeier, Stephen F., Gregory S. Gohn, Robert E. Weems, Robert L. Gelinas, and Meyer Rubin. "Geologic Evidence for Recurrent Moderate to Large Earthquakes near Charleston, South Carolina." *Science*, 227:408-411.

Obituary of Carlyle McKinley. Printed for private distribution by William A. Courtenay, 1904.

O'Brien, Joseph L. *A Chronicle History of Saint Patrick's Church, Charleston, South Carolina, 1837-1937.* Charleston, SC: John J. Furlong & Son, 1937.

Palm, Risa. "Improving Hazard Awareness." *Proceedings of Conference XX: A Workshop on The 1886 Charleston Earthquake and Its Implications for Today.* USGS Open File 83-843, 55-61.

Peters, Kenneth E. "Aiken's 'Rock of Ages.'" In *Historical Sketches on Aiken*, edited by William S. Brockington, Jr. and Judith T. VanSteenburg. Aiken, SC: Aiken Sesquicentennial Committee, 1985.

Peters, Kenneth E. "Disaster Relief Efforts Connected with the 1886 Charleston Earthquake." A Paper presented for the Fourth Citadel Conference on the South, Saturday, April 13, 1985. Charleston, SC: The Citadel, 1985.

Peters, Kenneth. E., and Robert. B. Herrmann, eds. *First-hand Observations of the Charleston Earthquake of August 31, 1886, and Other Earthquake Materials: Reports of W. J. McGee, Earle Sloan, Gabriel E. Manigault, Simon Newcomb, and Others.* Columbia, SC: Bulletin 41, South Carolina Geological Survey, 1986.

Phelps, W. Chris. *The Bombardment of Charleston, 1863-1865.* Gretna, LA: Pelican Publishing Co., 2002.

Pinckney, Paul. "Lesson Learned from the Charleston Quake. How the Southern City Was Rebuilt Finer Than Ever Within Four Years." *San Francisco Chronicle*, May 6, 1906.

Poppenheim, Mary B. *Southern Women at Vassar*. Columbia, SC: University of South Carolina Press, 2002.

Porter, Anthony Toomer. *Led On! Step by Step: Scenes from Clerical, Military, Educational, and Plantation Life in the South, 1828-1898. An Autobiography*. New York: G.P. Putnam Sons, 1898.

Poyas, Elizabeth Ann. *Days of Yore, or, Shadows of the Past*. Charleston, SC: W. G. Mazÿck, 1870.

Rabbitt, Mary C. *Minerals, Lands, and Geology for the Common Defence and General Welfare*. Vol. 2, 1879-1904. Washington, DC: USGS, U.S. Government Printing Office, 1980.

Ramsay, David. *The History of South-Carolina, from Its First Settlement in 1670, to the year 1808*. Vol. II. Newberry, South Carolina: W. J. Duffie, 1858.

Rankin, Douglas W., ed. *Studies Related to the Charleston, South Carolina Earthquake of 1886 – A Preliminary Report*. USGS Professional Paper 1028. Washington, DC: U. S. Government Printing Office, 1977.

Ravenel, Mrs. St. Julien. *Charleston: The Place and the People*. New York: The Macmillan Co., 1927.

Robinson, Andrew, and Pradeep Talwani. "Building damage at Charleston, South Carolina, associated with the 1886 earthquake." *Bulletin of the Seismological Society of America*, 73:633-652.

Rockwood, Charles Greene. "The Charleston Earthquake." *American Journal of Science*, 33:71-73.

Rodriguez, Pedro Pablo. "José Martí and the Charleston Earthquake," CubaNow.net, online at http://www.cubanow.net/global/loader.php?secc=5&cont=stories/num00/terr_Charleston.htm, May 10, 2006.

Rowe, Charles R. *Pages of History: 200 Years of the Post and Courier*. Charleston, SC: Evening Post Publishing Co., 2003.

Ruffman, Alan. "Potential for Large-Scale Submarine Slope Failure and Tsunami Generation Along the U.S. Mid-Atlantic Coast." *Geology*, 29:967.

Seeber, Leonardo, John. G. Armbruster, and G. A. Bellinger. "Large-scale patterns of seismicity before and after the 1886 South Carolina earthquake." *Geology*, 10:382-386.

Seeber, Leonardo, and John G. Armbruster. "The 1886-1889 Aftershocks of the Charleston, South Carolina Earthquake: A Widespread Burst of Seismicity." *Journal of Geophysical Research*, 92:2663-2696.

Selby, Julian A. *Memorabilia and Anecdotal Reminiscences of Columbia, S. C., and Incidents Connected Therewith*. Columbia, SC: The R.L. Bryan Co., 1905.

Shedlock, K. "Seismicity in South Carolina." *Seismological Research Letters*, 59:165-171.

Simkins, Frances Butler, and Robert Hilliard Woody. *South Carolina During Reconstruction*. Chapel Hill, NC: University of North Carolina Press, 1932.

Snowden, Yates, ed. "Earle Sloan." *The History of South Carolina*. New York: Lewis Publishing Co., 1920.

"The South Carolina Problem. The Epoch of Transition." *Scribner's Monthly*, June 1874, 129-160.

Southern Business Guide 1885-86... of the Southern States. Cincinnati, OH: The United States Central Publishing Co., 1885.

Sprunt, James. *Chronicles of the Cape Fear River, 1660-1916*. Raleigh, NC: Edwards & Broughton Printing Company, 1916.

Steinberg, Ted. *Acts of God: The Unnatural History of Natural Disaster in America*. London: Oxford University Press, 2003.

Stockdell, H. C. *Record of Earthquake Damages*. Atlanta, GA: Winham & Lester Publishers, 1886.

Stockton, Robert P. *The Great Shock: The Effects of the 1886 Earthquake on the Built Environment of Charleston, South Carolina.* Easley, SC: Southern Historical Press, 1986.

Stover, Carl W., and Jerry L. Coffman. *Seismicity of the United States, 1568-1989.* USGS Professional Paper 1527. Washington, DC: U.S. Government Printing Office, 1993.

The Summerville Committee. *Report of the Committee for the Relief of the Sufferers by the Earthquakes in Summerville and Vicinity.* Charleston, SC: The News and Courier Book Presses, 1986.

Taber, Stephen. "Seismic Activity in the Atlantic Coastal Plain near Charleston, South Carolina." *Bulletin of the Seismological Society of America*, 4:108-160.

Talwani, Pradeep. "Internally Consistent Pattern of Seismicity near Charleston, South Carolina." *Geology*, 10:654-658.

Talwani, Pradeep. Interview with Richard N. Côté, February 6, 2006, Columbia, SC.

Talwani, Pradeep. *Macroseismic Effects of the 1886 Charleston Earthquake.* Columbia, SC: Department of Geological Sciences, University of South Carolina, 2004.

Talwani, Pradeep, and Navin Sharma. "Reevaluation of the Magnitude of Three Destructive Aftershocks of the 1886 Charleston Earthquake." *Seismological Research Letters*, 70:360-367.

Talwani, Pradeep, and Robert A. Trenkamp. "Seismotectonics of the Charleston, South Carolina Region." Online at http://erp-web.er.usgs.gov/reports/annsum/vol40/ni/g0013/g0013.htm, November 25, 2005.

Talwani, Pradeep, and William T. Schaeffer. "Recurrence rates of large earthquakes in the South Carolina Coastal Plain based on paleoliquefaction data." *Journal of Geophysical Research*, 106:6621-6642.

Thomas, J. A. W. *A History of Marlboro County, with Traditions and Sketches of Numerous Families.* Atlanta, GA: The Foot & Davies Co., 1897.

Upchurch, Thomas Adams. "Senator John Tyler Morgan and the Genesis of Jim Crow Ideology, 1889-1891." *Alabama Review*, April 2004, 110-131.

"Up the Ashley and Cooper." *Harper's New Monthly Magazine*, December 1875, 1-24.

Vaughn, Thomas Wayland. "Memorial of Earle Sloan." *Bulletin of the Geological Society of America*, 40:57-61.

Visvanathan, T.R. *Earthquakes in South Carolina 1698-1975.* Bulletin 40. Columbia, SC: South Carolina Geological Survey, 1980.

Wallace, David Duncan. *The History of South Carolina.* New York: American Historical Society, 1934.

Ware, Lowry. *Associate Reformed Presbyterian Death and Marriage Notices, Volume II: 1866-1888.* Columbia, SC: South Carolina Magazine of Ancestral Research, 1998.

Wauchope, George Armstrong. *The Writers of South* Carolina. Columbia, SC: The State Co., 1910.

Way, William. *The History of Grace Church, Charleston South Carolina: The First Hundred Years.* Charleston, SC: Grace Episcopal Church, 1948.

Whitelaw, Robert N. S., and Alice F. Levkoff. *Charleston Come Hell or High Water.* Columbia, SC: University of South Carolina Press, 2002.

Wildwave, Willie. "Charleston Earthquake." copyright 1886 by E. J. H. Stecher. In *Fred Somers New Collection of 260 Songs.* Chicago, IL: Fred Somers, 1887.

INDEX

All places, streets, and institutions are in the city of Charleston, South Carolina, except as noted.